Scientific Computation

V. D. Liseikin

A Computational Differential Geometry Approach to Grid Generation

Second Edition

With 81 Figures
Including 3 Color Figures

Springer

Vladimir D. Liseikin
Russian Academy of Science
Institute of Computational Technologies
Pr. Lavrentyeva 6
630090 Novosibirsk, Russia
e-mail: lvd@ict.nsc.ru

ISSN 1434-8322
ISBN 978-3-642-07062-4 e-ISBN 978-3-540-34236-6

Springer is a part of Springer Science+Business Media
springer.com

Softcover reprint of the hardcover 2nd edition 2007

Cover design: *design & production* GmbH, Heidelberg

Preface to the Second Edition

This second edition of *A Computational Differential Geometry Approach to Grid Generation* is significantly expanded by new material that centers on the recent advances in grid generation technology based on the numerical solution of Beltrami and diffusion equations in monitor metrics. It gives a more detailed and practice-oriented description of the monitor metrics for providing the generation of adaptive, field-aligned, and balanced numerical grids. New finite-difference codes are described for generating both structured and unstructured surface and domain grids. Numerous applications of the codes for the generation of numerical grids with individual and balanced properties in surfaces and domains, in particular, in the tokamak-edge region are demonstrated. The new edition also boasts examples of the implementations of the grid generation codes in the codes for the numerical investigations of gas-dynamics and magnetized plasmas problems.

Grid technology, which has had a significant impact on the efficiency of numerical codes, remains a rapidly advancing field of computational physics and applied mathematics. New achievements are being added by the creation of more sophisticated techniques, modification of the available methods, and implementation of more subtle tools as well as the results of the theories of differential equations, calculus of variations, and Riemannian geometry in the formulation of grid models and analysis of grid properties.

The development of comprehensive differential and variational grid generation techniques reviewed in the monographs of J.F. Thompson, Z.U.A. Warsi, and C.W. Mastin, P. Knupp, and S. Steinberg, and V.D. Liseikin has been largely based on a popular concept in accordance with which a grid model realizing the required grid properties should be formulated through a linear combination of basic and control grid operators with weights. A typical, basic grid operator is the operator responsible for the well-posedness of the grid model and construction of unfolding grids, e.g., the Laplace equations (generalized Laplace equations referred also to as second-order Beltrami equations) or the function of grid smoothness, which produces fixed non-folding grids while grid clustering is controlled by source terms in differential grid formulations or by an adaptation function in variational models. However, such a formulation does not obey the fundamental invariance laws with respect to parameterizations of physical geometries and frequently results in cumbersome governing grid equations. Besides this, the choice of the weight and control functions for providing well-posedness, grid non-degeneracy, and adaptation is largely based on unreliable theoretical assumptions borrowed from one-dimensional models.

The current book revises this popular concept and pursues a more updated and somewhat revolutionary one based on the general fact that an arbitrary one-to-one, smooth multidimensional coordinate transformation deriving a numerical grid in a domain or on a surface is realized by a solution of a system of the Beltrami equations in a suitable monitor metric specified in the physical geometry. The system can be interpreted as the multidimensional equidistribution principle in which the monitor metric tensor is an extension of a scalar-valued weight function. With this interpretation for a mathematical model for generating grids in domains or on surfaces, one need only choose the Beltrami equations, without any complementary control operators that worsen the model, while the required grid properties are realized through the specification of suitable metric tensors.

Thus the single Beltrami mathematical model provides a real foundation for the solution of the challenging problem of the development of comprehensive grid generators. Consequently the efforts of research should be directed towards implementing this model into grid technology by developing approaches for formulating metrics in physical geometries and establishing necessary relations between them and the required grid properties for the purpose of setting up an adequate control of the grid quality by the choice of the suitable metric.

One natural approach for formulating metric tensors and corresponding tensor-valued weight functions is based on the notion of a monitor surface over the physical geometry that undergoes a gridding process. The monitor surface is defined as the graph of some (in general vector-valued) function that takes into account the behavior of the physical solution. This monitor surface, having an inherent metric tensor that can be considered as the very tensor-valued weight function, is suitable for generating adaptive grids with the use of a smoothness functional (which is the functional of energy) whose Euler–Lagrange equations are, in fact, equivalent to the Beltrami equations in the metric of the monitor surface. The resulting grid derived by this metric tends to cluster its nodes in the zones of the large gradient of the function. The approach for formulating the adaptive metric is readily extended to define more general monitor metrics in domains or on surfaces, thus turning them into Riemannian manifolds whose implementation in grid technology allows one to generate grids satisfying the most broad mesh quality requirements.

In order to control the required grid properties by the monitor metrics, one needs a knowledge of geometric characteristics of the monitor geometries and their relations to the resulting grid behavior. This knowledge can be attained with the aid of the theory of multidimensional differential geometry of Riemannian manifolds adjusted to the features of grid technology. The theory of multidimensional differential geometry is really one of the most promising branches of the pure mathematical field of science, capable of pushing grid technology to a more advanced level in its development. Indeed, many notions and characteristics of common surfaces, such as metric tensors, their invariants, first and second fundamental forms, curvatures and torsions of lines, the mean and Gauss curvatures, and Christoffel symbols, have already been used by many authors as natural elements in defining grid quality measures and formulating appropriate variational and differential grid techniques in a unified manner regardless of the geometry of the physical domains and surfaces. A theory of more general geometric objects, such as regular multidimensional surfaces and Riemannian manifolds implemented for generating grids with necessary properties, is expected to become a highly beneficial tool for boosting grid technology. The known relations and techniques of differential geometry also present an efficient means for transforming and modernizing the physical and grid equations into a suitable form. It is presumable that the science of differential geometry will play in numerical grid technology the same role played by the science of matrices in the theory of difference approximations of boundary value problems.

Therefore, there is a need for a monograph that is essentially aimed at providing deep and balanced insight into the fields of grid science, multidimensional geometry adjusted to grid technology, and up-to-date achievements of the applications of geometric tools to the creation of advanced grid techniques. With this background the reader will be able to formulate and develop well-posed grid models and algorithms and analyze grid properties with geometry related tools, thus taking part in the solution of the very challenging problem of the development of advanced comprehensive grid generators.

This monograph gives an account of the geometrization of popular comprehensive grid methods and presents an important extension to the methods related to the application of the technique of Riemannian manifolds to the formulation of grid equations by developing some procedures for the construction of monitor metric tensors. Contrary to classical geometric studies, which center on geometric features and characteristics of specified Riemannian manifolds, the problem of finding appropriate monitor metrics for producing grid systems with the required properties is somewhat an inverse problem of the creation of Riemannian manifolds with desirable geometric characteristics. In accordance with the concept of the inverse problem, the author of the monograph discusses rather thoroughly some new techniques aimed at the construction of special monitor metrics in physical geometries. The techniques are designed by generalizing the projection approach in which the monitor metric in an n-dimensional physical geometry is borrowed from a natural metric of the n-dimensional surface derived by a height monitor function over the geometry. This technology allows the required metric to be defined through the original metric of the physical geometry and certain vector-valued functions.

The book establishes and reviews some of the relations of the Riemannian geometry for the purpose of obtaining new equations with implemented metric characteristics aimed at facilitating the control of the generation of grids with the required properties. Taking advantage of the relations established, the author has converted the equations into a compact form convenient for numerical treatment via the available algorithms.

The technique of multidimensional differential geometry is also applied to study the qualitative effect of a general class of monitor metrics on the resulting mesh. For this purpose a new characteristic of grid clustering is formulated. Certain relations between this measurement and some geometric characteristics of grid hypersurfaces and the monitor functions forming the monitor metrics are established. The well-known results for grids generated by inverted Laplace equations about node-clustering near concave boundary segments of domains and node-rarefaction near convex boundary segments are, using these relations, extended to arbitrary boundary segments and to more general Beltrami equations in monitor metrics. On the basis of the established formulas, the monitor functions are readily estimated in the inverted diffusion and Beltrami grid equations to provide grid clustering or, if it is reasonable, grid rarefaction near arbitrary segments of physical geometries.

Some relations of the mean curvature of the monitor surfaces to the Beltrami equations for grid generation are exhibited. The book also includes a chapter devoted to the implementation of the comprehensive grid equations and the energy functional into numerical codes and to the application of the codes to the numerical solution of some gas-dynamics and plasma-related problems.

Since grid technology has widespread applications to nearly all field problems, this monograph will be useful for a broad range of readers, including teachers, students, and researchers as well as practitioners in applied mathematics, mechanics, biology, medicine, and physics interested in the numerical analysis of multidimensional field problems with complicated geometries and complex solutions.

The book is divided into two parts. Part I of the book gives a geometric background needed for the development of grid generators. The grid equations, codes, and applications are described in Part II.

Part I of the monograph includes Chaps. 1–4. Chapter 1 gives a general introduction to the subject of numerical grids and methods of their generation. Chapters 2–4 introduce the reader to multidimensional differential geometry for the purpose of better understanding those of its techniques that are suitable for the implementation into advanced grid generation technologies. The geometric implementation in grid technology pursued in the book assumes the development of robust techniques for producing appropriate monitor metrics over both physical domains and surfaces thus converting them into Riemannian manifolds. The metrics should guarantee generation of grids with the necessary properties through popular mathematical models.

Part II of the book is devoted to the implementation of geometric tools into the development of grid techniques and codes. It contains Chaps. 5–7. Chapter 5 deals with fundamental elliptic grid models formulated through the operators of Beltrami and diffusion and establishes compact formulas of monitor metrics. Two-dimensional Beltrami equations in the natural metric of a physical surface were originally proposed by Warsi for generating fixed grids on the surface. The ordinary Laplace equations that are the Beltrami equations in the Euclidean metric were applied to generate fixed grids in domains by Crowley and Winslow. One justification of the Beltramian operator is demonstrated in Chap. 5 by the proof of the statement that an arbitrary nondegenerate smooth transformation of a physical domain or surface is realized as a solution of the Dirichlet boundary value problem for the system of Beltrami grid equations in some appropriate metric. The chapter also discusses some variational and harmonic interpretations of the Beltrami equations, in particular, a variational approach for generating harmonic maps through the minimization of energy functionals, which was suggested by Dvinsky.

With the help of the geometric relations, established in Chap. 4, the grid equations introduced in Chap. 5 are transformed in Chap. 6 to equations in invariant forms with respect to independent logical variables. Special monitor metrics over two-dimensional surfaces are designed that result in simpler transformed equations, even in comparison with the equations that have been used for generating fixed grids. The chapter also establishes relations between the monitor functions and geometric characteristics of the Riemannian manifolds produced and the coordinate lines and surfaces generated by a corresponding mathematical model, for the purpose of realization of grid control through a suitable specification of the monitor functions.

Chapter 7 gives a description of some computational codes for generating grids with the numerical solution in the logical domain of the elliptic equations obtained in Chap. 6 by changing mutually dependent and independent variables in the original Beltrami and diffusion equations. Some numerical aspects related to the development of grid generation codes are reviewed, in particular, the application of layer-type functions to formulating monitor metrics and description of two techniques for generating smooth block-structured grids. Numerical results related to the application of the grid technology advocated in the book to some gas-dynamics and plasma problems are also exhibited in this chapter.

The book ends with a list of references.

Acknowledgement

The author is very grateful for helpful suggestions in geometry, algebra, and numerical techniques made by his colleagues, Professors Borisov, Churkin, Glasser, Kuzminov, Sharafutdinov, and Shvedov.

The author is also much obliged to the researchers who responded to his requests and sent files of their papers and of pictures for figures, namely

B.S. Azarenok (Figs. 7.10–7.14), A.A. Charakhch'yan (Figs. 7.8–7.9), and A.H. Glasser (Figs. 5.11–5.12). Figures 7.10–7.14 were published in Azarenok (2000, 2002), 7.8 in Charakhch'yan and Ivanenko (1997), 7.9 in Lomonosov, Frolova and Charakhch'yan (1997), and 5.10–5.11 in Glasser et al. (2005).

The work over the book was partly supported by the US Civilian Research & Development Foundation (CRDF): Award NO RU-M1-2579-NO-04. In particular, the efforts related to the development of grid generation codes, computing figures of grids, and preparing the text of the book in Latex code, made by I. Kitaeva, Yu. Likhanova, and D. Patrakhin, whom the author thanks very much, were remunerated by payments from the CRDF grant. Figures 5.6, 7.29 and 7.37 were published in Glasser, Liseikin and Kitaeva (2005), 5.4, 5.8, 7.32 and 7.34 in Glasser et al. (2005).

Novosibirsk, March 2006 *Vladimir D. Liseikin*

Contents

Part I Geometric Background to Grid Technology

Part I

Geometric Background to Grid Technology

The science of differential geometry is a powerful and helpful tool capable for significantly boosting the development of grid technology. In particular, many notions and characteristics of two-dimensional surfaces bounding physical domains such as metric tensors, their invariants, the first and second fundamental forms, curvatures and torsions of lines, mean and Gauss curvatures of surfaces, and Christoffel symbols have already been reviewed in the monographs of Warsi (1981), Thompson, Warsi, and Mastin (1985), Knupp and Steinberg (1993), and Liseikin (1999) as natural elements in defining grid quality measures and formulating appropriate variational and differential grid techniques in a unified manner regardless of the geometry of domains and surfaces.

The papers of Dvinski (1991) and Liseikin (1991) gave rise to the implementation of more general geometric objects such as multidimensional regular surfaces and Riemannian manifolds into two- and three-dimensional adaptive grid generation techniques. The known relations and techniques of the multidimensional differential geometry also present a useful tool for transforming and modernizing the physical and grid equations into a suitable form.

This part of the book, containing Chaps. 1–4, gives an introduction to the geometric and tensor theories necessary in advanced grid technology. Chapter 1 acquaints the reader with the most general aspects of grid generation. It expounds the most popular approaches and mapping techniques that have been developed for generating numerical grids in domains and on surfaces. The elementary theories of curves, multidimensional regular surfaces, and Riemannian manifolds are reviewed in Chaps. 2–4. These geometries play a crucial role in formulating comprehensive grid equations and in controlling grid properties in domains and on their boundaries. The metric tensors and basic geometry characteristics, in particular, the mean curvature of n-dimensional surfaces and Riemannian manifolds necessary for grid technology are discussed in detail.

1 Introductory Notions

This chapter gives an introduction to the subject of numerical grid technology. It delineates the notion of grid cells, requirements imposed on grid properties, and the most popular methods of a mapping approach.

1.1 Representation of Physical Geometries

The goal of advanced grid technology is the development of robust algorithms and codes for generating grids in arbitrary spatial domains. A spatial domain

$$X^3 \subset R^3\,, \quad X^3 = \{\mathbf{x} \in R^3\,, \mathbf{x} = (x^1, x^2, x^3)\}\,,$$

is defined by a specification of a number of two-dimensional boundary patches that bound it (Fig. 1.1). Let S^{x2} be some such patch that is a surface in R^3. There are two ways for describing points of the boundary surface S^{x2}: 1) implicit and 2) explicit. By the implicit way the points of the patch S^{x2} are found from some equation $F(x^1, x^2, x^3) = 0$. In the explicit approach the patch is specified by a nondegenerate parametrization

$$\mathbf{x}(\mathbf{s}) : S^2 \to X^3\,, \quad \mathbf{s} = (s^1, s^2)\,, \quad \mathbf{x}(\mathbf{s}) = [x^1(\mathbf{s}), x^2(\mathbf{s}), x^3(\mathbf{s})]\,, \tag{1.1}$$

which is a one-to-one map between a two-dimensional domain $S^2 \subset R^2$, referred to as a parametric domain, and S^{x2}.

In the same way there is specified a boundary curve S^{x1} of a patch: implicitly by two equations $F_1(x^1, x^2, x^3) = 0$ and $F_2(x^1, x^2, x^3) = 0$ or explicitly through a nondegenerate parametrization

$$\mathbf{x}(s) : [a, b] \to X^3\,, \quad \mathbf{x}(s) = [x^1(s), x^2(s), x^3(s)]\,, \tag{1.2}$$

between the parametric interval $[a, b]$ and S^{x1} (Fig. 1.1).

Analogous schemes (implicit and explicit) are applied to describing boundaries of two-dimensional domains.

Note the domain X^3 itself is readily represented in the form (1.1) and (1.2), as

$$\mathbf{x}(\mathbf{s}) : S^3 \to X^3\,, \quad \mathbf{s} = (s^1, s^2, s^3)\,, \tag{1.3}$$

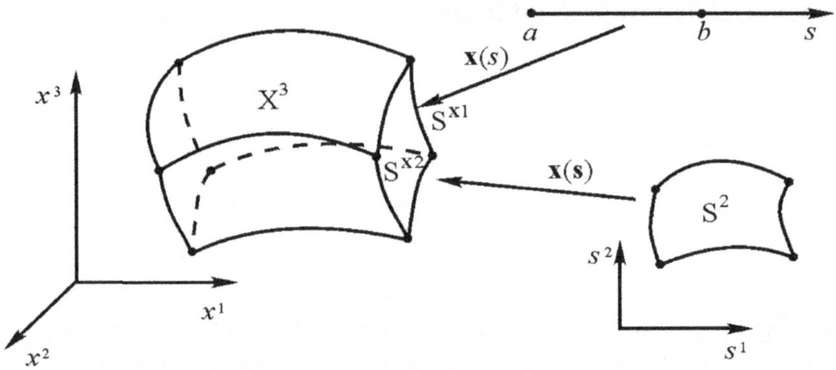

Fig. 1.1. Scheme for a representation of a domain as a curvilinear hexahedron

where S^3 is X^3 while $\mathbf{x}(\mathbf{s})$ is the identical transformation $\mathbf{x}(\mathbf{s}) \equiv \mathbf{s}$. However the domain may also have a general form of its representation (1.3) with the parametric domain S^3 different from X^3. The representations (1.1–1.3) give the full set of the explicit specifications of the domain X^3 and its boundary.

Of these two ways for describing physical geometries the modern grid technology relies only on the explicit specifications in the forms (1.1–1.3).

Thus the first inevitable step in the process of grid generation in a domain $X^3 \subset R^3$ is concluded with preparing a working place for the application of grid techniques, i.e. one must present an explicit representation of X^3 and the boundary of X^3 by choosing a set of surface patches covering this boundary and the parametrizations of these patches and their boundary curves.

If the domain X^3 is diffeomorphic to a three-dimensional cube then one of the typical approaches to the specification of X^3 concludes with considering X^3 in the form of a hexahedron with curvelinear faces, i.e. as a deformed cube (Fig. 1.1). Thus there are chosen 6 patches on the boundary of X^3 which represent 6 faces of the hexahedron. The intersection of two contiguous patches forms an edge of the hexahedron. As a result there are 12 edges. The intersections of contiguous edges form 8 vertices. The domain X^3, its faces, and edges should by specified by corresponding parametrizations (1.1–1.3).

In another consideration a domain X^3 may be viewed as a curvilinear tetrahedron, i.e. as a deformed three-dimensional simplex as in Fig. 1.2 (left). Consequently, the boundary patches are triangular surfaces while their parametric domains may naturally be triangles. In the similar way the domain X^3 may be interpreted as one more solid, for example, as a curvilinear prism (Fig. 1.2) (right). Analogously a two-dimensional domain X^2 or surface S^{x2} diffeomorphic to a square may be viewed as a curvilinear quadrilateral or triangle.

If a physical domain is not diffeomorphic to an n-dimesional cube then, for its representation, it is either divided into several subdomains called blocks,

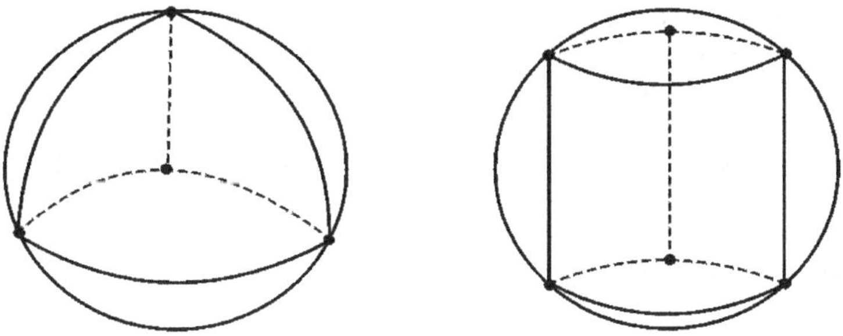

Fig. 1.2. Scheme for a representation of a globe as both a curvilinear tetrahedron and prism

each of which is diffeomorphic to the cube, or imaginary faces are introduced. Figure 1.3 demonstrates such representations of a two-dimensional ring.

Relying on the explicit representation of an arbitrary physical geometry a physical domain, surface, or curve can be considered in a unified manner, as a collection of geometric objects referred to as regular surfaces locally parametrized as

$$\mathbf{x}(\mathbf{s}) : S^n \to \mathrm{R}^{n+k},\ \mathbf{x} = (x^1, \dots, x^{n+k}),\ \mathbf{s} = (s^1, \dots, s^n),\ n \geq 1\ , \qquad (1.4)$$

where S^n is an n-dimensional parametric domain (an interval if $n = 1$), while $\mathbf{x}(\mathbf{s})$ is a smooth vector-valued function of rank n at all points $\mathbf{s} \in S^n$. We shall designate by S^{xn} the regular surface parametrized by (1.4). Note, when $k = 0$ then S^{xn} is a domain $Y^n \subset \mathrm{R}^n$.

We assume further throughout this book that we deal with an arbitrary geometry S^{xn} represented explicitly by the parametrization (1.4).

Using the specification (1.4) of the physical geometry S^{xn} allows one to generate grid points first on the parametric domain and then transforming them through the parametrization of S^{xn}. With such consideration the pro-

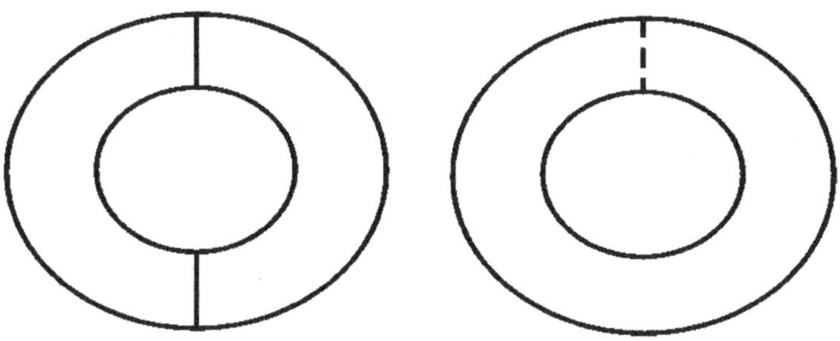

Fig. 1.3. Representations of a two-dimensional ring as both two and one curvilinear quadrilaterals

cess of grid generation can be carried out uniformly both for the boundary of a physical geometry and for its interior part. This scheme of grid generation leads to the natural requirement for grid techniques that the grids generated in the physical geometry for different parametrizations should be the same, i.e. the grid algorithms should be invariant of parametrizations of S^{xn}.

1.2 General Concepts Related to Grids

By a grid in a physical geometry S^{xn} there is understood an algorithmically described collection of elemental standard n-dimensional volumes covering or approximating the necessary area of the geometry. The standard volumes are referred to as grid cells. The cells have boundaries that are divided into a few segments which are $(n-1)$-dimensional cells. Therefore the cells can be formulated successively from one dimension to higher dimensions.

The boundary points of the one-dimensional cells are the cell vertices. These vertices are called the grid nodes.

This section discusses some general concepts related to grid cells and grids.

1.2.1 Grid Cells

In general the grid cells are small curvilinear volumes having simple standard shapes (Fig. 1.4 below). These curvilinear volumes are obtained by deforming reference cells. Such reference cells common in practical applications are demonstrated in Fig. 1.4 up.

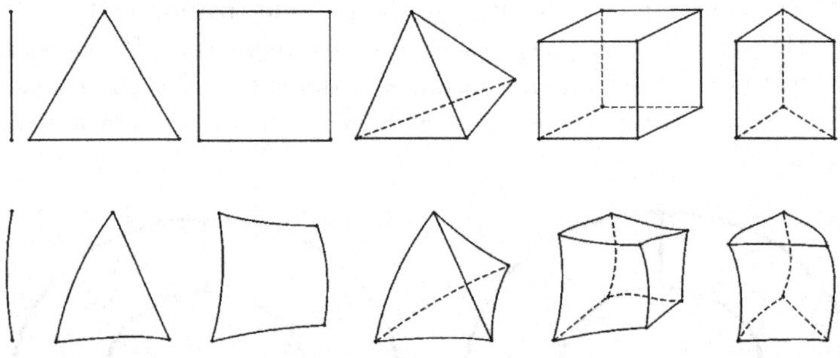

Fig. 1.4. Typical reference (up) and corresponding standard (below) grid cells

Reference Cells

In one dimension the reference cell is a closed straight segment, whose boundary is composed of two points referred to as the cell vertices.

A two-dimensional reference cell is a two-dimensional equilateral polygon whose boundary segments are one-dimensional reference cells referred to as the edges of the cell. Commonly, the two-dimensional reference cells are considered in the form of equilateral triangles or squares. The boundary of a triangular cell is composed of three edges, while the boundary of a square is represented by four edges. These edges are the one-dimensional reference cells.

By a reference three-dimensional cell there is meant a three-dimensional equilateral polyhedron whose boundary is partitioned into a finite number of two-dimensional cells called its faces. In practical applications, three-dimensional reference cells typically have the shape of either equilateral tetrahedrons or cubes. The boundary of a tetrahedral cell is composed of four triangular cells, while a cube is bounded by six squares. Some applications also use for three-dimensional reference cells the cells in the form of prisms having triangular top and bottom faces. Such prism has two triangular and three square faces, nine edges, and six vertices. Note bees also prefer cells in the form of prisms however with hexagonal top and bottom faces.

Standard cells

The edges and the faces of the reference cells are linear. The standard grid cells being deformed reference cells have, as a rule, nonequilateral edges and besides this they may be curvilinear. Thus, in general, the standard cells have the form of curves and curvilinear triangles, quadrilaterals, tetrahedrons, hexahedrons, and prisms as shown in Fig. 1.4 below.

The selection of the shapes shown in Fig. 1.4 to represent the reference and standard cells is justified, first, by their geometrical simplicity and, second, because the existing approaches for the numerical simulation of physical problems are largely based on approximations of partial differential equations using these elemental volumes. The choice of cell shape depends on the geometry and physics of the specific problem and on the method of solution. In particular, tetrahedrons (triangles in two dimensions) are well suited for finite-element methods, while hexahedrons are commonly used for finite-difference techniques.

The major advantage of hexahedral cells (quadrilaterals in two dimensions) is that their faces (or edges) may be aligned with the coordinate surfaces (or curves). In contrast, no coordinates can be aligned with tetrahedral meshes. However, strictly hexahedral meshes may be ineffective near boundaries with sharp corners.

Prismatic cells are generally placed near boundary surfaces which have previously been triangulated. The surface triangular cells serve as faces of prisms, which are grown out from these triangles. Prismatic cells are efficient for treating boundary layers, since they can be constructed with a high aspect ratio in order to resolve the layers, but without small angles, as would be the case for tetrahedral cells.

Triangular cells are the simplest two-dimensional volumes and can be produced from quadrilateral cells by constructing interior edges. Analogously, tetrahedral cells are the simplest three-dimensional volumes and can be derived from hexahedrons and prisms by constructing interior faces. The strength of triangular and tetrahedral cells is in their applicability to virtually any type of domain configuration. The drawback is that the integration of the physical equations becomes a few times more expensive with these cells in comparison with quadrilateral or hexahedral cells.

The vertices of the cells define grid points which approximate a physical geometry. Alternatively, the grid points in the physical geometry may have been generated previously by some other process. In this case the construction of the grid cells requires special techniques.

1.2.2 Requirements Imposed on Cells and Grids

The grid should discretize a physical geometry by displacing the standard cells in such a manner that the computation of the physical quantities on the geometry is carried out as efficiently as desired. The accuracy, which is one of the components of the efficiency of the computation, is influenced by a number of grid factors, such as grid size, grid topology, cell shape and size, and consistency of the grid with the geometry and with the solution. A general consideration of these grid factors is reviewed in this subsection.

Grid Size and Cell Size

The grid size is indicated by the number of grid points, while the cell size implies the maximum value of the lengths of the cell edges. Grid generation requires techniques which possess the intrinsic ability to increase the number of grid nodes. At the same time the edge lengths of the resulting cells should be reduced in such a manner that they approach zero as the number of nodes tends to infinity.

Small cells are necessary to obtain more accurate solutions and to resolve physical quantities on small scales, such as transition layers and turbulence. Also, the opportunity to increase the number of grid points and to reduce the size of the cells enables one to study the convergence rate of a numerical code and to improve the accuracy of the solution by multigrid approaches.

Cell and Grid Deformation

The cell deformation characteristics can be formulated as some measures of the departure of the cell from a reference one. Cells with low deformity are preferable from the point of view of simplicity and uniformity of the construction of the algebraic equations approximating the differential equations.

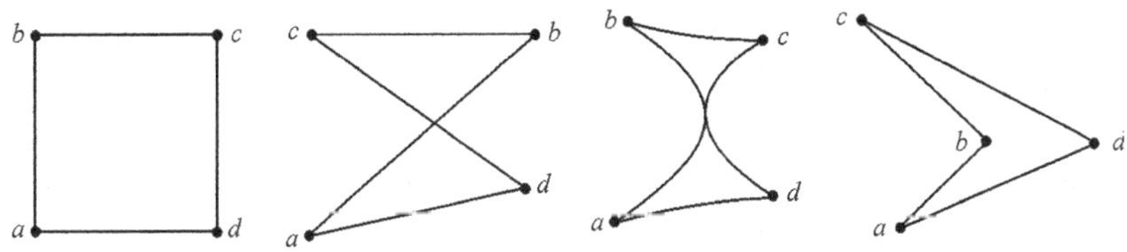

1 2 3 4

Fig. 1.5. Normal (*1*), singular (*2,3*), and badly deformed (*4*) quadrilateral cells

Typically, cell deformation is characterized through the aspect ratio, the angles between the cell edges, and the volume (area in two dimensions) of the cell.

The major requirement for the grid cells is that they must not be folded or degenerate at any points or lines, as demonstrated in Fig 1.5 (2,3,4). Unfolded cells are obtained from the reference cells by a one-to-one deformation. Commonly, the value of any grid generation method is judged by its ability to yield unfolded grids in regions with complex geometry. A tougher condition imposed on grid techniques also requires the generation of linear and convex cells only. The grid deformity is also characterized by the rate of the change of the geometrical features of contiguous cells. Grids whose neighboring cells do not change abruptly are referred to as smooth grids.

Grid Consistency

By a consistent grid or a consistent discretization there is meant a collection of n-dimensional strongly convex cells satisfying the following condition: if two different cells intersect, then the region of the intersection is a common face for both cells. This definition does not admit the fragments of discretizations depicted in Fig. 1.6 (2,3,4). If the union of the cells of the consistent discretization constitutes a simply connected n-dimensional geometry S^{xn}, i.e. a geometry which is homeomorphic to an n-dimensional cube, then, in accordance with the Euler theorem,

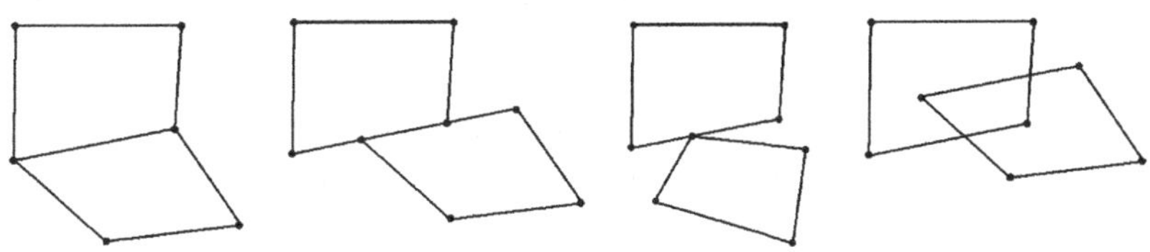

1 2 3 4

Fig. 1.6. Admitted (*1*) and nonadmitted (*2, 3, 4*) intersections of neighboring quadrilateral cells

$$\sum_{i=0}^{n-1}(-1)^i N_i = 1 + (-1)^{n-1} ,$$

where N_i, $i > 0$, is the number of i-dimensional boundary faces of the domain discretization, while N_0 is the number of boundary vertices. In particular, N_1 is the number of boundary edges. This relation can be used to verify the consistency of a generated grid.

Grid Organization

There also is a requirement on grids to have some organization of their nodes, faces, and cells, which is aimed at facilitating the procedures for formulating and solving the algebraic equations substituted for the differential equations. This organization should identify neighboring points and cells. The grid organization is especially important for that class of numerical methods whose procedures for obtaining the algebraic equations consist of substituting differences for derivatives. To a lesser degree, this organization is needed for finite-volume methods because of their inherent compatibility with irregular meshes.

Consistency with Geometry

The accuracy of the numerical solution of a partial differential equation and of the interpolation of a discrete function is considerably influenced by the degree of compatibility of the mesh with the shape of the physical geometry. First of all, the grid nodes must adequately approximate the original geometry, that is, the distance between any point of the physical geometry and the nearest grid node must not be too large. Moreover, this distance must approach zero when the grid size tends to infinity. This requirement of adequate geometry approximation by the grid nodes is indispensable for the accurate computation and interpolation of the solution over the whole physical geometry.

The second requirement for consistency of the grid with the geometry is concerned with the approximation of its boundary by the faces (edges in two dimensions) of the boundary grid cells, i.e. there is to be a sufficient number of nodes which can be considered as the boundary ones, so that a set of edges (in two dimensions) and cell faces (in three dimensions) formed by these nodes models efficiently the boundary. In this case, the boundary conditions may be applied more easily and accurately. If these points lie on the boundary of the geometry, then the grid is referred to as a boundary-fitting or boundary-conforming grid.

Consistency with Solution

It is evident that distribution of the grid points and the form of the grid cells should be dependent on the features of the physical solution. In particular, it is better to generate the cells in the shape of hexahedrons or prisms in boundary layers. Often, the grid points are aligned with some preferred directions, e.g. streamlines. Furthermore, a nonuniform variation of the solution requires clustering of the grid point in regions of high gradients, so that these areas of the physical geometry have finer resolution. Local grid clustering is needed because the uniform refinement of the entire domain may be very costly for multidimensional computations. It is especially true for problems whose solutions have localized regions of very rapid variation (layers). Without grid clustering in the layers, some important features of the solution can be missed, and the accuracy of the solution can be degraded. Problems with boundary and interior layers occur in many areas of application, for example in fluid dynamics, combustion, solidification, solid mechanics and wave propagation. Typically the locations where the high resolution is needed are not known beforehand but are found in the process of computation. Consequently, a suitable mesh, tracking the necessary features of the physical quantities, as the solution evolves, is required.

A local grid refinement is accomplished in two ways: (a) by moving a fixed number of grid nodes, with clustering of them in zones where this is necessary, and coarsening outside of these zones, and (b) by inserting new points in the zones of the physical geometry where they are needed. Local grid refinement in zones of large variation of the solution commonly results in the following improvements:

(1) the solution at the grid points is obtained more accurately;
(2) the solution is interpolated over the whole region more precisely;
(3) oscillations of the solution are eliminated;
(4) larger time steps can be taken in the process of computing solutions of time-dependent problems.

Independence of Parametrizations of Geometries

If a grid algorithm uses parametrizations of a physical geometry S^{xn} in the process of grid generation then, inevitably, this algorithm should be independent of the choice of a parametrization. To clarify this we consider one popular equidistribution approach for generating grids on curves (Fig. 1.7). Let a curve S^{x1} in R^n be specified by two parametrizations

$$\mathbf{x}_1(s) : [0,1] \to R^n , \quad \mathbf{x}_1 = (x_1^1, \dots , x_1^n) , \tag{1.5}$$

and

$$\mathbf{x}_2(t) : [0,1] \to R^n , \quad \mathbf{x}_2(t) = \mathbf{x}_1[s(t)] , \tag{1.6}$$

where

$$s(t) : [0,1] \to [0,1]$$

is a smooth one-to-one function connecting these parametrizations. The popular universal approach, based on the parametrization (1.5), for generating

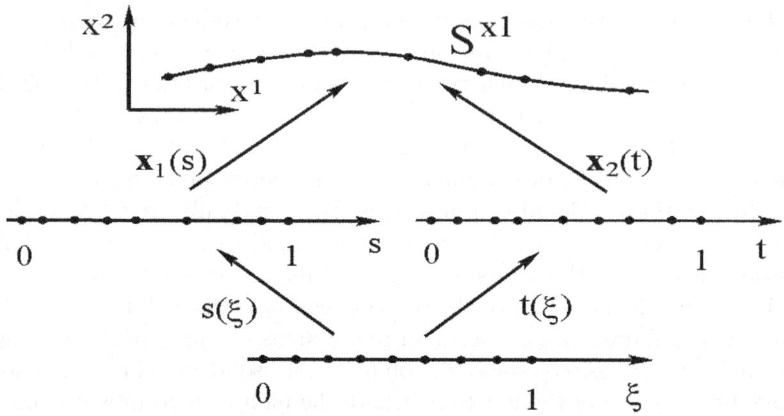

Fig. 1.7. Illustration for independence of grid generation of the choice of parametrizations

grid nodes on S^{x1}, uses a solution of the following two-point boundary value problem

$$\begin{aligned} &\frac{\mathrm{d}}{\mathrm{d}\xi}\Big[\frac{\mathrm{d}s}{\mathrm{d}\xi}w_1(s)\Big] = 0\,, \quad 0<\xi<1\,, \\ &s(0)=0\,, \quad s(1)=1\,, \end{aligned} \tag{1.7}$$

where $w(s) > 0$ is some function called a weight function. If $s(\xi)$ is a solution of this problem then the grid nodes $\mathbf{x}_i$, $i = 0, 1, \ldots, N$, on the curve S^{x1}, obtained by the method, are defined as follows:

$$\mathbf{x}_i = \mathbf{x}_1[s(ih)]\,, \quad i = 0, \ldots, N\,, \quad h = 1/N\,.$$

Let now the parametrization $\mathbf{x}_2(t)$ specified by (1.6) be used in the problem (1.7) with some weight function $w_2(t)$ for the generation of the same grid nodes on S^{x1}. In this case the function $t_1(\xi)$ for which

$$\mathbf{x}_2(t_1(ih)) = \mathbf{x}_1[s(t_1(ih))] = \mathbf{x}_i = \mathbf{x}_1(s(ih))\,, \quad i = 0, 1, \ldots, n\,,$$

must coincide with $t[s(\xi)]$, where $s(\xi)$ is the solution of (1.7), while $t(s)$ is the inverse of $s(t)$. Therefore the function $t_1(\xi)$ is a solution of the boundary value problem

$$\begin{aligned} &\frac{\mathrm{d}}{\mathrm{d}\xi}\Big[\frac{\mathrm{d}t_1}{\mathrm{d}\xi}\frac{\mathrm{d}s}{\mathrm{d}t}w_1[s(t_1)]\Big] = 0\,, \quad 0<\xi<1\,, \\ &t_1(0) = 0\,, \quad t_1(1) = 1\,. \end{aligned}$$

Since the weight functions $w_1(s)$ and $w_2(t)$ for defining the same grid nodes on S^{x1} by the model (1.7) through the parametrizations (1.5) and (1.6), respectively, are not independent but they should be connected by the following relation

$$w_2(t) = w_1[s(t)]\frac{ds}{dt} . \tag{1.8}$$

If this relation is not satisfied, the grid nodes obtained with the help of the solution of (1.7) may vary for different parametrizations of S^{x1}.

It appears that if we take for the weight functions related to the parametrization (1.5) and (1.6) the corresponding functions $w_1(s) = \sqrt{g^s}$ and $w_2(t) = \sqrt{g^t}$ where

$$g^s = \frac{d\mathbf{x}_1}{ds} \cdot \frac{d\mathbf{x}_1}{ds} \quad \text{and} \quad g^t = \frac{d\mathbf{x}_2}{dt} \cdot \frac{d\mathbf{x}_2}{dt}$$

is the covariant metric tensor of S^{x1} in the coordinate s and t, respectively, then the equation (1.8) is held since there is valid the obvious equation

$$g^t = g^s\left(\frac{ds}{dt}\right)^2 .$$

The consideration given for the curvilinear curve also is actual for surfaces. As well as in the one-dimensional case, the application of the elements of the metric tensors of n-dimensional surfaces allows one to formulate grid equations which produce the same grid nodes for different surface parametrizations.

Compatibility with Numerical Methods

The locations of the zones of local refinement are also dependent on the numerical approximation to the physical equations. In particular, the areas of high solution error require more refined grid cells. However, the error is estimated through the derivatives of the solution and the size of the grid cells. Thus, ultimately, the grid point locations are to be defined in accordance with the derivatives of the solution.

In general, numerical methods for solving partial differential equations can be divided into two classes: methods based on direct approximations of the derivatives in the differential equation and methods that approximate the solution of the continuum differential equation by linear combinations of trial functions. Finite-difference methods belong to the first class. This difference in methods has a direct impact on the construction of the numerical grid. For the finite-difference methods it is desirable to locate the grid points along directions of constant coordinates in the physical region in order to provide a natural approximation of the derivatives: on the other hand, the methods in the second class that approximate the solution with trial functions do not impose such a restriction on the grid, since the approximate derivatives are obtained after substitution of the approximate solution.

1.3 Grid Generation Models

There are two fundamental classes of grid popular in the numerical solution of boundary value problems in multidimensional regions: structured and unstructured. These classes differ in the way in which the mesh points are locally organized. In the most general sense, this means that if the local organization of the grid points and the form of the grid cells do not depend on their position but are defined by a general rule, the mesh is considered as structured. When the connection of the neighboring grid nodes varies from point to point, the mesh is called unstructured. As a result, in the structured case the connectivity of the grid is implicitly taken into account, while the connectivity of unstructured grids must be explicitly described by an appropriate data structure procedure.

Detailed descriptions of the most popular structured methods and their theoretical and logical justifications and numerical implementations were given in the monographs by Thompson, Warsi, and Mastin (1985), Knupp and Steinberg (1993), and Liseikin (1999, 2004). Particular issues concerned with the generation of one-dimensional moving grids, the stretching technique for the numerical solution of singularly perturbed equations, nonstationary grid techniques, and equidistribution methods for wave propagation problems were considered in the books by Alalykin et al. (1970), Liseikin (2001a), Zegeling (1993), and Khakimzyanov et al. (2001), respectively.

A considerable number of general structured grid generation methods were reviewed in surveys by Thompson, Warsi, and Mastin (1982), Thompson (1984a, 1996), Eiseman (1985), Liseikin (1991b), and Thompson and Weatherill (1993).

Adaptive structured grid methods were first surveyed by Anderson (1983) and Thompson (1984b, 1985). Then a series of surveys on general adaptive methods was presented by Eiseman (1987), Hawken, Gottlieb, and Hansen (1991), Liseikin (1996), and Baker (1997). Adaptive techniques for moving grids were described by Hedstrom, Rodrigue (1982) and Zegeling (1993).

Methods for unstructured grids were reviewed by Thacker (1980), Ho-Le (1988), Shephard et al. (1988), Baker (1995, 1997), Field (1995), Carey (1997), George and Borouchaki (1998), and Krugljakova et al. (1998). An exhaustive survey of both structured and unstructured techniques has been given by Thompson and Weatherill (1993).

The two fundamental classes of mesh give rise to three additional subdivisions of grid types: block-structured, overset, and hybrid. These kinds of mesh possess to some extent the features of both structured and unstructured grids, thus occupying an intermediate position between the purely structured and unstructured grids.

The multiblock strategy for generating grids around complicated shapes was originally proposed by Lee et al. (1980); however, the idea of using different coordinates in different subregions of the domain can be traced back to Thoman and Szewezyk (1969). Some of the first applications of block-structured grids to the numerical solution of three-dimensional fluid-flow problems in realistic configurations were demonstrated by Rizk and Ben-Shmuel (1985), Sorenson (1986), Atta, Birchelbaw, and Hall (1987), and Belk and Whitfield (1987).

The overset grid approach was introduced by Atta and Vadyak (1982), Berger and Oliger (1983), Benek, Steger, and Dougherty (1983), Miki and Takagi (1984), and Benek, Buning, and Steger (1985). The concept of blocks with a continuous alignment of grid lines across adjacent block boundaries was described by Weatherill and Forsey (1984) and Thompson (1987). Thomas (1982) and Eriksson (1983) applied the concept of continuous line slope, while a discontinuity in slope was discussed by Rubbert and Lee (1982).

A shape recognition technique based on an analysis of a physical domain and an interactive construction of a computational domain with a similar geometry was proposed by Takahashi and Shimizu (1991) and extended by Chiba et al. (1998). The embedding technique was considered by Albone and Joyce (1990) and Albone (1992).

1.3.1 Mapping Approach

The process of grid generation on the physical geometry S^{xn} locally represented by (1.4) is generally carried out by the mapping approach that concludes with finding an intermediate transformation

$$\mathrm{s}(\boldsymbol{\xi}) : \Xi^n \to S^n , \quad \boldsymbol{\xi} = (\xi^1, \dots , \xi^n) \tag{1.9}$$

from a suitable computational (logical) domain Ξ^n to the parametric domain S^n. Consequently the mesh points on S^{xn} are generated as images through

$$\mathbf{x}[\mathrm{s}(\boldsymbol{\xi})] : \Xi^n \to R^{n+k}$$

of the nodes of a reference grid in Ξ^n which can be either structured or unstructured (see Figs. 1.8–1.10).

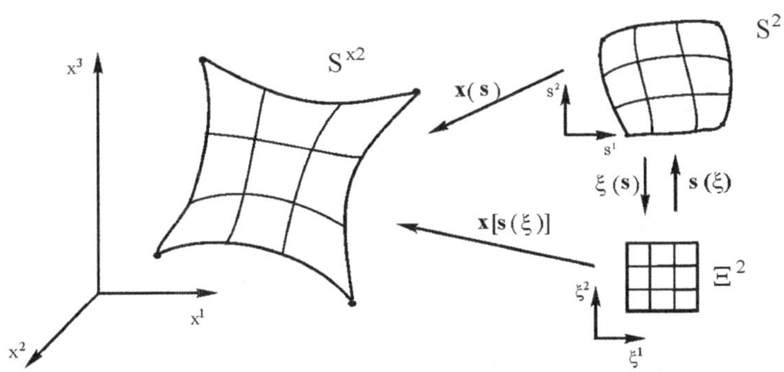

Fig. 1.8. Illustration for quadrilateral grid generation by the mapping approach

Realization of Grid Requirements

The notion of using an intermediate transformation to generate a mesh is very helpful. The idea is to choose a computational domain Ξ^n with a simpler geometry than that of the parametric domain S^n and then to find a transformation $\boldsymbol{s}(\boldsymbol{\xi})$ between these domains which eliminates the need for a complicated mesh when approximating the physical quantities. That is, if the computational area and the transformation are well chosen, the transformed physical boundary value problem can be accurately represented by a small number of mesh points. Emphasis is placed on a small number of points, because any transformed problem (provided only that the transformation is nonsingular) may be accurately approximated with a sufficiently fine, simple mesh. In practice, there will be a trade-off between the difficulty of finding the transformation and the number of grid nodes required to find the solution to a given accuracy.

The idea of using mappings to generate grids is extremely appropriate for finding the conditions that the grid must satisfy for obtaining accurate solutions of partial differential equations in the physical geometry S^{xn}, because these conditions can be readily defined in terms of the transformations. For example, the grid requirements described in Sect. 1.3.2 are readily formulated through the transformation concept.

Since a solution which is a linear function is computed accurately at the grid points and is approximated accurately over the whole region, an attractive possible method for generating grids is to find a transformation $\boldsymbol{s}(\boldsymbol{\xi})$ such that the solution is linear in Ξ^n. Though in practice this requirement for the transformation is not attained even theoretically (except in the case of strongly monotonic univariate functions), it is useful in the sense of an ideal that the developers of grid generation techniques should bear in mind. One modification of this requirement which can be practically realized consists of the requirement of a local linearity of the solution in Ξ^n.

The requirements imposed on the grid and the cell size are realized by the construction of a uniform grid in Ξ^n and a smooth function $\boldsymbol{s}(\boldsymbol{\xi})$. The grid cells are not folded if $\boldsymbol{s}(\boldsymbol{\xi})$ is a one-to-one mapping. Consistency with the geometry is satisfied with a transformation $\boldsymbol{x}(\boldsymbol{\xi})$ that maps the boundary of Ξ^n onto the boundary of S^n. Grid concentration in zones of large variation of a function $\boldsymbol{u}(\boldsymbol{x})$ is accomplished with a mapping $\boldsymbol{x}[\boldsymbol{s}(\boldsymbol{\xi})]$ which provides nearly uniform variations of the function $\boldsymbol{u}[\boldsymbol{x}(\boldsymbol{\xi})]$ at all points of the domain Ξ^n.

Coordinate Grids

Among grids, coordinate grids in which the nodes and cell faces are defined by the intersection of lines and surfaces of a coordinate system in S^{xn} are very popular in finite-difference methods. The range of values of this system defines a computation region Ξ^n in which the cells of the uniform grid are rectangular n-dimensional parallelepipeds, and the coordinate values define the function $\boldsymbol{x}[\boldsymbol{s}(\boldsymbol{\xi})] : \Xi^n \to S^{xn}$.

In the case S^{xn} is a domain $X^n \subset R^n$ the simplest of such grids are the Cartesian grids obtained by the intersection of the Cartesian coordinates in X^n. The cells of these grids are rectangular parallelepipeds (rectangles in two dimensions). The use of Cartesian coordinates avoids the need to transform the physical equations. However, the nodes of the Cartesian grid do not coincide with the curvilinear boundary, which leads to difficulties in implementing the boundary conditions with second-order accuracy.

Boundary-Conforming Grids

An important subdivision of grids is the boundary-fitted or boundary-conforming grids. These grids are obtained from one-to-one transformations $\boldsymbol{s}(\boldsymbol{\xi})$ which map the boundary of the domain Ξ^n onto the boundary of S^n.

The most popular of these, for finite-difference methods, have become the coordinate boundary-fitted grids whose points are formed by intersection of the coordinate lines, while the boundary of S^n is composed of a finite number of coordinate surfaces (lines in two dimensions) $\xi^i = \xi^i_0$. Consequently, in this case the computation region Ξ^n is a rectangular domain, the boundaries of which are determined by $(n-1)$-dimensional coordinate planes in R^n, and the uniform grid in Ξ^n is the Cartesian grid. Thus the physical geometry S^{xn} is represented as a deformation of a rectangular domain and the generated grid as a deformed lattice. These grids give a good approximation to the boundary of S^{xn} and are therefore suitable for the numerical solution of problems with boundary singularities, such as those with boundary layers in which the solution depends very much on the accuracy of the approximation of the boundary conditions.

The requirements imposed on boundary-conforming grids are naturally satisfied with the intermediate coordinate transformations $\boldsymbol{s}(\boldsymbol{\xi})$.

The algorithm for the organization of the nodes of boundary-fitted coordinate grids consists of the trivial identification of neighboring points by incrementing the coordinate indices, while the cells are curvilinear hexahedrons. This kind of grid is very suitable for algorithms with parallelization.

Its design makes it easy to increase or change the number of nodes as required for multigrid methods or in order to estimate the convergence rate and error, and to improve the accuracy of numerical methods for solving boundary value problems.

With boundary-conforming grids there is no necessity to interpolate the boundary conditions of the problem, and the boundary values of the region can be considered as input data to the algorithm, so automatic codes for grid generation can be designed for a wide class of regions and problems.

In the case of unsteady problems the most direct way to set up a moving grid is to do it via a coordinate transformation. These grids do not require a complicated data structure, since they are obtained from uniform grids in

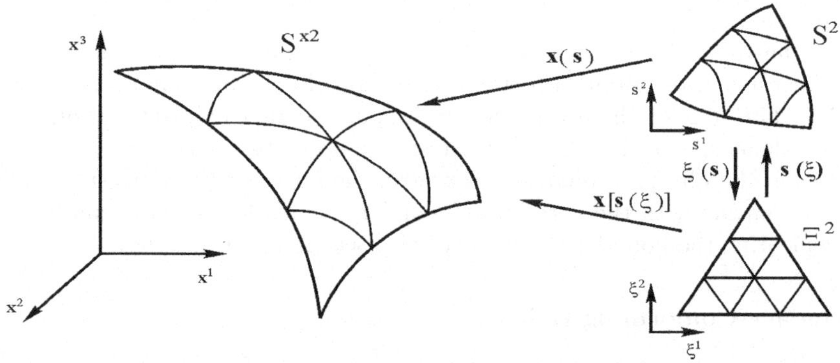

Fig. 1.9. Illustration for triangular grid generation by the mapping approach

simple fixed domains such as rectangular ones, where the grid data structure remains intact.

Shape of Computational Domains

The idea of the mapping approach is to transform a complex physical geometry S^{xn} to a simpler domain Ξ^n with the help of the parametrization $\boldsymbol{x}[\boldsymbol{s}(\boldsymbol{\xi})]$. The region Ξ^n in (1.9), which is called the computational or logical domain, can be either rectangular (Fig. 1.8) or of a different shape matching qualitatively the shape of the physical geometry; in particular, it can be triangular for $n = 2$ (Fig. 1.9) or tetrahedral for $n = 3$. Using such approach, a numerical solution of a partial differential equation in a physical region of arbitrary shape can be carried out in a standard computational domain, and codes can be developed that require only changes in the input.

Shape of a Reference Grid

The cells of the reference grid in the computational domain Ξ^n can be rectangular or of a different shape. Schematic illustration of grid cells is presented in Fig. 1.1. Note that regions in the form of curvilinear triangles, such as that shown in Fig. 1.9, may be more suitable for gridding by triangular cells than by quadrilateral ones.

The triangular grid such as in Fig. 1.9 can also be obtained by mapping a rectangle Ξ^2 onto the triangular parametric domain S^2 provided the reference grid in Ξ^2 is not a structured one but it is obtained from a uniform grid in the triangle T^2 which undergoes a deformation $\boldsymbol{\xi}(\mathbf{t})$ (Fig. 1.10). This deformation is the inverse of the contraction $\boldsymbol{t}(\boldsymbol{\xi})$ of the rectangle Ξ^2 along the horizontal lines to transform it to the triangle T^2.

Analogous scheme with a nonstructured reference grid in the logical domain Ξ^3 having the shape of a three-dimensional rectangular parallelepiped can be applied to the generation of tetrahedral grids.

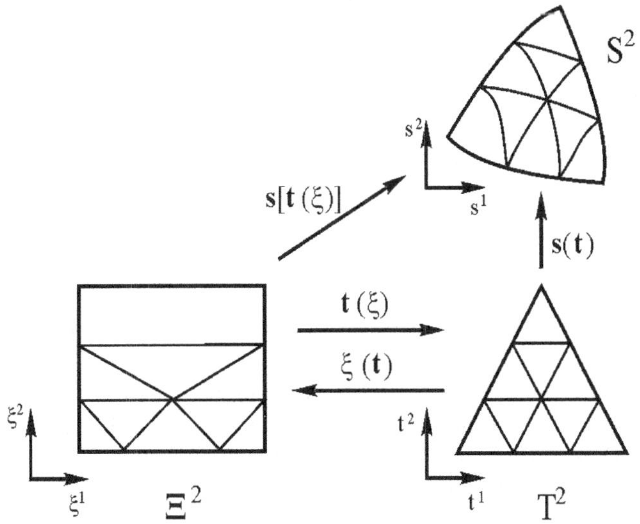

Fig. 1.10. Illustration for generating triangular cells by the mapping approach with a nonstructured reference grid

1.3.2 Requirements Imposed on Mathematical Models

The chief practical difficulty facing methods for gridding general physical geometries is that of formulating satisfactory techniques which can realize the user's requirements. Grid generation technology should develop methods that can help in handling problems with multiple variables, each varying over many orders of magnitude. These methods also should be capable of generating grids whose node displacement is independent of parametrizations of a physical geometry. The methods should incorporate specific control tools, with simple and clear relationships between these control tools and characteristics of the grid such as mesh spacing, skewness, smoothness, and aspect ratio, in order to provide a reliable way to influence the efficiency of the computation. And finally, the methods should be computationally efficient and easy to code. Thus with the mapping approach a mathematical model for generating grids through intermediate transformations between the computational and parametric domains, which is claimed to be a foundation of a robust comprehensive grid generator, should satisfy the following fundamental properties:

1. well-posedness of the mathematical problem formulated by this model for the grid generator;
2. independence of the grid construction of a parametrization of the geometry;
3. malleability to a numerical implementation into an automatic code;
4. existence of a straightforward means for efficient control of the grid quality;
5. ability to obtain in a unified manner the domain and surface grids required in practice.

A number of techniques for generating intermediate transformation have been developed. Every method has its strengths and its weaknesses. Therefore, there is also the question of how to choose the most efficient method for the solution of any specific problem, taking into account the geometrical complexity, the computing cost for generating the grid, the grid structure, and other factors.

The goal of the development of these methods is to provide effective and acceptable grid generation systems.

The most efficient numerical grids are boundary-conforming grids. The generation of these grids can be performed by a number of approaches and techniques. Many of these methods are specifically oriented to the generation of grids for the finite-difference method.

A boundary-fitted coordinate grid in the physical geometry S^{xn} is commonly generated first on the chosen edges of S^{xn}, then on its faces, and finally in its interior. Thus at each step the intermediate transformation $\boldsymbol{s}(\boldsymbol{\xi})$ in the mapping approach is known at the boundary of the corresponding logical domain and this boundary map is extended from the boundary to the domain interior. This process is analogous to the interpolation of a function from a boundary or to the solution of the Dirichlet boundary value problem. On this basis there have been developed three basic groups of methods for finding the intermediate transformation $\boldsymbol{s}(\boldsymbol{\xi}) : \Xi^n \to S^n$ provided there exists a boundary transformation

$$\partial \mathbf{s}(\boldsymbol{\xi}) : \partial \Xi^n \to \partial S^n \ .$$

These methods are

(1) algebraic methods, which use various forms of interpolation or special functions;
(2) differential methods, based mainly on the solution of elliptic, parabolic, and hyperbolic equations in a selected computational domain Ξ^n;
(3) variational methods, based on optimization of grid quality properties.

1.3.3 Algebraic Methods

In the algebraic approach the intermediate transformation $\boldsymbol{s}(\boldsymbol{\xi})$ which extends the boundary mapping $\partial \mathbf{s}(\boldsymbol{\xi}) : \partial \Xi^n \to \partial S^n$ found on the previous step is commonly computed through the formulas of transfinite interpolation. There are two types of transfinite interpolations popular with grid generation: Lagrange and Hermite.

In particular the three-dimensional Lagrange intermediate transformation $\boldsymbol{s}(\boldsymbol{\xi})$ between the interior of the unit cube $0 \le \xi^i \le 1$, $i = 1,2,3$, and the parametric domain S^3 is defined by the following recursive formula

$$\begin{aligned} \boldsymbol{F}_1(\boldsymbol{\xi}) &= \alpha_0^1(\xi^1)\boldsymbol{s}(0,\xi^2,\xi^3) + \alpha_1^1(\xi^1)\boldsymbol{s}(1,\xi^2,\xi^3)\,, \\ \boldsymbol{F}_2(\boldsymbol{\xi}) &= \boldsymbol{F}_1(\boldsymbol{\xi}) + \alpha_0^2(\xi^2)\boldsymbol{s}(\xi^1,0,\xi^3) + \alpha_1^2(\xi^2)\boldsymbol{s}(\xi^1,1,\xi^3)\,, \\ \boldsymbol{s}(\boldsymbol{\xi}) &= \boldsymbol{F}_2(\boldsymbol{\xi}) + \alpha_0^3(\xi^3)\boldsymbol{s}(\xi^1,\xi^2,0) + \alpha_1^3(\xi^3)\boldsymbol{s}(\xi^1,\xi^2,1)\,, \end{aligned} \tag{1.10}$$

where the univariate functions $\alpha_k^i(\xi)$, $i = 1,2,3$, $k = 0,1$, referred to as blending functions, are subject to the relations of consistency

$$\begin{aligned} \alpha_0^i(0) &= 1\,, \quad \alpha_0^i(1) = 0\,, \\ \alpha_1^i(0) &= 0\,, \quad \alpha_1^i(1) = 1\,. \end{aligned}$$

Analogous formulas are held for the Hermite interpolation that matches at the points of the boundary of Ξ^n the values of both the function $\boldsymbol{s}(\boldsymbol{\xi})$ and its first derivatives in the directions normal to the boundary segments. A detailed review of the Lagrange and Hermit techniques for generating algebraic grids is presented in the monograph of Liseikin (1999).

Algebraic methods are simple; they enable the grid to be generated rapidly and the spacing and slope of the coordinate lines to be controlled by the tangential derivatives at the boundary points and blending coefficients in the transfinite interpolation formulas. However, in regions of complicated shape the coordinate surfaces obtained by algebraic methods can become degenerate or the cells can overlap or cross the boundary. Moreover, they basically preserve the features of the boundary surfaces, in particular, discontinuties. Besides this the algebraic methods of transfinite interpolation do not guarantee the independence of grid nodes displacement on parametrizations of a physical geometry.

Algebraic approaches are commonly used to generate grids in regions with smooth boundaries that are not highly deformed, or as an initial approximation in order to start the iterative process of an elliptic grid solver.

The construction of intermediate transformations through the formulas of transfinite interpolation was originally formulated by Gordon and Hall (1973) and Gordon and Thiel (1982). The Hermite interpolation was presented by Smith (1982).

The multisurface method which allows for the specification of the intermediate transformation $\mathbf{s}(\boldsymbol{\xi})$ at the points of some interior sections of the physical domain was described by Eiseman (1980). The blending functions were implicitly derived from global and/or local interpolants which result from an expression for the tangential derivative spanning between the exterior boundary surfaces. A two-boundary technique was introduced by Smith

(1981). It is based on the description of two opposite boundary surfaces, tangential derivatives on the boundary surfaces which are used to compute normal derivatives, and Hermite cubic blending functions.

The construction of some special blending functions aimed at grid clustering at the boundaries of physical geometries was performed by Eriksson (1982), Smith and Eriksson (1987), and Liseikin (1998a,b). A detailed description of various forms of blending functions was presented in monographs by Thompson, Warsi, and Mastin (1985) and Liseikin (1999).

1.3.4 Differential Methods

For gridding geometries with arbitrary boundaries, differential methods based on the solution of elliptic and parabolic equations are commonly used. Such equations generate smooth grids, allow for full specification of grid nodes on the boundary of a physical geometry, does not propagate boundary singularities into its interior, have less danger of producing cell overlapping, and can be solved efficiently using many well-developed codes. The use of parabolic and elliptic systems enables orthogonal and clustering coordinate lines to be constructed, while, in many cases, the maximum principle, which is typical for these systems, ensures that the intermediate transformations are nondegenerate. Elliptic equations are also used to smooth algebraic or unstructured grids.

Elliptic Equations

Originally the most popular elliptic equations with differential grid approaches were the generalized Poisson equations formulated with respect to the components $\xi^i(\mathbf{s})$ of the transformation

$$\boldsymbol{\xi}(\boldsymbol{s}) : S^n \to \Xi^n \tag{1.11}$$

that is the inverse of the intermediate transformation (1.9). The equations for generating grids of the physical geometry S^{xn} include coefficients defined by the elements

$$g_{ij}^{xs} = \boldsymbol{x}_{s^i} \cdot \boldsymbol{x}_{s^j} \ , \quad i,j = 1,\dots,n \ ,$$

of the covariant metric tensor of S^{xn} in the parametric coordinates $s^1,\dots,s^n$. A general form of these generalized Poisson equations is as follows:

$$\Delta_B[\xi^i] = P^i(\boldsymbol{s}) \ , \quad i = 1,\dots,n \ . \tag{1.12}$$

Here Δ_B is the operator of Beltrami defined at a function $f(\boldsymbol{s})$ by the formula

$$\Delta_B[f] = \frac{1}{\sqrt{g^{xs}}} \sum_{j=1}^{n} \frac{\partial}{\partial s^j} \left(\sqrt{g^{xs}} \sum_{k=1}^{n} g_{sx}^{jk} \frac{\partial f(\mathrm{s})}{\partial s^k} \right) , \tag{1.13}$$

where $g^{xs} = \det(g^{xs}_{ij})$, while g^{jk}_{sx} is the (jk)th element of the contravariant metric tensor of S^{xn} in the parametric coordinates $s^1, \dots, s^n$. The elements g^{ij}_{sx}, $i,j = 1, \dots, n$, comprise a matrix that is the inverse of the matrix formed by the elements g^{xs}_{ij}, $i,j = 1, \dots, n$. The terms $P^i(\boldsymbol{s})$ in (1.12) referred to as source terms or control functions are introduced to control the grid behavior.

A particular form of (1.13) for generating grids in a domain $S^n \subset R^n$ is given by the Poisson equations

$$\sum_{j=1}^{n} \frac{\partial^2 \xi^i}{\partial s^j \partial s^j} = P^i(\boldsymbol{s}) \, , \quad i = 1, \dots, n \, . \tag{1.14}$$

The intermediate transformation $\boldsymbol{s}(\boldsymbol{\xi}) = [s^1(\boldsymbol{\xi}), \dots, s^n(\boldsymbol{\xi})]$ for generating grids on a physical geometry S^{xn} is found from the solution of the Dirichlet boundary value problem for the transformed equations obtained from (1.12) by changing mutually dependent and independent variables. These equations are of the form

$$\sum_{i,j=1}^{n} g^{ij}_{\xi x} \frac{\partial^2 s^k}{\partial \xi^i \partial \xi^j} = \Delta_B[s^k] - \sum_{i=1}^{n} P^i \frac{\partial s^k}{\partial \xi^i} \, , \quad k = 1, \dots, n \, , \tag{1.15}$$

where $g^{ij}_{\xi x}$ is the (ij)th element of the contravariant metric tensor of S^{xn} in the grid coordinates $\xi^1, \dots, \xi^n$.

A two-dimensional Laplace system which implied the parametric coordinates to be solutions in the logical domain Ξ^2 was introduced by Godunov and Prokopov (1967), Barfield (1970), and Amsden and Hirt (1973). A general two-dimensional elliptic system for generating structured grids was considered by Chu (1971). A two-dimensional system (1.14) with $P^i(\mathrm{s}) \equiv 0$ using the logical coordinates ξ^i as dependent variables was proposed by Crowley (1962) and Winslow (1967).

Godunov and Prokopov (1972) obtained a system of the Poisson type (1.14) assuming that its solution is a composition of conformal and stretching transformations. The general Poisson system (1.14) was justified by Thompson, Thames, and Mastin (1974) and Tompson, Warsi, and Mastin (1985) in their monograph.

The algorithm aimed at grid clustering at a boundary and forcing grid lines to intersect the boundary in a nearly normal fashion through the source terms of the Poisson system (1.14) was developed by Steger and Sorenson (1979), Visbal and Knight (1982), and White (1990). Thomas and Middlecoff (1980) described a procedure to control the local angle of intersection between transverse grid lines and the boundary through the specification of the control functions. Control of grid spacing and orthogonality was performed by Tamamidis and Assanis (1991) by introducing a distortion function (the ratio of the diagonal metric elements) into the system of Poisson equations. Warsi (1982) replaced the source terms P^i in (1.14) by $g^{ii}P^i$ (i fixed) to improve the numerical behavior of the grid generator. As a result the modified system acquired the property of satisfying the maximum principle.

The technique based on setting to zero the off-diagonal elements of the elliptic system (1.15) was proposed by Lin and Shaw (1991) to generate nearly orthogonal grids.

The use of generalized Laplace equations to generate surface grids was proposed by Warsi (1981), in analogy with the widely utilized Laplace grid generator of Crowley (1962) and Winslow (1967). Warsi (1990) has also justified these equations by using some fundamental results of differential geometry.

A surface grid generation scheme that uses a quasi-two-dimensional elliptic system, obtained by projecting the inverted three-dimensional Laplace system, to generate grids on smooth surfaces analytically specified by the equation $z = f(x, y)$ was proposed by Thomas (1982). The method was extended and updated by Takagi et al. (1985) and Warsi (1986) for arbitrary curved surfaces using a parametric surface representation. An adaptive surface grid technique based on control functions in (1.12) and parametric specifications was also considered by Lee and Loellbach (1989).

Since 1991 a new elliptic approach for controlling grid properties is being developed. By this approach the task of grid adaptation, instead of source terms $P^i(\mathbf{s})$ in (1.12), is put on monitor metrics in Beltrami equations. Namely, as an elliptic model there are used the Beltrami equations

$$\Delta_B[\xi^i] = 0, \quad i = 1, \dots, n, \tag{1.16}$$

in a monitor metric, where the operator Δ_B is of the form (1.13), however the contravariant metric elements are not obliged to be the elements of the physical geometry S^{xn}. The equations (1.16) are the Euler-Lagrange equations for the functional of energy. The solutions of these equations are referred to as harmonic transformations.

Liseikin (1991a, 1992) used the elliptic system (1.16) derived from a variational principle to produce n-dimensional harmonic coordinate transformations which generate both uniform and adaptive grids on surfaces. The harmonic mapping approach was also used by Arina and Casella (1991) to derive a surface elliptic system. The conformal mapping technique for generating surface grids was formulated by Khamayseh and Mastin (1996). In the papers of Liseikin (2001b, 2002a,b, 2003, 2005) there have been designed monitor metrics which provide the generation through the system (1.16) both adaptive, field-aligned, and balanced numerical grids.

Hyperbolic Equations

The most known hyperbolic equations are the first order partial differential equations of the Cauchy–Riemann type. In practice, two-dimensional hyperbolic equations with respect to the intermediate transformation $\mathbf{s}(\boldsymbol{\xi})$ have the following form

$$A\mathbf{s}_{\xi^1} + B\mathbf{s}_{\xi^2} = \mathbf{f}, \tag{1.17}$$

where A and B some matrices. These equations are simpler then nonlinear elliptic equations (1.15) and enable marching methods to be used and an orthogonal system of coordinates to be constructed, while grid adaptation can be performed using the coefficients of the equations. However, methods based on the solution of hyperbolic equations are not always mathematically correct and they are not applicable to regions in which the complete boundary node distribution is strictly specified. Therefore hyperbolic methods are mainly used for simple regions which have several lateral faces for which no special nodal distribution is required. The marching procedure for the solution of hyperbolic equations allows one to decompose only the boundary geometry in such a way that neighboring boundary grids overlap.

The first systematic analysis of the use of two-dimensional hyperbolic equations to generate grids was made by Starius (1977) and Steger and Chaussee (1980), although hyperbolic grid generation can be traced back to McNally (1972). This system was generalized by Cordova and Barth (1988). They developed a two-dimensional hyperbolic system with an angle-control source term which allows one to constrain a grid with more than one boundary. A combination of grids using the hyperbolic technique of Steger and Chaussee (1980), which starts from each boundary segment was generated by Jeng and Shu (1995). The extension to three dimensions was performed by Steger and Rizk (1985), Chan and Steger (1992), and Tai, Chiang, and Su (1996), who introduced grid smoothing as well.

The hyperbolic approach based on grid orthogonality was extended to surfaces by Steger (1991). An analogous technique for generating overset surface grids was presented by Chan and Buning (1995).

Parabolic Equations

The parabolic grid approach lies between the elliptic and hyperbolic ones.

The two-dimensional parabolic grid generation equation where the marching direction is ξ^2 may be written in the following form:

$$\mathbf{s}_{\xi^2} = A_1 \mathbf{s}_{\xi^1\xi^1} - B_1 \mathbf{s} + \mathbf{P} , \tag{1.18}$$

where A_1 and B_1 are matrix coefficients, and $\mathbf{P}$ is a source vector-valued function that contains the information about the outer boundary configuration. Analogously, the three-dimensional parabolic equations may be written as follows:

$$\mathbf{s}_{\xi^3} = \sum_{i=1}^{2} A_i \mathbf{s}_{\xi^i\xi^i} - B_1 \mathbf{s} + \mathbf{P} . \tag{1.19}$$

The generation of grids based on a parabolic scheme approximating the inverted Poisson equations was first proposed for two-dimensional grids by Nakamura (1982). A variation of the method of Nakamura was developed by Noack (1985) to use in space-marching solutions to the Euler equations. Extensions of this parabolic technique to generate solution adaptive grids were performed by Edwards (1985) and Noack and Anderson (1990).

Hybrid Grid Generation Scheme

The combination of the hyperbolic and parabolic schemes into a single scheme is attractive because it can use the advantages of both schemes. These advantages are; first, it is a noniterative scheme; second, the orthogonality of the grid near the initial boundary is well controlled; and third, the outer boundary can be prescribed.

A hybrid grid generation scheme in two dimensions for the particular marching direction ξ^2 can be derived by combining hyperbolic and parabolic equations, in particular, as the sum of equations (1.17) multiplied by B^{-1} and (1.18) with weights α and $1-\alpha$, respectively:

$$\alpha(B^{-1}A\mathbf{s}_{\xi^1} + \mathbf{s}_{\xi^2}) + (1-\alpha)(\mathbf{s}_{\xi^2} - A_1\mathbf{s}_{\xi^1\xi^1} + B_1\mathbf{s})$$

$$= \alpha B^{-1}\boldsymbol{f} + (1-\alpha)\mathbf{P}\,. \tag{1.20}$$

The parameter α can be changed as desired to control the proportions of the two methods. If α approaches 1, the system (1.20) becomes the hyperbolic grid system, while if α approaches zero it becomes the parabolic grid system. In practical applications α is set to 1, when the grid generation starts from the initial boundary curve $\xi^2 = 0$, but it gradually decreases and approaches zero when the grid reaches the outer boundary.

An analogous combination can be used to generate three-dimensional grids through a hybrid of parabolic and hyperbolic equations.

A combination of hyperbolic and parabolic schemes that uses the advantages of the two but eliminates the drawbacks of each was proposed by Nakamura and Suzuki (1987).

1.3.5 Variational Methods

Variational methods are widely used to generate grids which are required to satisfy several critical properties, e. g., mesh concentration in areas needing high resolution of the physical solution, mesh alignment to some prescribed vector fields, mesh nondegeneracy, smoothness, uniformity, and near-orthogonality that cannot be realized simultaneously with algebraic or differential techniques. Variational methods take into account the conditions imposed on the grid by constructing special functionals defined on a set of smooth or discrete transformations. A compromise grid, with properties close to those required, is obtained with the optimum transformation for a combination of these functionals.

The major task of the variational approach to grid generation is to describe all basic measures of the desired grid features in an appropriate functional form and to formulate a combined functional that provides a well-posed minimization problem. These functionals can provide mathematical feedback in an automatic grid procedure.

Commonly, in the calculus of variations, any functional over some admissible set of functions $\boldsymbol{f} : D^n \to R^m$ is defined by the integral

$$I[\boldsymbol{f}] = \int_{D^n} G(\boldsymbol{f}) \mathrm{d}V \, , \tag{1.21}$$

where D^n is a bounded n-dimensional domain, and $G(\boldsymbol{f})$ is some operator specifying, for each vector-valued function $\boldsymbol{f} : D^n \to R^m$, a scalar function $G(\boldsymbol{f}) : D^n \to R$. The admissible set is composed of those functions $\boldsymbol{f}$ which satisfy a prescribed boundary condition

$$\boldsymbol{f} \mid_{\partial D^n} = \boldsymbol{\phi}$$

and for which the integral (1.21) is limited.

In the application of the calculus of variations to grid generation this set of admissible functions is a set of sufficiently smooth invertible coordinate transformations (1.11) between the parametric domain S^n and the computational domain Ξ^n or, vice versa, a set of sufficiently smooth invertible intermediate transformations (1.9) from the computational domain Ξ^n onto the parametric region S^n. The integral (1.21) is defined over the domain S^n or Ξ^n, respectively.

In grid generation applications the operator G is commonly chosen as a combination of weighted local grid characteristics which are to be optimized. The choice depends, of course, on what is expected from the grid. Some forms of the weight functions and both local and integral grid characteristics were formulated in a monograph of Liseikin (1999) through the transformations (1.9) or (1.11) and their first and second derivatives. Therefore, for the purpose of grid generation, it can be supposed that the most widely acceptable formula for the operator G in (1.21) is one which is derived from some expressions containing the first and second derivatives of the coordinate transformations. Thus it is generally assumed that the functional (1.21), depending on the transformation $\boldsymbol{\xi}(\boldsymbol{s})$, is of the form

$$I[\boldsymbol{\xi}] = \int_{S^n} G(\mathrm{s}, \boldsymbol{\xi}, \boldsymbol{\xi}_{s^i}, \boldsymbol{\xi}_{s^i s^j}) \mathrm{d}\mathrm{s} \, ,$$

where G is a smooth function of its variables.

Analogously, the functional (1.21) formulated over a set of invertible intermediate transformations $\mathrm{s}(\boldsymbol{\xi})$ has the form

$$I[\mathrm{s}] = \int_{\Xi^n} G_1(\boldsymbol{\xi}, \mathrm{s}, \mathrm{s}_{\xi^i}, \mathrm{s}_{\xi^i \xi^j}) \mathrm{d}\boldsymbol{\xi} \, .$$

In one popular approach the functional formulated with respect to the intermediate mapping $\mathrm{s}(\boldsymbol{\xi})$ has the following form

$$I[\mathrm{s}] = \int_{\Xi^n} (\sqrt{g^{m\xi}} \sum_{i,j=1}^{n} g^{ij}_{\xi m} g^{s\xi}_{ij}) \mathrm{d}\boldsymbol{\xi} \, , \tag{1.22}$$

where $g_{\xi m}^{ij}$, $i,j = 1,\dots,n$, are the elements of the contravariant tensor in the logical coordinates $\xi^1,\dots,\xi^n$ of a monitor metric $g_{ij}^{m\xi}$ imposed on Ξ^n, $g^{m\xi} = \det(g_{ij}^{m\xi})$, while $g_{ij}^{s\xi}$ is the covariant Euclidean metric tensor of S^n in the coordinates $\xi^1,\dots,\xi^n$. This functional was proposed for $n = 2$ by Godunov and Prokopov (1967) for generating conformal and quasi-conformal grids in S^2. In their consideration the elements $g_{ij}^{m\xi}$, $i,j = 1,2$, of the monitor metric should be dependent on $\boldsymbol{\xi}$ and some, in general vector-valued parameter $\mathbf{r}$. Belinskii et al. (1975) and Godunov, Romenskii, and Chumakov (1990) discussed the same two-dimensional functional of the form (1.22) with the following monitor metric introduced in Ξ^2

$$g_{ij}^{m\xi} = \begin{pmatrix} e^{2p(\boldsymbol{\xi})} & e^{p(\boldsymbol{\xi})+q(\boldsymbol{\xi})}\cos[\alpha(\boldsymbol{\xi}) - \beta(\boldsymbol{\xi})] \\ e^{p(\boldsymbol{\xi})+q(\boldsymbol{\xi})}\cos[\alpha(\boldsymbol{\xi}) - \beta(\boldsymbol{\xi})] & e^{2q(\boldsymbol{\xi})} \end{pmatrix},$$

where the functions $p(\boldsymbol{\xi})$, $q(\boldsymbol{\xi})$, $\alpha(\boldsymbol{\xi})$, and $\beta(\boldsymbol{\xi})$ are subject to the restrictions

$$p(\boldsymbol{\xi}) - q(\boldsymbol{\xi}) = \ln\sqrt{g_{11}^{s\xi}/g_{22}^{s\xi}}\,,$$

$$\alpha(\boldsymbol{\xi}) - \beta(\boldsymbol{\xi}) = \arccos\Big(g_{12}^{s\xi}/\sqrt{g_{11}^{s\xi}g_{22}^{s\xi}}\Big)\,.$$

The grid approach based on the minimization of the functional (1.22) for $n = 2$ was also used by Chumakov and Chumakov (1998) for generating quasi-isometric grids by introducing in Ξ^2 a monitor metric borrowed from the metric of a surface of a constant Gauss curvature.

Note the functional (1.22) is twice the energy functional of the function $\mathbf{s}(\boldsymbol{\xi}) : \Xi^n \to S^n$ where Ξ^n is endowed by the monitor metric $g_{ij}^{m\xi}$, while S^n has the Euclidean metric.

The twice energy functional of the function $\boldsymbol{\xi}(\mathbf{s}) : S^n \to \Xi^n$ between S^n with an imposed monitor metric $g_{ij}^{\mathbf{s}}$ and Ξ^n with the Euclidean metric for generating adaptive grids was considered by Dvinsky (1991) and Liseikin (1991a). This functional has the following form

$$I[\boldsymbol{\xi}] = \int_{S^n}\Big(\sqrt{g^{\mathbf{s}}}\sum_{i,j,k=1}^{n} g_{\mathbf{s}}^{ij}\frac{\partial\xi^k}{\partial s^i}\frac{\partial\xi^k}{\partial s^j}\Big)d\mathbf{s}\,, \tag{1.23}$$

where $g^{\mathbf{s}} = \det(g_{ij}^{\mathbf{s}})$, $g_{\mathbf{s}}^{ij}$, $i,j = 1,\dots,n$, are the elements of the contravariant monitor metric tensor introduced in S^n. The Euler-Lagrange equations for the functional (1.23) are equivalent to the Beltrami equations (1.16).

The functionals are used to control and realize various grid properties. This is carried out by combining these functionals with weights in the form

$$I = \sum_i \lambda_i I_i\,, \qquad i = 1,\cdots,k\,. \tag{1.24}$$

Here λ_i, $i = 1, \cdots, k$, are specified parameters which determine the individual contribution of each functional I_i to I. The ranges of the parameters λ_i controlling the relative contributions of the functionals can be defined readily when the functionals I_i are dimensionally homogeneous. However, if they are dimensionally inhomogeneous, then the selection of a suitable value for λ_i presents some difficulties. A common rule for selecting the parameters λ_i involves making each component $\lambda_i I_i$ in (1.24) of a similar scale by using a dimensional analysis.

The most common practice in forming the combination (1.24) uses both the functionals of adaptation to the physical solution and the functionals of grid regularization. The first reason for using such a strategy is connected with the fact that the process of adaptation can excessively distort the form of the grid cells. The distortion can be prevented by functionals which impede cell deformation. These functionals are ones which control grid skewness, smoothness, and conformality. The second reason for using the regularization functionals is connected with the natural requirement for the well-posedness of the grid generation process. This requirement is achieved by the utilization of convex functionals in variational grid generators. The convex functionals are represented by energy-type functionals (1.23) producing harmonic maps and by the functionals of conformality.

The various functionals provide broad opportunities to control and realize the required grid properties, though problems still remain; these require more detailed studies of all properties of the functionals. The knowledge of these properties will allow one to utilize the functionals as efficient tools to generate high-quality grids.

Liseikin and Yanenko (1977), Danaev, Liseikin, and Yanenko (1978), Ghia, Ghia, and Shin (1983), Brackbill and Saltzman (1982), Bell and Shubin (1983), Huang, Ren, and Russell (1994), and Huang (2001) have each used the variational principle for grid adaptation. The variational formulation of grid properties was described by Warsi and Thompson (1990).

The functional measuring the alignment of the two-dimensional grid with a specified vector field was formulated by Giannakopoulos and Engel (1988). The extension of this approach to three dimensions was discussed by Brackbill (1993). A variational method optimizing cell aspect ratios was presented and analyzed by Mastin (1992). A dimensionally homogeneous functional of two-dimensional grid skewness was proposed by Steinberg and Roache (1986).

The introduction of the volume-weighted functional was originally proposed in two dimensions by Yanenko, Danaev, and Liseikin (1977).

The approach of determining functionals which depend on invariants of orthogonal transformations of the metric tensor $g_{ij}^{s\xi}$, to ensure that the problems are well-posed and to obtain more compact formulas for the Euler–Lagrange equations, was proposed by Jacquotte (1987). In his paper, the grids were constructed through functionals obtained by modeling different elastic and plastic properties of a deformed body.

The possibility of using harmonic function theory to provide a general framework for developing multidimensional mesh generators was discussed by Dvinsky (1991) and Liao (1991). A survey of the theory of harmonic mappings was published by Eells and Lenaire (1988). The interpretation of the functional of diffusion as a version of the energy functional was presented by Brackbill (1993). A detailed description of the properties of the functional (1.23) and the application of its Beltrami equations (1.16) to the grid development was presented in the monograph of Liseikin (2004).

1.4 Comprehensive Codes

A comprehensive grid generation code is an effective system for generating structured and unstructured grids, as well as hybrid and overset combinations, in arbitrary physical geometries. The development of such codes is a considerable problem in its own right. The present comprehensive grid generation codes developed for the solution of multidimensional problems have to incorporate combinations of block-structured, hybrid, and overset grid methods and are still rather cumbersome, rely on interactive tools, and take too many man-hours to generate a complicated grid. Efforts to increase the efficiency and productivity of these codes are mainly being conducted in two interconnected research areas.

The first, the "array area", is concerned with the automation of those routine processes of grid generation which require interactive tools and a great deal of human time and effort. Some of these are:

(1) the decomposition of a domain into a set of contiguous or overlapping blocks consistent with the distinctive features of the domain geometry, the singularities of the physical medium and the sought-for solution, and the computer architecture;
(2) numbering the set of blocks, their faces, and their edges with a connectivity hierarchy and determining the order in which the grids are constructed in the blocks and their boundaries;
(3) choosing the grid topology and the requirements placed on the qualitative and quantitative characteristics of the internal and boundary grids and on their communication between the blocks;
(4) selecting appropriate methods to satisfy the requirements put on the grid in accordance with a particular geometry and solution;
(5) assessment and enhancement of grid quality.

The second, more traditional, "methods area" deals with developing new, more reliable, and more elaborate methods for generating, adapting, and smoothing grids in domains in a unified manner, irrespective of the geometry of the domain or surface and of the qualitative and quantitative characteristics the grids should possess, so that these methods, when incorporated in the comprehensive codes, should ease the bottlenecks of the array area, in particular, by enabling a considerable reduction of the number of blocks required. The contents of the current monograph is aimed at advocating one such method based on the use of the Beltramian operator with respect to monitor metrics and the theory of multidimensional differential geometry. This method is a natural extention of the approaches proposed by Crowley (1962), Winslow (1967), Godunov and Prokopov (1967), Warsi (1981), Dvinski (1991), and Liseikin (1991). Some recent results of the method have been presented in the papers of Liseikin (2001b, 2002a, 2002b, 2003, 2005).

The overall purpose of the development of the comprehensive grid generation codes is to create a system which enables one to generate grids in a "black box" mode without or with only a slight human interaction. Currently, however, the user has to take active role and be occupied in the grid generation process. The user has to make conclusions about qualitative properties of the grid and undertake corrective measures when necessary. The present codes include significant measures to increase the productivity of such human activity, namely, graphical interactive systems and user-friendly interfaces. Efforts to eliminate the "human component" of the codes are directed towards developing new techniques, in particular, new grid generation methods and automated block decomposition techniques.

The first comprehensive grid codes were described by Holcomb (1987), Thompson (1987), Thomas, Bache, and Blumenthal (1990), Widhopf et al. (1990), and Steinbrenner, Chawner, and Fouts (1990). These codes have stimulated the development of updated ones, reviewed by Thompson (1996). This paper also describes the current domain decomposition techniques developed by Shaw and Weatherill (1992), Stewart (1992), Dannenhoffer (1995), Wulf and Akrag (1995), Schonfeld, Weinerfelt, and Jenssen (1995), and Kim and Eberhardt (1995). The first attempts to overcome the problem of domain decomposition were discussed by Andrews (1988), Georgala and Shaw (1989), Allwright (1989), and Vogel (1990).

2 General Coordinate Systems in Domains

We consider here some notions and relations connected with smooth invertible coordinate transformations of the physical region $X^n \subset R^n$ from the parametric domain $\Xi^n \subset R^n$:

$$\mathbf{x}(\boldsymbol{\xi}) : \Xi^n \to X^n \ , \quad \boldsymbol{\xi} = (\xi^1, \cdots, \xi^n) \ , \quad \mathbf{x} = (x^1, \cdots, x^n) \ .$$

If Ξ^n is a standard logical domain then, in accordance with Chap. 1, such coordinate transformations are used to generate grids in X^n. Here and later R^n presents the Euclidean space with the Cartesian basis of an orthonormal system of vectors $\mathbf{e}_1, \cdots, \mathbf{e}_n$, i.e.

$$\mathbf{e}_i \cdot \mathbf{e}_j = \delta^i_j \ , \quad i, j = 1, \ldots, n \ ,$$

where δ^i_j is the Kronecker symbol:

$$\delta^i_j = 0 \quad \text{if} \quad i \neq j \ , \quad \delta^i_j = 1 \quad \text{if} \quad i = j \ .$$

Thus the position of a point $\mathbf{x}$ in R^n is determined unequivocally by the expansion

$$\mathbf{x} = x^1 \mathbf{e}_1 + \cdots + x^n \mathbf{e}_n \ .$$

The values x^i, $i = 1, \cdots, n$, are called the Cartesian coordinates of the point $\mathbf{x}$. Each coordinate transformation $\mathbf{x}(\boldsymbol{\xi})$ defines, in the domain X^n, new coordinates $\xi^1, \cdots, \xi^n$ which are called the curvilinear coordinates.

2.1 Jacobi Matrix

The matrix

$$J = \left(\frac{\partial x^i}{\partial \xi^j}\right) , \quad i, j = 1, \cdots, n \ ,$$

is referred to as the Jacobi matrix, and its Jacobian is designated by J:

$$J = \det\left(\frac{\partial x^i}{\partial \xi^j}\right) , \quad i, j = 1, \cdots, n \ .$$

The inverse transformation to the coordinate mapping $\mathbf{x}(\boldsymbol{\xi})$ is denoted by

$$\boldsymbol{\xi}(\mathbf{x}) : X^n \to \Xi^n\ , \quad \boldsymbol{\xi}(\mathbf{x}) = (\xi^1(\mathbf{x}), \dots, \xi^n(\mathbf{x}))\ .$$

This transformation can be considered analogously as a mapping introducing a curvilinear coordinate system $x^1, \cdots, x^n$ in the domain $\Xi^n \subset R^n$. It is obvious that the inverse to the matrix $\jmath$ is

$$\jmath^{-1} = \left(\frac{\partial \xi^i}{\partial x^j}\right)\ , \quad i, j = 1, \cdots, n\ ,$$

and consequently

$$\det\left(\frac{\partial \xi^i}{\partial x^j}\right) = \frac{1}{J}\ , \quad i, j = 1, \cdots, n\ .$$

In the case of two-dimensional space the elements of the matrices $(\partial x^i/\partial \xi^j)$ and $(\partial \xi^i/\partial x^j)$ are connected by

$$\frac{\partial \xi^i}{\partial x^j} = (-1)^{i+j} \frac{\partial x^{3-j}}{\partial \xi^{3-i}} \Big/ J\ , \quad i, j = 1, 2\ ,$$

$$\frac{\partial x^i}{\partial \xi^j} = (-1)^{i+j} J \frac{\partial \xi^{3-j}}{\partial x^{3-i}}\ , \quad i, j = 1, 2\ . \tag{2.1}$$

Similar relations between the elements of the corresponding three-dimensional matrices have the form

$$\frac{\partial \xi^i}{\partial x^j} = \frac{1}{J}\left(\frac{\partial x^{j+1}}{\partial \xi^{i+1}} \frac{\partial x^{j+2}}{\partial \xi^{i+2}} - \frac{\partial x^{j+1}}{\partial \xi^{i+2}} \frac{\partial x^{j+2}}{\partial \xi^{i+1}}\right)\ , \quad i, j = 1, 2, 3\ ,$$

$$\frac{\partial x^i}{\partial \xi^j} = J\left(\frac{\partial \xi^{j+1}}{\partial x^{i+1}} \frac{\partial \xi^{j+2}}{\partial x^{i+2}} - \frac{\partial \xi^{j+1}}{\partial x^{i+2}} \frac{\partial \xi^{j+2}}{\partial x^{i+1}}\right)\ , \quad i, j = 1, 2, 3\ , \tag{2.2}$$

where for each superscript or subscript index, say l, $l \pm 3$ is equivalent to l. With this condition the sequence of indices $(l, l+1, l+2)$ is a cyclic permutation of $(1, 2, 3)$ and vice versa; the indices of a cyclic sequence (i, j, k) satisfy the relation $j = i + 1,\ k = i + 2$.

2.2 Coordinate Lines, Tangential Vectors, and Grid Cells

The value of the function $\mathbf{x}(\boldsymbol{\xi}) = [x^1(\boldsymbol{\xi}), \dots, x^n(\boldsymbol{\xi})]$ in the Cartesian basis $\mathbf{e}_1, \cdots, \mathbf{e}_n$, i.e.

$$\mathbf{x}(\boldsymbol{\xi}) = x^1(\boldsymbol{\xi})\mathbf{e}_1 + \ldots + x^n(\boldsymbol{\xi})\mathbf{e}_n\ , \quad \boldsymbol{\xi} = (\xi^1, \dots, \xi^n)\ , \tag{2.3}$$

is a position vector for every $\boldsymbol{\xi} \in \Xi^n$. If one variable ξ^i varies and the others ξ^j, $j \neq i$, are kept constant then the function (2.3) depends upon a single

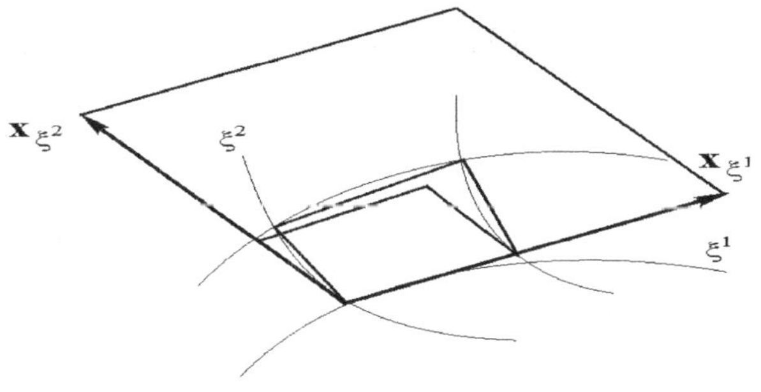

Fig. 2.1. Basic and contracted parallelograms and corresponding grid cell

parameter and therefore describes a curve. This curve is referred to as the ξ^i curvilinear coordinate line.

The vector-valued function $\mathbf{x}(\boldsymbol{\xi})$ generates the nodes, edges, faces, etc. of the cells of the coordinate grid in the domain X^n. Each edge of the cell corresponds to a coordinate line ξ^i for some i and is defined by the vector

$$\Delta_i \mathbf{x} = \mathbf{x}(\boldsymbol{\xi} + h_i \mathbf{e}_i) - \mathbf{x}(\boldsymbol{\xi}) ,$$

where h_i is the step size of the uniform grid in the ξ^i direction in the logical domain Ξ^n. We have

$$\Delta_i \mathbf{x} = h_i \mathbf{x}_{\xi^i} + \mathbf{t} ,$$

where

$$\mathbf{x}_{\xi^i} = \left(\frac{\partial x^1}{\partial \xi^i}, \cdots, \frac{\partial x^n}{\partial \xi^i}\right) \tag{2.4}$$

is the vector tangential to the coordinate curve ξ^i, and $\mathbf{t}$ is a residual vector whose length does not exceed the following quantity:

$$\frac{1}{2} \max |\mathbf{x}_{\xi^i \xi^i}| (h_i)^2 , \quad i = 1, \ldots, n .$$

So the cells in the domain X^n whose edges are formed by the vectors $h_i \mathbf{x}_{\xi^i}$, $i = 1, \cdots, n$, are approximately the same as those obtained by mapping the uniform coordinate cells in the computational domain Ξ^n with the transformation $\mathbf{x}(\boldsymbol{\xi})$. The tangential vectors $\mathbf{x}_{\xi^i}$, $i = 1, \cdots, n$, form a parallelepiped (parallelogram when n=2) referred to as a basic parallelepiped. Thus the uniformly contracted basic parallelepiped spanned by the tangential vectors $\mathbf{x}_{\xi^i}$, $i = 1, \cdots, n$, represents to a high order of accuracy with respect to h_i the cell of the coordinate grid at the corresponding point in X^n (see Fig. 2.1 for $n = 2$). In particular, for the length l_i of the ith grid edge we have

$$l_i = h_i |\mathbf{x}_{\xi^i}| + O(h_i^2) , \quad i = 1, \ldots, n .$$

The volume V_h (area in two dimensions) of the cell is expressed as follows:

$$V_h = \prod_{i=1}^{n} h_i V + O(\prod_{i=1}^{n} h_i \sum_{j=1}^{n} h_j) ,$$

where V is the volume of the n-dimensional basic parallelepiped determined by the tangential vectors $\mathbf{x}_{\xi^i}$, $i = 1, \cdots, n$.

The tangential vectors $\mathbf{x}_{\xi^i}$, $i = 1, \cdots, n$, are called the base covariant vectors since they comprise a vector basis. The sequence $\mathbf{x}_{\xi^1}, \cdots, \mathbf{x}_{\xi^n}$ of the tangential vectors has a right-handed orientation if the Jacobian of the transformation $\mathbf{x}(\boldsymbol{\xi})$ is positive. Otherwise, the base vectors $\mathbf{x}_{\xi^i}$ have a left-handed orientation.

The operation of the dot product on the tangential vectors produces elements of the covariant metric tensor. These elements generate the coefficients that appear in the transformed grid equations. Besides this, the metric elements play a primary role in studying and formulating various geometric characteristics of the grid cells in domains.

2.3 Coordinate Surfaces and Normal Vectors

If one curvilinear coordinate, say ξ^i, is fixed ($\xi^i = \xi^i_0$) then the position function (2.3) describes an $(n-1)$-dimensional surface which is called the coordinate hypersurface. Thus the coordinate hypersurface is defined by the equation $\xi^i = \xi^i_0$; i.e. along the surface all of the coordinates $\xi^1, \cdots, \xi^n$ except ξ^i are allowed to vary.

The inverse transformation

$$\boldsymbol{\xi}(\mathbf{x}) = [\xi^1(\mathbf{x}), \ldots, \xi^n(\mathbf{x})]$$

to (2.3) yields for each fixed i the base contravariant vector

$$\boldsymbol{\nabla}\xi^i = \left(\frac{\partial \xi^i}{\partial x^1}, \cdots, \frac{\partial \xi^i}{\partial x^n}\right) , \tag{2.5}$$

which is the gradient of $\xi^i(\mathbf{x})$ with respect to the Cartesian coordinates $x^1, \cdots, x^n$. The set of the vectors $\boldsymbol{\nabla}\xi^i$, $i = 1, \cdots, n$, is called the set of base contravariant vectors.

Similarly, as the tangential vectors relate to the coordinate curves, the contravariant vectors $\boldsymbol{\nabla}\xi^i$, $i = 1, \cdots, n$, are connected with their respective coordinate hypersurfaces (curves in two dimensions). Indeed for all of the tangent vectors $\mathbf{x}_{\xi^j}$ to the coordinate lines on the surface $\xi^i = \xi^i_0$ we have the obvious identity

$$\mathbf{x}_{\xi^j} \cdot \boldsymbol{\nabla}\xi^i = 0 , \quad i \neq j ,$$

and thus the vector $\boldsymbol{\nabla}\xi^i$ is a normal to the coordinate hypersurface $\xi^i = \xi^i_0$. Therefore the vectors $\boldsymbol{\nabla}\xi^i$, $i = 1, \cdots, n$, are also called the normal base vectors.

Since

$$\mathbf{x}_{\xi^i} \cdot \boldsymbol{\nabla}\xi^i = 1$$

for each fixed $i = 1, \cdots, n$, the vectors $\mathbf{x}_{\xi^i}$ and $\boldsymbol{\nabla}\xi^i$ intersect each other at an angle θ which is less than $\pi/2$. Now, taking into account the orthogonality of the vector $\boldsymbol{\nabla}\xi^i$ to the hypersurface $\xi^i - \xi^i_0$, we find that these two vectors $\mathbf{x}_{\xi^i}$ and $\boldsymbol{\nabla}\xi^i$ are directed to the same side of the coordinate hypersurface (curve in two dimensions). An illustration of this fact in two dimensions is given in Fig. 2.2. The length of any normal base vector $\boldsymbol{\nabla}\xi^i$ is linked to the distance d_i between the corresponding opposite boundary segments (joined by the vector $\mathbf{x}_{\xi^i}$) of the n-dimensional basic parallelepiped formed by the base tangential vectors, namely,

$$d_i = 1/|\boldsymbol{\nabla}\xi^i|\,, \quad |\boldsymbol{\nabla}\xi^i| = \sqrt{\boldsymbol{\nabla}\xi^i \cdot \boldsymbol{\nabla}\xi^i}\,, \quad i = 1, \ldots, n\,.$$

To prove this relation we recall that the vector $\boldsymbol{\nabla}\xi^i$ is a normal to all of the vectors $\mathbf{x}_{\xi^j}$, $j \neq i$, and therefore to the boundary segments of the parallelepiped formed by these $n-1$ vectors. Hence, the unit normal vector $\mathbf{n}_i$ to these segments is expressed by

$$\mathbf{n}_i = \boldsymbol{\nabla}\xi^i/|\boldsymbol{\nabla}\xi^i|\,.$$

Now, taking into account that

$$d_i = \mathbf{x}_{\xi^i} \cdot \mathbf{n}_i\,,$$

we readily obtain

$$d_i = \mathbf{x}_{\xi^i} \cdot \boldsymbol{\nabla}\xi^i/|\boldsymbol{\nabla}\xi^i| = 1/|\boldsymbol{\nabla}\xi^i|\,. \square$$

Let l_i denote the distance between a grid point on the coordinate hypersurface $\xi^i = c$ and the nearest point on the neighboring coordinate hypersurface $\xi^i = c + h$; then

$$l_i = hd_i + O(h^2) = h/|\boldsymbol{\nabla}\xi^i| + O(h^2)\,.$$

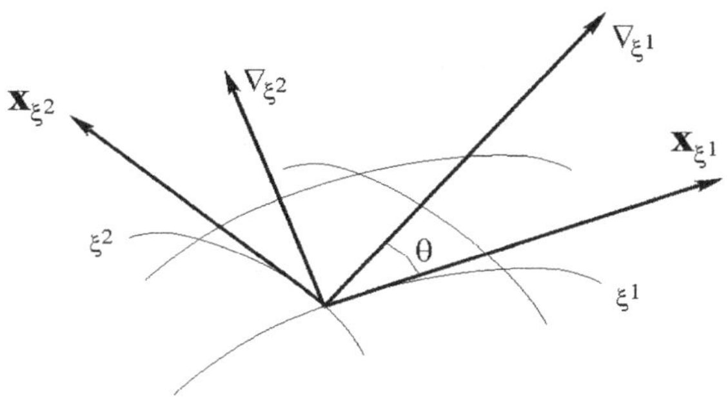

Fig. 2.2. Base tangential and normal vectors in two dimensions

This equation shows that the inverse length of the normal vector $\nabla\xi^i$ multiplied by h represents with high accuracy the distance between the corresponding faces of the coordinate cells in the domain X^n.

Note that the volume of the parallelepiped spanned by the tangential vectors equals J, so we find that the volume of the n-dimensional parallelepiped defined by the normal vectors $\nabla\xi^i$, $i = 1, \cdots, n$, is equal to $1/J$. Thus both the base normal vectors $\nabla\xi^i$ and the base tangential vectors $\mathbf{x}_{\xi^i}$ have the same right-handed or left-handed orientation.

If the coordinate system $\xi^1, \cdots, \xi^n$ is orthogonal, i.e.

$$\mathbf{x}_{\xi^i} \cdot \mathbf{x}_{\xi^j} = p(\boldsymbol{\xi})\delta^i_j\,, \quad i,j = 1, \cdots, n\,, \quad p(\boldsymbol{\xi}) \neq 0$$

then for each fixed $i = 1, \cdots, n$ the vector $\nabla\xi^i$ is parallel to $\mathbf{x}_{\xi^i}$.

2.4 Representation of Vectors Through the Base Vectors

If there are n independent vectors $\mathbf{a}_1, \cdots, \mathbf{a}_n$ of the Euclidean space R^n then any vector $\mathbf{b}$ with components $b^1, \cdots, b^n$ in the Cartesian basis $\mathbf{e}_1, \cdots, \mathbf{e}_n$ is represented through the vectors $\mathbf{a}_i$, $i = 1, \cdots, n$, by

$$\mathbf{b} = a^{ij}(\mathbf{b} \cdot \mathbf{a}_j)\mathbf{a}_i, \quad i,j = 1, \cdots, n\,, \tag{2.6}$$

where a^{ij} are the elements of the matrix (a^{ij}) which is the inverse of the tensor (a_{ij}),

$$a_{ij} = \mathbf{a}_i \cdot \mathbf{a}_j\,, \quad i,j = 1, \cdots, n\,, \tag{2.7}$$

(Fig. 2.3).

It is assumed in (2.6) and later, unless otherwise noted, a popular geometric index convention that a summation is carried out over repeated indices in a product or single term, namely, a sign $\sum$ is understood whenever an index is repeated in the aforesaid cases. The components of the vector $\mathbf{b}$ in the natural basis of the tangential vectors $\mathbf{x}_{\xi^i}$, $i = 1, \cdots, n$, are called contravariant. Let them be denoted by $\overline{b}^i$, $i = 1, \cdots, n$. Thus

$$\mathbf{b} = \overline{b}^1 \mathbf{x}_{\xi^1} + \cdots + \overline{b}^n \mathbf{x}_{\xi^n}\,.$$

Assuming in (2.6) $\mathbf{a}_i = \mathbf{x}_{\xi^i}$, $i = 1, \cdots, n$, we obtain

$$\overline{b}^i = a^{mj}\left(b^k \frac{\partial x^k}{\partial \xi^j}\right)\frac{\partial x^i}{\partial \xi^m}\,, \quad i,j,k,m = 1, \cdots, n\,, \tag{2.8}$$

where $b^1, \cdots, b^n$ are the components of the vector $\mathbf{b}$ in the Cartesian basis $\mathbf{e}_1, \cdots, \mathbf{e}_n$. Since (2.7)

$$a_{ij} = \mathbf{x}_{\xi^i} \cdot \mathbf{x}_{\xi^j} = \frac{\partial x^k}{\partial \xi^i}\frac{\partial x^k}{\partial \xi^j}\,, \quad i,j,k = 1, \cdots, n\,,$$

we have

$$a^{ij} = \frac{\partial \xi^i}{\partial x^k}\frac{\partial \xi^j}{\partial x^k}, \quad i,j,k = 1,\cdots,n .$$

Therefore, from (2.8),

$$\overline{b}^i = \frac{\partial \xi^i}{\partial x^k}\frac{\partial \xi^j}{\partial x^k} b^m \frac{\partial x^m}{\partial \xi^j} = b^j \frac{\partial \xi^i}{\partial x^j}, \quad i,j,k,m = 1,\cdots,n ,$$

or, using the dot product notation

$$\overline{b}^i = \mathbf{b}\cdot\nabla\xi^i , \quad i = 1,\cdots,n . \tag{2.9}$$

Thus, in this case (2.6) has the form

$$\mathbf{b} = (\mathbf{b}\cdot\nabla\xi^i)\mathbf{x}_{\xi^i} , \quad i = 1,\cdots,n . \tag{2.10}$$

For example, the normal base vector $\nabla\xi^i$ is expanded through the base tangential vectors $\mathbf{x}_{\xi^j}$, $j = 1,\cdots,n$, by the following formula:

$$\nabla\xi^i = (\nabla\xi^i\cdot\nabla\xi^k)\mathbf{x}_{\xi^k} = \frac{\partial \xi^i}{\partial x^j}\frac{\partial \xi^k}{\partial x^j}\mathbf{x}_{\xi^k} , \quad i,j,k, = 1,\cdots,n . \tag{2.11}$$

Analogously, a component $\overline{b}_i$ of the vector $\mathbf{b}$ in the basis $\nabla\xi^i$, $i = 1,\cdots,n$, is expressed by the formula

$$\overline{b}_i = b^j\frac{\partial x^j}{\partial \xi^i} = \mathbf{b}\cdot\mathbf{x}_{\xi^i} , \quad i,j = 1,\cdots,n , \tag{2.12}$$

and consequently

$$\mathbf{b} = \overline{b}_i\nabla\xi^i = (\mathbf{b}\cdot\mathbf{x}_{\xi^i})\nabla\xi^i, \quad i = 1,\cdots,n . \tag{2.13}$$

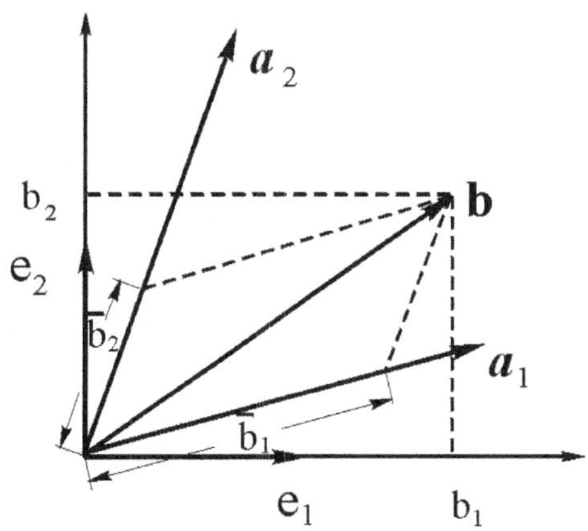

Fig. 2.3. Expansion of the vector **b** in the vectors $\mathbf{a}_1$ and $\mathbf{a}_2$

These components $\bar{b}_i$, $i = 1, \cdots, n$, of the vector $\mathbf{b}$ are called covariant. In particular, the base tangential vector $\mathbf{x}_{\xi^i}$ is expressed through the base normal vectors $\nabla \xi^j$, $j = 1, \cdots, n$, as follows:

$$\mathbf{x}_{\xi^i} = (\mathbf{x}_{\xi^i} \cdot \mathbf{x}_{\xi^j}) \nabla \xi^j = \frac{\partial x^k}{\partial \xi^i} \frac{\partial x^k}{\partial \xi^j} \nabla \xi^j \,, \quad i, j, k = 1, \cdots, n \,. \tag{2.14}$$

2.5 Metric Tensors

Many grid generation algorithms, in particular those based on the calculus of variations, are typically formulated in terms of fundamental features of coordinate transformations and the corresponding mesh cells. These features are compactly described with the use of the metric notation, which is discussed in this section.

2.5.1 Covariant Metric Tensor

The matrix

$$(g_{ij}) \,, \quad i, j = 1, \cdots, n \,,$$

whose elements g_{ij} are the dot products of the pairs of the basic tangential vectors $\mathbf{x}_{\xi^i}$, $i = 1, \ldots, n$,

$$g_{ij} = \mathbf{x}_{\xi^i} \cdot \mathbf{x}_{\xi^j} = \frac{\partial x^k}{\partial \xi^i} \frac{\partial x^k}{\partial \xi^j} \,, \quad i, j, k = 1, \cdots, n \,, \tag{2.15}$$

is called a fundamental or covariant metric tensor of the domain X^n in the coordinates $\xi^1, \cdots, \xi^n$.

Geometrically, each diagonal element g_{ii} of the matrix (g_{ij}) is the length of the tangent vector $\mathbf{x}_{\xi^i}$ squared:

$$g_{ii} = |\mathbf{x}_{\xi^i}|^2 \,, \quad i = 1, \cdots, n \,, \ i \text{ fixed} \,.$$

Also,

$$g_{ij} = |\mathbf{x}_{\xi^i}| |\mathbf{x}_{\xi^j}| \cos\theta = \sqrt{g_{ii}} \sqrt{g_{jj}} \cos\theta \,, \quad i, j = 1, \cdots, n \,, \ i, j \text{ fixed} \,,$$

where θ is the angle between the tangent vectors $\mathbf{x}_{\xi^i}$ and $\mathbf{x}_{\xi^j}$. As a reminder, the notification "fixed" in these expressions for g_{ii} and g_{ij} means that the subscripts ii and jj are fixed, i.e. here the summation over the repeated indices is not carried out.

We designate by g the Jacobian of the covariant matrix (g_{ij}). It is evident that

$$(g_{ij}) = \jmath \jmath^T \,, \quad i, j = 1, \ldots, n \,,$$

and hence

$$J^2 = g \ .$$

The covariant metric tensor is a symmetric matrix, i.e. $g_{ij} = g_{ji}$. If a coordinate system at a point $\boldsymbol{\xi}$ is orthogonal then the tensor (g_{ij}) has a simple diagonal form at this point. Note that these advantageous properties are in general not possessed by the Jacobi matrix $(\partial x^i / \partial \xi^j)$ from which the covariant metric tensor (g_{ij}) is defined.

2.5.2 Line Element

Let P be the point of R^n whose curvilinear coordinates are $\xi^1, \dots, \xi^n$ and let Q be a neighboring point with the curvilinear coordinates $\xi^1 + \mathrm{d}\xi^1, \dots, \xi^n + \mathrm{d}\xi^n$. Then the Cartesian coordinates of these points are

$$x^1(\boldsymbol{\xi}), \dots, x^n(\boldsymbol{\xi}) \ , \quad \boldsymbol{\xi} = (\xi^1, \dots, \xi^n)$$

and

$$x^1(\boldsymbol{\xi} + \mathrm{d}\boldsymbol{\xi}), \dots, x^n(\boldsymbol{\xi} + \mathrm{d}\boldsymbol{\xi}) \ , \quad \mathrm{d}\boldsymbol{\xi} = (\mathrm{d}\xi^1, \dots, \mathrm{d}\xi^n) \ ,$$

respectively. The infinitesimal distance PQ denoted by ds is called the element of length or the line element. In the Cartesian coordinates the line element is the length of the diagonal of the elementary parallelepiped whose edges are $\mathrm{d}x^1, \dots, \mathrm{d}x^n$, where

$$\mathrm{d}x^i = x^i(\boldsymbol{\xi} + \mathrm{d}\boldsymbol{\xi}) - x^i(\boldsymbol{\xi}) = \frac{\partial x^i}{\partial \xi^j} \mathrm{d}\xi^j + o(|d\boldsymbol{\xi}|) \ , \quad i, j = 1, \dots, n \ ,$$

(see Fig. 2.4). Therefore

$$\mathrm{d}s = \sqrt{(\mathrm{d}x^1)^2 + \ldots + (\mathrm{d}x^n)^2} = \sqrt{\mathbf{dx} \cdot \mathbf{dx}} \ ,$$

where

$$\mathbf{dx} = \mathbf{x}(\boldsymbol{\xi} + \mathrm{d}\boldsymbol{\xi}) - \mathbf{x}(\boldsymbol{\xi}) = \mathbf{x}_{\xi^i} \mathrm{d}\xi^i + o(|\mathrm{d}\boldsymbol{\xi}|) \ , \quad i = 1, \dots, n \ ,$$

and we readily find that the expression for ds in the curvilinear coordinates is as follows:

$$\mathrm{d}s = \sqrt{\mathbf{x}_{\xi^i} \mathrm{d}\xi^i \cdot \mathbf{x}_{\xi^j} \mathrm{d}\xi^j + o(|\mathrm{d}\boldsymbol{\xi}|)} = \sqrt{g_{ij} \mathrm{d}\xi^i \mathrm{d}\xi^j + o(|\mathrm{d}\boldsymbol{\xi}|)} \ , \quad i, j = 1, \cdots, n \ .$$

Thus the length s of the curve in X^n, prescribed by the parametrization

$$\mathbf{x}[\boldsymbol{\xi}(t)] : [a, b] \to X^n \ ,$$

is computed by the formula

$$s = \int_a^b \sqrt{g_{ij} \frac{\mathrm{d}\xi^i}{\mathrm{d}t} \frac{\mathrm{d}\xi^j}{\mathrm{d}t}} \mathrm{d}t \ , \quad i, j = 1, \dots, n \ . \tag{2.16}$$

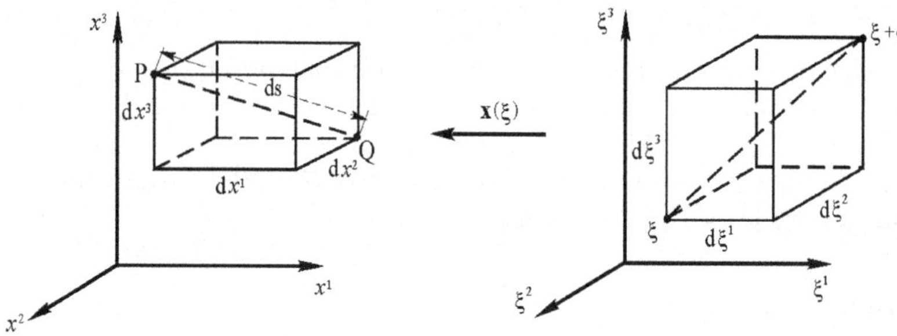

Fig. 2.4. Illustration for the line element

2.5.3 Contravariant Metric Tensor

The contravariant metric tensor of the domain X^n in the coordinates $\xi^1, \cdots, \xi^n$ is the matrix

$$(g^{ij}) , \quad i,j = 1, \cdots, n ,$$

inverse to (g_{ij}), i.e.

$$g_{ij}g^{jk} = \delta_i^k , \quad i,j,k = 1, \cdots, n . \qquad (2.17)$$

Therefore

$$\det(g^{ij}) = \frac{1}{g} , \quad i,j = 1, \ldots, n .$$

It is easily shown that (2.17) is satisfied if and only if

$$g^{ij} = \nabla\xi^i \cdot \nabla\xi^j = \frac{\partial \xi^i}{\partial x^k}\frac{\partial \xi^j}{\partial x^k} , \quad i,j,k = 1, \cdots, n , \qquad (2.18)$$

where $\nabla\xi^l$, $l = 1, \ldots, n$, is the base normal vector determined by (2.5). Thus, each diagonal element g^{ii} (where i is fixed) of the matrix (g^{ij}) is the square of the length of the vector $\nabla\xi^i$:

$$g^{ii} = |\nabla\xi^i|^2 , \quad i = 1, \ldots, n , \ i \text{ fixed} . \qquad (2.19)$$

Geometric Interpretation

Now we discuss the geometric meaning of a diagonal element g^{ii} with a fixed index i, say g^{11}, of the matrix (g^{ij}). Let us consider a three-dimensional coordinate transformation $\mathbf{x}(\boldsymbol{\xi}) : \Xi^3 \to X^3$. Its tangential vectors $\mathbf{x}_{\xi^1}$, $\mathbf{x}_{\xi^2}$, $\mathbf{x}_{\xi^3}$ at some point P form the basic parallelepiped whose edges are these vectors (Fig. 2.5). For the distance d_1 between the opposite faces of the parallelepiped which are defined by the vectors $\mathbf{x}_{\xi^2}$ and $\mathbf{x}_{\xi^3}$, we have

$$d_1 = \mathbf{x}_{\xi^1} \cdot \mathbf{n}_1 ,$$

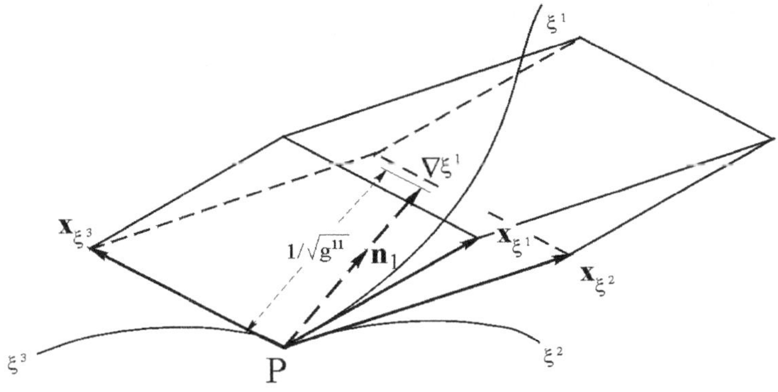

Fig. 2.5. Geometric meaning of the diagonal contravariant metric element g^{11}

where $\mathbf{n}_1$ is the unit normal to the plane spanned by the vectors $\mathbf{x}_{\xi^2}$ and $\mathbf{x}_{\xi^3}$. It is clear, that

$$\nabla_{\xi^1} \cdot \mathbf{x}_{\xi^j} = 0 , \quad j = 2, 3 ,$$

and hence the unit normal $\mathbf{n}_1$ is parallel to the normal base vector ∇_{ξ^1}. Thus we obtain

$$\mathbf{n}_1 = \nabla_{\xi^1}/|\nabla_{\xi^1}| = \nabla_{\xi^1}/\sqrt{g^{11}} .$$

Therefore

$$d_1 = \nabla_{\xi^1} \cdot \nabla_{\xi^1}/\sqrt{g^{11}} = 1/\sqrt{g^{11}} ,$$

and consequently

$$g^{11} = 1/(d_1)^2 .$$

Analogous relations are valid for g^{22} and g^{33}, i.e. in three dimensions the diagonal element g^{ii} for a fixed i means the inverse square of the distance d_i between those faces of the basic parallelepiped which are connected by the vector $\mathbf{x}_{\xi^i}$. In two-dimensional space the element g^{ii} (where i is fixed) is the inverse square of the distance between the edges of the basic parallelogram defined by the tangential vectors $\mathbf{x}_{\xi^1}$ and $\mathbf{x}_{\xi^2}$.

The same interpretation of g^{ii} is valid for general multidimensional coordinate transformations:

$$g^{ii} = 1/(d_i)^2, \quad i = 1, \cdots, n , \tag{2.20}$$

where the index i is fixed, and d_i is the distance between those $(n-1)$-dimensional faces of the n-dimensional parallelepiped which are linked by the tangential vector $\mathbf{x}_{\xi^i}$.

2.5.4 Relations Between Covariant and Contravariant Elements

Now, in analogy with (2.1) and (2.2), we write out very convenient formulas for natural relations between the contravariant elements g^{ij} and the covariant ones g_{ij} in two and three dimensions.

For $n = 2$,

$$g^{ij} = (-1)^{i+j} \frac{g_{3-i\;3-j}}{g} ,$$

$$g_{ij} = (-1)^{i+j} g g^{3-i\;3-j} , \quad i,j = 1,2 , \qquad (2.21)$$

where the indices i, j on the right-hand side of the relations (2.21) are fixed, i.e. summation over the repeated indices is not carried out here. For $n = 3$ we have

$$g^{ij} = \frac{1}{g}(g_{i+1\;j+1}\, g_{i+2\;j+2} - g_{i+1\;j+2}\, g_{i+2\;j+1}) ,$$

$$g_{ij} = g(g^{i+1\;j+1}\, g^{i+2\;j+2} - g^{i+1\;j+2}\, g^{i+2\;j+1}) , \quad i,j = 1,2,3 , \quad (2.22)$$

with the convention that any index, say l, is identified with $l \pm 3$, so, for instance, $g_{45} = g_{12}$.

We also note that, in accordance with the expressions (2.15), (2.18) for g_{ij} and g^{ij}, respectively, the relations (2.11) and (2.14) between the basic vectors $\mathbf{x}_{\xi^i}$ and $\boldsymbol{\nabla}\xi^j$ can be written in the form

$$\mathbf{x}_{\xi^i} = g_{ik} \boldsymbol{\nabla} \xi^k ,$$

$$\boldsymbol{\nabla} \xi^i = g^{ik} \mathbf{x}_{\xi^k} , \quad i,k = 1,\cdots,n . \qquad (2.23)$$

So the first derivatives $\partial x^i/\partial \xi^j$ and $\partial \xi^k/\partial x^m$ of the transformations $\mathbf{x}(\boldsymbol{\xi})$ and $\boldsymbol{\xi}(\mathbf{x})$, respectively, are connected through the metric elements:

$$\frac{\partial x^i}{\partial \xi^j} = g_{mj} \frac{\partial \xi^m}{\partial x^i} ,$$

$$\frac{\partial \xi^i}{\partial x^j} = g^{mi} \frac{\partial x^j}{\partial \xi^m} , \quad i,j,m = 1,\cdots,n . \qquad (2.24)$$

2.6 Cross Product

In addition to the dot product there is another important operation on three-dimensional vectors. This is the cross product, $\times$, which for any two vectors $\mathbf{a} = (a^1, a^2, a^3)$, $\mathbf{b} = (b^1, b^2, b^3)$ expanded in the Cartesian vector basis $(\mathbf{e}_1,\ \mathbf{e}_2,\ \mathbf{e}_3)$ of the Euclidean space R^3 is expressed as the determinant of a matrix:

$$\mathbf{a} \times \mathbf{b} = \det \begin{pmatrix} \mathbf{e}_1 & \mathbf{e}_2 & \mathbf{e}_3 \\ a^1 & a^2 & a^3 \\ b^1 & b^2 & b^3 \end{pmatrix} . \qquad (2.25)$$

Thus

$$\mathbf{a} \times \mathbf{b} = (a^2b^3 - a^3b^2,\ a^3b^1 - a^1b^3,\ a^1b^2 - a^2b^1)\ ,$$

or, with the previously assumed convention in three dimensions of the identification of any index j with $j \pm 3$,

$$\mathbf{a} \times \mathbf{b} = (a^{i+1}b^{i+2} - a^{i+2}b^{i+1})\mathbf{e}_i\ , \quad i = 1, 2, 3\ . \tag{2.26}$$

We will now state some facts connected with the cross product operation.

2.6.1 Geometric Meaning

We can readily see that $\mathbf{a} \times \mathbf{b} = \mathbf{0}$ if the vectors $\mathbf{a}$ and $\mathbf{b}$ are parallel. Also, from (2.26) we find that $\mathbf{a} \cdot (\mathbf{a} \times \mathbf{b}) = 0$ and $\mathbf{b} \cdot (\mathbf{a} \times \mathbf{b}) = 0$, i.e. the vector $\mathbf{a} \times \mathbf{b}$ is orthogonal to each of the vectors $\mathbf{a}$ and $\mathbf{b}$. Thus, if these vectors are not parallel then

$$\mathbf{a} \times \mathbf{b} = \alpha |\mathbf{a} \times \mathbf{b}| \mathbf{n}\ , \tag{2.27}$$

where $\alpha = 1$ or $\alpha = -1$ and $\mathbf{n}$ is a unit normal vector to the plane determined by the vectors $\mathbf{a}$ and $\mathbf{b}$.

Now we show that the length of the vector $\mathbf{a} \times \mathbf{b}$ equals the area of the parallelogram formed by the vectors $\mathbf{a}$ and $\mathbf{b}$, i.e.

$$|\mathbf{a} \times \mathbf{b}| = |\mathbf{a}||\mathbf{b}| \sin\theta\ , \tag{2.28}$$

where θ is the angle between the two vectors $\mathbf{a}$ and $\mathbf{b}$. To prove (2.28) we first note that

$$|\mathbf{a}|^2|\mathbf{b}|^2 \sin^2\theta = |\mathbf{a}|^2|\mathbf{b}|^2(1 - \cos^2\theta) = |\mathbf{a}|^2|\mathbf{b}|^2 - (\mathbf{a} \cdot \mathbf{b})^2\ .$$

We have, further,

$$\begin{aligned} |\mathbf{a}|^2|\mathbf{b}|^2 - |\mathbf{a} \cdot \mathbf{b}|^2 &= \Big(\sum_{i=1}^{3} a^i a^i\Big)\Big(\sum_{j=1}^{3} b^j b^j\Big) - \Big(\sum_{k=1}^{3} a^k b^k\Big)^2 \\ &= \sum_{k=1}^{3} [(a^l)^2(b^m)^2 + (a^m)^2(b^l)^2 - 2a^l b^l a^m b^m] \\ &= \sum_{k=1}^{3} (a^l b^m - a^m b^l)^2\ , \end{aligned}$$

where (k, l, m) are cyclic, i.e. $l = k+1,\ m = k+2$ with the convention that $j+3$ is equivalent to j for any index j. According to (2.26) the quantity $a^l b^m - a^m b^l$ for the cyclic sequence (k, l, m) is the kth component of the vector $\mathbf{a} \times \mathbf{b}$, so we find that

$$|\mathbf{a}||\mathbf{b}| \sin^2\theta = |\mathbf{a}|^2|\mathbf{b}|^2 - |\mathbf{a} \cdot \mathbf{b}|^2 = |\mathbf{a} \times \mathbf{b}|^2\ , \tag{2.29}$$

what proves (2.28). □

Thus we obtain the result that if the vectors $\mathbf{a}$ and $\mathbf{b}$ are not parallel then the vector $\mathbf{a} \times \mathbf{b}$ is orthogonal to the parallelogram formed by these vectors and its length equals the area of the parallelogram. Therefore the three vectors $\mathbf{a}$, $\mathbf{b}$ and $\mathbf{a} \times \mathbf{b}$ are independent in this case and represent a base vector system in the three-dimensional space R^3. Moreover, the vectors $\mathbf{a}$, $\mathbf{b}$ and $\mathbf{a} \times \mathbf{b}$ form a right-handed triad since $\mathbf{a} \times \mathbf{b} \neq \mathbf{0}$, and consequently the Jacobian of the matrix determined by $\mathbf{a}$, $\mathbf{b}$, and $\mathbf{a} \times \mathbf{b}$ is positive; it equals

$$(\mathbf{a} \times \mathbf{b}) \cdot (\mathbf{a} \times \mathbf{b}) = (\mathbf{a} \times \mathbf{b})^2 .$$

2.6.2 Relation to Volumes

Let $\mathbf{c} = (c^1, c^2, c^3)$ be one more vector. The volume V of the parallelepiped whose edges are the vectors $\mathbf{a}$, $\mathbf{b}$ and $\mathbf{c}$ equals the area of the parallelogram formed by the vectors $\mathbf{a}$ and $\mathbf{b}$ multiplied by the modulas of the dot product of the vector $\mathbf{c}$ and the unit normal $\mathbf{n}$ to the parallelogram. Thus

$$V = |\mathbf{a} \times \mathbf{b}||\mathbf{n} \cdot \mathbf{c}|$$

and from (2.27) we obtain

$$V = |(\mathbf{a} \times \mathbf{b}) \cdot \mathbf{c}| . \tag{2.30}$$

Taking into account (2.26), we obtain

$$(\mathbf{a} \times \mathbf{b}) \cdot \mathbf{c} = c^1(a^2b^3 - a^3b^2) + c^2(a^3b^1 - a^1b^3) + c^3(a^1b^2 - a^2b^1) .$$

The right-hand side of this equation is the Jacobian of the matrix whose rows are formed by the vectors $\mathbf{a}$, $\mathbf{b}$, and $\mathbf{c}$, i.e.

$$(\mathbf{a} \times \mathbf{b}) \cdot \mathbf{c} = \det \begin{pmatrix} a^1 & a^2 & a^3 \\ b^1 & b^2 & b^3 \\ c^1 & c^2 & c^3 \end{pmatrix} . \tag{2.31}$$

From this equation we readily obtain

$$(\mathbf{a} \times \mathbf{b}) \cdot \mathbf{c} = \mathbf{a} \cdot (\mathbf{b} \times \mathbf{c}) = (\mathbf{c} \times \mathbf{a}) \cdot \mathbf{b} .$$

Thus the volume of the parallelepiped determined by the vectors $\mathbf{a}$, $\mathbf{b}$, and $\mathbf{c}$ equals the Jacobian of the matrix formed by the components of these vectors. In particular, we obtain that the Jacobian of a three-dimensional coordinate transformation $\mathbf{x}(\boldsymbol{\xi})$ is expressed as follows:

$$J = \mathbf{x}_{\xi^1} \cdot (\mathbf{x}_{\xi^2} \times \mathbf{x}_{\xi^3}) . \tag{2.32}$$

2.6.3 Relation to Base Vectors

Applying the operation of the cross product to two base tangential vectors $\mathbf{x}_{\xi^l}$ and $\mathbf{x}_{\xi^m}$, we find that the vector $\mathbf{x}_{\xi^l} \times \mathbf{x}_{\xi^m}$ is a normal to the coordinate surface $\xi^i = \xi^i_0$ with (i, l, m) cyclic. The base normal vector $\nabla \xi^i$ is also orthogonal to the surface and therefore it is a scalar multiple of $\mathbf{x}_{\xi^l} \times \mathbf{x}_{\xi^m}$, i.e.

$$\nabla \xi^i = c(\mathbf{x}_{\xi^l} \times \mathbf{x}_{\xi^m}) .$$

Multiplying this equation for a fixed i by $\mathbf{x}_{\xi^i}$, using the operation of the dot product, we obtain, using (2.32),

$$1 = c\, J ,$$

and therefore

$$\nabla \xi^i = \frac{1}{J}(\mathbf{x}_{\xi^l} \times \mathbf{x}_{\xi^m}) . \tag{2.33}$$

Thus the elements of the three-dimensional contravariant metric tensor (g^{ij}) which are computed by the dot product of the normal base vectors (formula (2.18)) can also be found through the tangential vectors $\mathbf{x}_{\xi^i}$ by the formula

$$g^{ij} = \nabla \xi^i \cdot \nabla \xi^j = \frac{1}{g}(\mathbf{x}_{\xi^{i+1}} \times \mathbf{x}_{\xi^{i+2}}) \cdot (\mathbf{x}_{\xi^{j+1}} \times \mathbf{x}_{\xi^{j+2}}) , \quad i, j = 1, 2, 3 .$$

Analogously, every base vector $\mathbf{x}_{\xi^i}$, $i = 1, 2, 3$, is expressed by the tensor product of the normal base vectors $\nabla \xi^j$, $j = 1, 2, 3$:

$$\mathbf{x}_{\xi^i} = J(\nabla \xi^l \times \nabla \xi^k), \quad i = 1, 2, 3 , \tag{2.34}$$

where $l = i + 1$, $k = i + 2$, and m is equivalent to $m + 3$ for any index m. Accordingly we have, taking into account (2.15),

$$g_{ij} = g(\nabla \xi^{i+1} \times \nabla \xi^{i+2}) \cdot (\nabla \xi^{j+1} \times \nabla \xi^{j+2}) , \quad i, j = 1, 2, 3 .$$

Using the relations (2.33) and (2.34) in (2.30), we also obtain

$$\frac{1}{J} = \nabla \xi^1 \cdot \nabla \xi^2 \times \nabla \xi^3 . \tag{2.35}$$

Thus the volume of the parallelepiped formed by the base normal vectors $\nabla \xi^1$, $\nabla \xi^2$, and $\nabla \xi^3$ is the modulus of the inverse of the Jacobian J of the transformation $\mathbf{x}(\boldsymbol{\xi})$.

2.7 Relations Concerning Second Derivatives

The elements of the covariant and contravariant metric tensors are defined by the dot products of the base tangential and normal vectors, respectively.

These elements are suitable for describing the internal features of the cells such as the lengths of the edges, the areas of the faces, their volumes, and the angles between the edges and the faces. However, as they are derived from the first derivatives of the coordinate transformation $\mathbf{x}(\boldsymbol{\xi})$, the direct use of the metric elements is not sufficient for the description of the dynamic features of the grid (e.g. curvature), which reflect changes between adjacent cells. This is because the formulation of these grid features relies not only on the first derivatives but also on the second derivatives of $\mathbf{x}(\boldsymbol{\xi})$. Therefore there is a need to study relations connected with the second derivatives of the coordinate parametrizations.

This section presents some notations and formulas which are concerned with the second derivatives of the components of the coordinate transformations. These notations and relations will be used to describe the curvatures of the coordinate lines and surfaces.

2.7.1 Christoffel Symbols of Domains

The edge of a grid cell in the ξ^i direction can be represented with high accuracy by the base vector $\mathbf{x}_{\xi^i}$ contracted by the factor h_i, which represents the step size of a uniform grid in Ξ^n in the ξ^i direction. Therefore the local change of the edge in the ξ^j direction is characterized by the derivative of $\mathbf{x}_{\xi^i}$ with respect to ξ^j, i.e. by $\mathbf{x}_{\xi^i\xi^j}$.

Since the second derivatives may be used to formulate quantitative measures of the grid, we describe these vectors $\mathbf{x}_{\xi^i\xi^j}$ through the base tangential and normal vectors using certain three-index quantities known as Christoffel symbols. The Christoffel symbols are commonly used in formulating measures of the mutual interaction of the cells and in formulas for differential equations.

Let us denote by Γ^k_{ij} the kth contravariant component of the vector $\mathbf{x}_{\xi^i\xi^j}$ in the base tangential vectors $\mathbf{x}_{\xi^k}$, $k = 1, \cdots, n$. The superscript k in this designation relates to the base vector $\mathbf{x}_{\xi^k}$ and the subscript ij corresponds to the mixed derivative with respect to ξ^i and ξ^j. Thus

$$\mathbf{x}_{\xi^i\xi^j} = \Gamma^k_{ij}\, \mathbf{x}_{\xi^k} , \quad i, j, k = 1, \cdots, n , \tag{2.36}$$

and consequently

$$\frac{\partial^2 x^p}{\partial \xi^j \partial \xi^k} = \Gamma^m_{kj} \frac{\partial x^p}{\partial \xi^m} , \quad j, k, m, p = 1, \cdots, n . \tag{2.37}$$

Multiplying these equations by $\partial \xi^i / \partial x^p$ gives

$$\Gamma^i_{kj} = \frac{\partial^2 x^p}{\partial \xi^k \partial \xi^j} \frac{\partial \xi^i}{\partial x^p} , \quad i, j, k, p = 1, \cdots, n , \tag{2.38}$$

or in vector form,

$$\Gamma^i_{kj} = \mathbf{x}_{\xi^k\xi^j} \cdot \boldsymbol{\nabla}\xi^i \, . \tag{2.39}$$

The quantities Γ^i_{kj} are called the space Christoffel symbols of the second kind and the expression (2.36) is a form of the Gauss relation representing the second derivatives of the position vector $\mathbf{x}(\boldsymbol{\xi})$ through the tangential vectors $\mathbf{x}_{\xi^i}$.

Analogously, the components of the second derivatives of the position vector $\mathbf{x}(\boldsymbol{\xi})$ expanded in the base normal vectors $\boldsymbol{\nabla}\xi^i$, $i = 1, \cdots, n$, are referred to as the space Christoffel symbols of the first kind. The mth component of the vector $\mathbf{x}_{\xi^k\xi^j}$ in the base vectors $\boldsymbol{\nabla}\xi^i$, $i = 1, \cdots, n$, is denoted by $[kj, m]$. Thus, according to (2.12),

$$[kj, m] = \mathbf{x}_{\xi^k\xi^j} \cdot \mathbf{x}_{\xi^m} = \frac{\partial^2 x^l}{\partial\xi^k \partial\xi^j} \frac{\partial x^l}{\partial\xi^m} \, , \quad j, k, l, m = 1, \cdots, n \, , \tag{2.40}$$

and consequently

$$\mathbf{x}_{\xi^k\xi^j} = [kj, m] \boldsymbol{\nabla}\xi^m \, . \tag{2.41}$$

So, in analogy with (2.37), we obtain

$$\frac{\partial^2 x^l}{\partial\xi^j \partial\xi^k} = [kj, m] \frac{\partial\xi^m}{\partial x^i} \, , \quad i, j, k, m = 1, \cdots, n \, . \tag{2.42}$$

Multiplying (2.40) by g^{im} and summing over m we find that the space Christoffel symbols of the first and second kind are connected by the following relation:

$$\Gamma^i_{kj} = g^{im} [kj, m] \, , \quad i, j, k, m = 1, \cdots, n \, . \tag{2.43}$$

Conversely, from (2.38) and (2.40),

$$[kj, m] = g_{ml} \Gamma^l_{kj} \, , \quad j, k, l, m = 1, \cdots, n \, . \tag{2.44}$$

The space Christoffel symbols of the first kind $[kj, m]$ can be expressed through the first derivatives of the covariant elements g_{ij} of the metric tensor (g_{ij}) by the following readily verified formula:

$$[kj, m] = \frac{1}{2} \Big(\frac{\partial g_{jm}}{\partial\xi^k} + \frac{\partial g_{km}}{\partial\xi^j} - \frac{\partial g_{kj}}{\partial\xi^m} \Big) \, , \quad i, j, k, m = 1, \cdots, n \, . \tag{2.45}$$

Thus, taking into account (2.43), we see that the space Christoffel symbols of the second kind Γ^i_{kj} can be written in terms of metric elements and their first derivatives. In particular, in the case of an orthogonal coordinate system ξ^i, we obtain from (2.43, 2.45)

$$\Gamma^i_{kj} = \frac{1}{g} g^{ii} \Big(\frac{\partial g_{ii}}{\partial\xi^k} + \frac{\partial g_{ii}}{\partial\xi^j} - \frac{\partial g_{kj}}{\partial\xi^i} \Big) \, , \quad i, j, k = 1, \cdots, n \, .$$

Here the index i is fixed, i.e. the summation over i is not carried out.

2.7.2 Differentiation of the Jacobian

Of critical importance in establishing relations between geometric characteristics is the formula for differentiation of the Jacobian of a coordinate transformation $\mathbf{x}(\boldsymbol{\xi})$

$$\frac{\partial J}{\partial \xi^k} \equiv J \frac{\partial^2 x^i}{\partial \xi^k \partial \xi^m} \frac{\partial \xi^m}{\partial x^i} \equiv J \frac{\partial}{\partial x^i}\left(\frac{\partial x^i}{\partial \xi^k}\right) \equiv J \mathrm{div}_x \frac{\partial \mathbf{x}}{\partial \xi^k} ,$$

$$i, k, m = 1, \cdots, n . \tag{2.46}$$

In accordance with (2.38), this identity can also be expressed through the space Christoffel symbols of the second kind Γ^i_{kj} by

$$\frac{\partial J}{\partial \xi^k} = J \Gamma^i_{ik} , \quad i, k = 1, \cdots, n ,$$

with the summation convention over the repeated index i.

In order to prove the identity (2.46) we note that in the case of an arbitrary matrix (a_{ij}) the first derivative of its Jacobian is obtained by the process of differentiating the first row (the others are left unchanged), then performing the same operation on the second row, and so on with all of the rows of the matrix. The summation of the Jacobians of the matrices derived in such a manner gives the first derivative with respect to ξ^k of the Jacobian of the original matrix (a_{ij}). Thus

$$\frac{\partial}{\partial \xi^k} \det(a_{ij}) = \frac{\partial a_{im}}{\partial \xi^k} G^{im} , \quad i, j, k, m = 1, \cdots, n , \tag{2.47}$$

where G^{im} is the cofactor of the element a_{im}. For the Jacobi matrix $(\partial x^i / \partial \xi^j)$ of the coordinate transformation $\mathbf{x}(\boldsymbol{\xi})$ we have

$$G^{im} = J \frac{\partial \xi^m}{\partial x^i} , \quad i, j = 1, \cdots, n .$$

Therefore, applying (2.47) to the Jacobi matrix, we obtain (2.46). □

2.7.3 Basic Identity

The identity (2.46) implies the extremely important relation

$$\frac{\partial}{\partial \xi^j}\left(J \frac{\partial \xi^j}{\partial x^i}\right) \equiv 0 , \quad i, j = 1, \cdots, n , \tag{2.48}$$

which leads to specific forms of new dependent variables for conservation-law equations. To prove (2.48) we first note that

$$\frac{\partial^2 \xi^p}{\partial x^k \partial x^j} \frac{\partial x^l}{\partial \xi^p} = - \frac{\partial^2 x^l}{\partial \xi^p \partial \xi^m} \frac{\partial \xi^m}{\partial x^k} \frac{\partial \xi^p}{\partial x^j} .$$

Multiplying this equation by $\partial\xi^i/\partial x^l$ and summing over l, we obtain a formula representing the second derivative $\partial^2\xi^i/\partial x^k\partial x^m$ of the functions $\xi^i(\mathbf{x})$ through the second derivatives $\partial^2 x^m/\partial\xi^l\partial\xi^p$ of the functions $x^m(\boldsymbol{\xi})$, $m = 1,\cdots,n$:

$$\begin{aligned}\frac{\partial^2\xi^i}{\partial x^k\partial x^m} &= -\frac{\partial^2 x^p}{\partial\xi^l\partial\xi^j}\frac{\partial\xi^j}{\partial x^k}\frac{\partial\xi^l}{\partial x^m}\frac{\partial\xi^i}{\partial x^p}\\ &= -\Gamma^i_{lj}\frac{\partial\xi^l}{\partial x^m}\frac{\partial\xi^j}{\partial x^k}\,, \quad i,j,k,l,m,p = 1,\cdots,n\,.\end{aligned} \tag{2.49}$$

Now, using this relation and the formula (2.46) for differentiation of the Jacobian in the identity

$$\frac{\partial}{\partial\xi^j}\Big(J\frac{\partial\xi^j}{\partial x^i}\Big) = \frac{\partial J}{\partial\xi^j}\frac{\partial\xi^j}{\partial x^i} + J\frac{\partial^2\xi^j}{\partial x^i\partial x^k}\frac{\partial x^k}{\partial\xi^j}\,,$$

we obtain

$$\begin{aligned}\frac{\partial}{\partial\xi^j}\Big(J\frac{\partial\xi^j}{\partial x^i}\Big) &= J\frac{\partial^2 x^k}{\partial\xi^p\partial\xi^j}\frac{\partial\xi^p}{\partial x^k}\frac{\partial\xi^j}{\partial x^i} - J\frac{\partial^2 x^p}{\partial\xi^l\partial\xi^m}\frac{\partial\xi^m}{\partial x^i}\frac{\partial\xi^l}{\partial x^k}\frac{\partial\xi^j}{\partial x^p}\frac{\partial x^k}{\partial\xi^j}\\ &= J\frac{\partial^2 x^k}{\partial\xi^p\partial\xi^j}\frac{\partial\xi^p}{\partial x^k}\frac{\partial\xi^j}{\partial x^i} - J\frac{\partial^2 x^p}{\partial\xi^l\partial\xi^m}\frac{\partial\xi^l}{\partial x^p}\frac{\partial\xi^m}{\partial x^i} = 0\,,\end{aligned}$$

$$i,j,k,l,m,p = 1,\quad ,n\,,$$

i.e. (2.48) has been proved. □

The identity (2.48) is obvious when $n = 1$ or $n = 2$. For example, for $n = 2$ we have from (2.1)

$$J\frac{\partial\xi^j}{\partial x^i} = (-1)^{i+j}\frac{\partial x^{3-i}}{\partial\xi^{3-j}}\,, \quad i,j = 1,2\,,$$

with fixed indices i and j, and therefore

$$\frac{\partial}{\partial\xi^j}\Big(J\frac{\partial\xi^j}{\partial x^i}\Big) = (-1)^{i+1}\Big(\frac{\partial}{\partial\xi^1}\frac{\partial x^{3-i}}{\partial\xi^2} - \frac{\partial}{\partial\xi^2}\frac{\partial x^{3-i}}{\partial\xi^1}\Big) = 0\,, \quad i,j = 1,2\,.$$

An inference of (2.48) for $n = 3$ also follows from the differentiation of the cross product of the base tangential vectors $\mathbf{r}_{\xi^i}$, $i = 1,2,3$. Taking into account (2.26), we readily obtain the following formula for the differentiation of the cross product of two three-dimensional vector-valued functions $\mathbf{a}$ and $\mathbf{b}$:

$$\frac{\partial}{\partial\xi^i}(\mathbf{a}\times\mathbf{b}) = \frac{\partial}{\partial\xi^i}\mathbf{a}\times\mathbf{b} + \mathbf{a}\times\frac{\partial}{\partial\xi^i}\mathbf{b}\,, \quad i = 1,2,3\,.$$

With this formula we obtain

$$\sum_{i=1}^{3} \frac{\partial}{\partial \xi^i}(\mathbf{x}_{\xi^j} \times \mathbf{x}_{\xi^k}) = \sum_{i=1}^{3} \mathbf{x}_{\xi^j \xi^i} \times \mathbf{x}_{\xi^k} + \sum_{i=1}^{3} \mathbf{x}_{\xi^j} \times \mathbf{x}_{\xi^k \xi^i} , \tag{2.50}$$

where the indices $(i,\ j,\ k)$ are cyclic, i.e. $j = i+1,\ k = i+2$, m is equivalent to $m+3$. For the last summation of the above formula, we obtain

$$\sum_{i=1}^{3} \mathbf{x}_{\xi^j} \times \mathbf{x}_{\xi^k \xi^i} = \sum_{i=1}^{3} \mathbf{x}_{\xi^k} \times \mathbf{x}_{\xi^i \xi^j} .$$

Therefore, from (2.50),

$$\sum_{i=1}^{3} \frac{\partial}{\partial \xi^i}(\mathbf{x}_{\xi^j} \times \mathbf{x}_{\xi^k}) = 0 ,$$

since

$$\mathbf{x}_{\xi^i} \times \mathbf{x}_{\xi^j \xi^k} = -\mathbf{x}_{\xi^j \xi^k} \times \mathbf{x}_{\xi^i}$$

and (2.33) implies (2.48) for $n = 3$.

3 Geometry of Curves

3.1 Curves in Multidimensional Space

3.1.1 Definition

Commonly, a curve in the n-dimensional Euclidean space R^n is the locus of points of R^n whose positions are represented by a vector-valued position function $\mathbf{r}$ of a single parameter, say φ,

$$\mathbf{r}(\varphi) : [a, b] \to R^n \ , \quad \mathbf{r}(\varphi) = [x^1(\varphi), \dots, x^n(\varphi)] \ , \tag{3.1}$$

(see Fig. 3.1). The position function $\mathbf{r}(\varphi)$ is referred to as a parametrization of the curve. Note each curve can be given parametric representation in an infinity of ways, namely, as

$$\mathbf{r}[\varphi(\psi)] : [a_1, b_1] \to [a, b] \ ,$$

where

$$\psi : [a_1, b_1] \to [a, b]$$

is an arbitrary univariate one-to-one monotone function.

It is assumed that the parametrization (3.1) is $p \geq 1$ times continuously differentiable with respect to φ and $\mathbf{r}_\varphi \neq \mathbf{0}$ for all $\varphi \in [a, b]$. In our considerations we will use the designation S^{r1} for the curve with the parametrization $\mathbf{r}(\varphi)$. In this chapter we discuss the important measures of the local curve quality known as curvature and torsion. These measures are derived by some manipulations of basic curve vectors using the operations of the dot and cross products.

3.1.2 Basic Curve Vectors

Tangent Vector

The first derivative of the parametrization $\mathbf{r}(\varphi)$ is a tangential vector

$$\mathbf{r}_\varphi = \left(\frac{\mathrm{d}x^1}{\mathrm{d}\varphi}, \dots, \frac{\mathrm{d}x^n}{\mathrm{d}\varphi}\right)$$

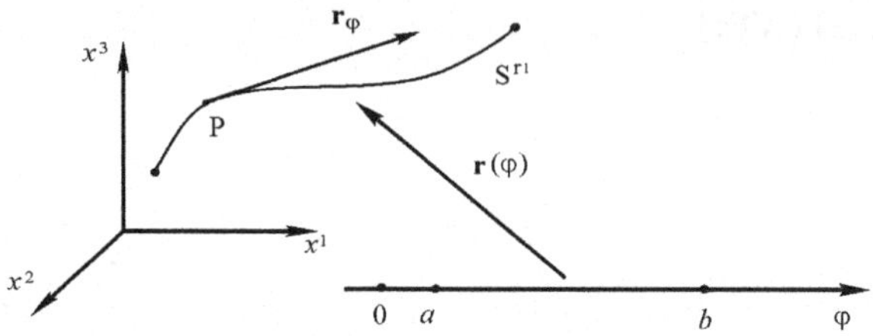

Fig. 3.1. Scheme of a curve representation in R^3

to the curve S^{r1} (Fig. 3.1). The quantity

$$g^{r\varphi} = \mathbf{r}_\varphi \cdot \mathbf{r}_\varphi = \frac{\mathrm{d}x^i}{\mathrm{d}\varphi}\frac{\mathrm{d}x^i}{\mathrm{d}\varphi}\,, \quad i = 1, \dots, n\,, \tag{3.2}$$

is the metric tensor of the curve and its square root is the length of the tangent vector $\mathbf{r}_\varphi$. Accordingly the length l of the curve S^{r1} is computed from the integral

$$l = \int_a^b \sqrt{g^{r\varphi}}\mathrm{d}\varphi\,.$$

The most important notions related to curves are connected with the arc length parameter s defined by the equation

$$s(\varphi) = \int_a^\varphi \sqrt{g^{r\varphi}}\mathrm{d}\varphi\,. \tag{3.3}$$

The vector $\mathrm{d}\mathbf{r}[\varphi(s)]/\mathrm{d}s$, where $\varphi(s)$ is the inverse of $s(\varphi)$, is a tangent vector designated by $\mathbf{t}$. Using (3.3) we obtain

$$\mathbf{t} = \frac{\mathrm{d}}{\mathrm{d}s}\mathbf{r}[\varphi(s)] = \frac{\mathrm{d}\varphi}{\mathrm{d}s}\mathbf{r}_\varphi = \frac{1}{\sqrt{g^{r\varphi}}}\mathbf{r}_\varphi\,. \tag{3.4}$$

Therefore $\mathbf{t}$ is the unit tangent vector.

Principal Normal Vector

Any nonzero vector, which is orthogonal to the tangent vector, is called a normal vector to the curve. Thus a vector $\mathbf{v}$ is normal to S^{r1} if $\mathbf{v} \cdot \mathbf{t} = 0$. Since (3.4) $\mathbf{t}$ is a unit vector, i.e.

$$\mathbf{t} \cdot \mathbf{t} = 1\,,$$

and if we take the derivative with respect to s of this equation, we obtain

$$\mathbf{t}_s \cdot \mathbf{t} = 0\,.$$

This means that if $\mathbf{t}_s \neq \mathbf{0}$ then this vector is normal to the curve S^{r1}. The vector $\mathbf{t}_s$ is called the principal normal of S^{r1}. Let $\mathbf{n}$ be a unit vector that is co-directional with $\mathbf{t}_s$. Then

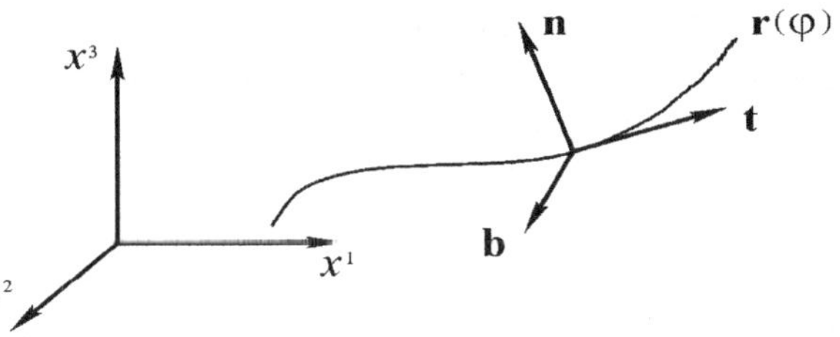

Fig. 3.2. Base curve vectors

$$\mathbf{n} = \frac{1}{k}\mathbf{t}_s \,, \quad k = (\mathbf{t}_s \cdot \mathbf{t}_s)^{1/2} \,. \tag{3.5}$$

The magnitude k is called the curvature at the point in question, while the quantity $\rho = 1/k$ is called the radius of curvature of the curve.

Using the identity $\mathbf{r}_\varphi = \sqrt{g^{r\varphi}}\mathbf{t}$ obtained from (3.4) we find by virtue of (3.2) and (3.3),

$$\begin{aligned} \mathbf{r}_{\varphi\varphi} &= \frac{1}{\sqrt{g^{r\varphi}}}(\mathbf{r}_{\varphi\varphi} \cdot \mathbf{r}_\varphi)\mathbf{t} + g^{r\varphi}\mathbf{t}_s \\ &= \frac{1}{\sqrt{g^{r\varphi}}}(\mathbf{r}_{\varphi\varphi} \cdot \mathbf{r}_\varphi)\mathbf{t} + g^{r\varphi}k\mathbf{n} \,. \end{aligned} \tag{3.6}$$

The identity (3.6) is an analog of the Gauss relations (2.36). This identity shows that the vector $\mathbf{r}_{\varphi\varphi}$ lies in the $\mathbf{t} - \mathbf{n}$ plane.

3.2 Curves in Three-Dimensional Space

3.2.1 Basic Vectors

In three dimensions we can apply the operation of the cross product to the basic tangential and normal vectors of a curve. The vector $\mathbf{b} = \mathbf{t} \times \mathbf{n}$ is a unit vector which is orthogonal to both $\mathbf{t}$ and $\mathbf{n}$. It is called the binormal vector. From (3.6) we find that $\mathbf{b}$ is orthogonal to $\mathbf{r}_{\varphi\varphi}$.

The three vectors ($\mathbf{t}$, $\mathbf{n}$, $\mathbf{b}$) form a right-handed triad (Fig. 3.2). Note that if the curve lies in a plane, then the vectors $\mathbf{t}$ and $\mathbf{n}$ lie in the plane as well and $\mathbf{b}$ is a constant unit vector normal to the plane. The vectors $\mathbf{t}$, $\mathbf{n}$, and $\mathbf{b}$ are connected by the Serret–Frenet equations

$$\begin{aligned} \frac{d\mathbf{t}}{ds} &= k\mathbf{n} \,, \\ \frac{d\mathbf{n}}{ds} &= -k\mathbf{t} + \tau\mathbf{b} \,, \\ \frac{d\mathbf{b}}{ds} &= -\tau\mathbf{n} \,, \end{aligned} \tag{3.7}$$

where the coefficient τ is called the second curvature or torsion of the curve. The first equation of the system (3.7) is taken from (3.5). The second and third equations are readily obtained from the formula (2.6) by replacing the $\mathbf{b}$ in (2.6) by the vectors on the left-hand side of (3.7), while the vectors $\mathbf{t}$, $\mathbf{n}$, and $\mathbf{b}$ substitute for $\mathbf{e}_1$, $\mathbf{e}_2$, and $\mathbf{e}_3$, respectively. The vectors $\mathbf{t}$, $\mathbf{n}$, and $\mathbf{b}$ constitute an orthonormal basis, i.e.

$$a_{ij} = a^{ij} = \delta^i_j\,, \quad i,j = 1,2,3\,,$$

where, in accordance with Sect. 2.4 $a_{ij} = \mathbf{e}_i \cdot \mathbf{e}_j$, and (a^{ij}) is the inverse of the tensor (a_{ij}). Now, using (2.6) we obtain

$$\frac{\mathrm{d}\mathbf{n}}{\mathrm{d}s} = \Big(\frac{\mathrm{d}\mathbf{n}}{\mathrm{d}s}\cdot\mathbf{t}\Big)\mathbf{t} + \Big(\frac{\mathrm{d}\mathbf{n}}{\mathrm{d}s}\cdot\mathbf{n}\Big)\mathbf{n} + \Big(\frac{\mathrm{d}\mathbf{n}}{\mathrm{d}s}\cdot\mathbf{b}\Big)\mathbf{b} = -k\mathbf{t} + \Big(\frac{\mathrm{d}\mathbf{n}}{\mathrm{d}s}\cdot\mathbf{b}\Big)\mathbf{b}\,,$$

since $\mathbf{n}_s \cdot \mathbf{t} = -\mathbf{n}\cdot\mathbf{t}_s$, $\mathbf{n}_s\cdot\mathbf{n} = 0$. Thus we obtain the second equation of (3.7) with $\tau = \mathbf{n}_s \cdot \mathbf{b}$. Analogously we obtain the last equation of (3.7) by expanding the vector $\mathbf{b}_s$ through $\mathbf{t}$, $\mathbf{n}$, and $\mathbf{b}$ using the relation (2.6):

$$\frac{\mathrm{d}\mathbf{b}}{\mathrm{d}s} = \Big(\frac{\mathrm{d}\mathbf{b}}{\mathrm{d}s}\cdot\mathbf{t}\Big)\mathbf{t} + \Big(\frac{\mathrm{d}\mathbf{b}}{\mathrm{d}s}\cdot\mathbf{n}\Big)\mathbf{n} + \Big(\frac{\mathrm{d}\mathbf{b}}{\mathrm{d}s}\cdot\mathbf{b}\Big)\mathbf{b} = -\Big(\frac{\mathrm{d}\mathbf{n}}{\mathrm{d}s}\cdot\mathbf{b}\Big)\mathbf{n} = -\tau\mathbf{n}\,,$$

as $\mathbf{b}_s\cdot\mathbf{t} = -\mathbf{b}\cdot\mathbf{t}_s = 0$, $\mathbf{b}_s\cdot\mathbf{b} = 0$. □

Note the formula (3.7) also has the following form

$$\frac{\mathrm{d}\mathbf{p}}{\mathrm{d}s} = \begin{pmatrix} 0 & k & 0 \\ -k & 0 & \tau \\ 0 & -\tau & 0 \end{pmatrix}\mathbf{p}\,,$$

where $\mathbf{p} = (\mathbf{t},\mathbf{n},\mathbf{b})^T$. We also find from this expression

$$\begin{pmatrix} \dfrac{\mathrm{d}\mathbf{r}}{\mathrm{d}s} \\ \dfrac{\mathrm{d}^2\mathbf{r}}{\mathrm{d}s} \\ \dfrac{\mathrm{d}^3\mathbf{r}}{\mathrm{d}s} \end{pmatrix} = \begin{pmatrix} 1 & 0 & 0 \\ 0 & k & 0 \\ -k^2 & k' & k\tau \end{pmatrix}\mathbf{p}\,,$$

where $'$ is the first derivative with respect to s.

3.2.2 Curvature

A very important characteristic of a curve which is related to grid generation is the curvature k. This quantity is used as a measure of coordinate line bending.

One way to compute the curvature is to multiply (3.5) by $\mathbf{n}$ using the dot product operation. As from (3.3–3.5),

$$k\mathbf{n} = \frac{\mathrm{d}\mathbf{t}}{\mathrm{d}s} = \frac{1}{\sqrt{g^{r\varphi}}}\frac{\mathrm{d}}{\mathrm{d}\varphi}\left(\frac{1}{\sqrt{g^{r\varphi}}}\mathbf{r}_\varphi\right) = \frac{1}{g^{r\varphi}}\mathbf{r}_{\varphi\varphi} - \frac{1}{(g^{r\varphi})^2}(\mathbf{r}_\varphi \cdot \mathbf{r}_{\varphi\varphi})\mathbf{r}_\varphi \,,$$

the result is

$$k = \frac{1}{g^{r\varphi}}\mathbf{r}_{\varphi\varphi} \cdot \mathbf{n} \,. \tag{3.8}$$

The vector $\mathbf{n}$ is independent of the curve parametrization, and therefore we find from (3.6) and (3.8) that k is an invariant of parametrizations of the curve.

In two dimensions,

$$\mathbf{n} = \frac{1}{\sqrt{g^{r\varphi}}}(-x^2_\varphi, x^1_\varphi) \,,$$

therefore in this case we obtain from (3.8),

$$k^2 = \frac{(x_\varphi y_{\varphi\varphi} - y_\varphi x_{\varphi\varphi})^2}{[(x_\varphi)^2 + (y_\varphi)^2]^3} \tag{3.9}$$

with the convention $x = x^1$, $y = x^2$. In particular, when the curve in R^2 is defined by a function $u = u(x)$, we obtain from (3.9), assuming $\mathbf{r}(\varphi) = [\varphi, u(\varphi)]$, $\varphi = x$,

$$k^2 = (u_{xx})^2/[1 + (u_x)^2]^3 \,. \tag{3.10}$$

In the case of three-dimensional space the curvature k can also be computed from the relation obtained by multiplying (3.6) by $\mathbf{r}_\varphi$ with the cross product operation:

$$\mathbf{r}_\varphi \times \mathbf{r}_{\varphi\varphi} = g^{r\varphi}k(\mathbf{r}_\varphi \times \mathbf{n}) = (g^{r\varphi})^{3/2}k\mathbf{b} \,.$$

Thus we obtain

$$k^2 = \frac{|\mathbf{r}_\varphi \times \mathbf{r}_{\varphi\varphi}|^2}{(g^{r\varphi})^3} \tag{3.11}$$

and, consequently, from (2.26)

$$k^2 = \frac{(x^1_\varphi x^2_{\varphi\varphi} - x^2_\varphi x^1_{\varphi\varphi})^2 + (x^2_\varphi x^3_{\varphi\varphi} - x^3_\varphi x^2_{\varphi\varphi})^2 + (x^3_\varphi x^1_{\varphi\varphi} - x^1_\varphi x^3_{\varphi\varphi})^2}{[(x^1_\varphi)^2 + (x^2_\varphi)^2 + (x^3_\varphi)^2]^3} \,. \tag{3.12}$$

3.2.3 Torsion

Another important quality measure of curves in three-dimensional space is the torsion τ. This quantity is suitable for measuring the rate of twisting of the lines of coordinate grids.

In order to figure out the value of τ for a curve in R^3 we use the last relation in (3.7), which yields

$$\tau = -\frac{\mathrm{d}\mathbf{b}}{\mathrm{d}s} \cdot \mathbf{n} \,.$$

As $\mathbf{b} = \mathbf{t} \times \mathbf{n}$, we obtain

$$\frac{\mathrm{d}\mathbf{b}}{\mathrm{d}s} = \frac{\mathrm{d}\mathbf{t}}{\mathrm{d}s} \times \mathbf{n} + \mathbf{t} \times \frac{\mathrm{d}\mathbf{n}}{\mathrm{d}s} = \mathbf{t} \times \frac{\mathrm{d}\mathbf{n}}{\mathrm{d}s},$$

since $\mathrm{d}\mathbf{t}/\mathrm{d}s = k\mathbf{n}$. Thus

$$\tau = \left(-\mathbf{t} \times \frac{\mathrm{d}\mathbf{n}}{\mathrm{d}s}\right) \cdot \mathbf{n}. \tag{3.13}$$

From (3.3–3.5) we have the following obvious relations for the basic vectors $\mathbf{t}$ and $\mathbf{n}$ in terms of the parametrization $\mathbf{r}(\varphi)$ and its derivatives:

$$\mathbf{t} = \frac{1}{\sqrt{g^{r\varphi}}}\mathbf{r}_\varphi,$$

$$\mathbf{n} = \frac{1}{k}\frac{\mathrm{d}\mathbf{t}}{\mathrm{d}s} = \frac{1}{k}\left(\frac{1}{g^{r\varphi}}\mathbf{r}_{\varphi\varphi} - \frac{\mathbf{r}_\varphi \cdot \mathbf{r}_{\varphi\varphi}}{(g^{r\varphi})^2}\mathbf{r}_\varphi\right),$$

$$\frac{\mathrm{d}\mathbf{n}}{\mathrm{d}s} = \frac{1}{k}\left(\frac{1}{(g^{r\varphi})^{3/2}}\mathbf{r}_{\varphi\varphi\varphi} - 2\frac{\mathbf{r}_\varphi \cdot \mathbf{r}_{\varphi\varphi}}{(g^{r\varphi})^2}\mathbf{r}_{\varphi\varphi}\right.$$
$$\left. - \frac{\mathrm{d}}{\mathrm{d}\varphi}\left(\frac{\mathbf{r}_\varphi \cdot \mathbf{r}_{\varphi\varphi}}{(g^{r\varphi})^2}\right)\mathbf{r}_\varphi - \frac{1}{k}\frac{\mathrm{d}k}{\mathrm{d}s}\mathbf{n}\right). \tag{3.14}$$

Thus

$$\mathbf{t} \times \frac{\mathrm{d}\mathbf{n}}{\mathrm{d}s} = \frac{1}{k(g^{r\varphi})^2}\mathbf{r}_\varphi \times \mathbf{r}_{\varphi\varphi\varphi} - 2\frac{\mathbf{r}_\varphi \cdot \mathbf{r}_{\varphi\varphi}}{k(g^{r\varphi})^{5/2}}\mathbf{r}_\varphi \times \mathbf{r}_{\varphi\varphi} - \frac{1}{k^2\sqrt{g^{r\varphi}}}\frac{\mathrm{d}k}{\mathrm{d}s}\mathbf{r}_\varphi \times \mathbf{n}.$$

As $(\mathbf{a} \times \mathbf{b}) \cdot \mathbf{a} = (\mathbf{a} \times \mathbf{b}) \cdot \mathbf{b} = 0$ for arbitrary vectors $\mathbf{a}$ and $\mathbf{b}$, we obtain from (3.13, 3.14)

$$\tau = -\frac{1}{k^2(g^{r\varphi})^3}(\mathbf{r}_\varphi \times \mathbf{r}_{\varphi\varphi\varphi}) \cdot \mathbf{r}_{\varphi\varphi} = \frac{1}{k^2(g^{r\varphi})^3}(\mathbf{r}_\varphi \times \mathbf{r}_{\varphi\varphi}) \cdot \mathbf{r}_{\varphi\varphi\varphi}. \tag{3.15}$$

And, using (2.31), we also find

$$\tau = \frac{1}{k^2(g^{r\varphi})^3} \det \begin{pmatrix} \frac{\mathrm{d}x^1}{\mathrm{d}\varphi} & \frac{\mathrm{d}x^2}{\mathrm{d}\varphi} & \frac{\mathrm{d}x^1}{\mathrm{d}\varphi} \\ \frac{\mathrm{d}^2x^1}{\mathrm{d}\varphi^2} & \frac{\mathrm{d}^2x^2}{\mathrm{d}\varphi^2} & \frac{\mathrm{d}^2x^3}{\mathrm{d}\varphi^2} \\ \frac{\mathrm{d}^3x^1}{\mathrm{d}\varphi^3} & \frac{\mathrm{d}^3x^2}{\mathrm{d}\varphi^3} & \frac{\mathrm{d}^3x^3}{\mathrm{d}\varphi^3} \end{pmatrix}. \tag{3.16}$$

4 Multidimensional Geometry

The notion of a curve in R^n is readily extended to a notion of an n-dimensional surface in R^{n+l}, $l \geq 0$.

A regular n-dimensional surface of class $C^m(m \geq 1)$ is the point set in real $(n+l)$-dimensional space R^{n+l} locally represented by some parametric n-dimensional domain S^n and a parametrization

$$\mathbf{r}(\mathbf{s}) : S^n \to R^{n+l} \ , \ \mathbf{r}(\mathbf{s}) = [r^1(\mathbf{s}), \ldots, r^{n+l}(\mathbf{s})] \ , \ \mathbf{s} = (s^1, \ldots, s^n) \ , \tag{4.1}$$

such that all partial derivatives of $\mathbf{r}(\mathbf{s})$ of order m are continuous in S^n and the rank of the matrix $(\partial r^i / \partial s^j)$, $i = 1, \ldots, n+l$, $j = 1, \ldots, n$, equals n at each point of S^n. The vector equation (4.1) is called a parametric equation of the n-dimensional surface while the variables s^i, $i = 1, \ldots, n$, are referred to as curvilinear coordinates on the surface. We shall use the designation S^{rn} for the surface represented by (4.1). In accordance with the definition a curve is meant as a one-dimensional surface.

In grid generation methods regular n-dimensional surfaces are typical objects as boundaries of the domains under consideration, coordinate hypersurfaces, and monitor surfaces specified to generate adaptive meshes. The advanced grid technology also requires the application of the theories of more sophisticated geometries, namely, Riemannian manifolds which generalize regular surfaces. These geometries have real potential to provide efficient means to control the qualitative properties of grids and develop advanced grid technologies.

This chapter gives an introduction to the theory of multidimensional surfaces and Riemannian manifolds.

4.1 Tangent and Normal Vectors and Tangent Plane

Let $\mathbf{s}(t) : [a, b] \to S^n$ be a representation of a curve in S^n. Then the parametrization

$$\mathbf{r}[\mathbf{s}(t)] : [a, b] \to R^{n+l} \tag{4.2}$$

represents a curve in the n-dimensional surface S^{rn} specified by (4.1). The tangent vector to this curve forms a tangent vector to the surface S^{rn}. The

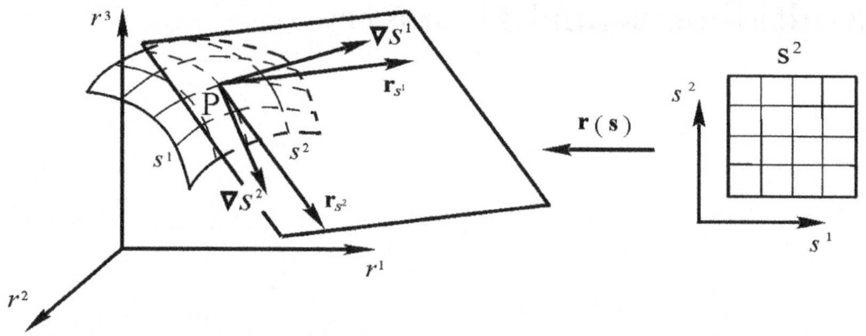

Fig. 4.1. Base tangent and normal vectors to coordinate lines

set of all tangent vectors at a point on the surface S^{rn} forms the tangent n-dimensional plane to the surface at this point.

Analogously to the definition of the coordinate line in space (section 2.2) there is defined the s^ith coordinate line in the surface S^{rn} as a curve represented by the following vector-valued function dependent upon a variable φ in the capacity of s^i:

$$\begin{aligned} &\mathbf{r}[\mathbf{s}^i(\varphi)] : [a,b] \to R^{n+l} , \\ &\mathbf{s}^i(\varphi) = (s_0^1, \dots, s_0^{i-1}, \varphi, s_0^{i+1}, \dots, s_0^n) , \quad i \text{ fixed} , \end{aligned} \tag{4.3}$$

here $\mathbf{r}(\mathbf{s})$ is the function from (4.1), $\mathbf{s}^i(\varphi) \in S^n$, $\varphi \in [a,b]$, the constants s_0^j, $j \neq i$ are fixed.

Each s^ith coordinate line defines one basic tangent vector along this curve

$$\mathbf{r}_{s^i} = \frac{\partial \mathbf{r}}{\partial s^i} , \quad i = 1, \dots , n ,$$

assuming in (4.3) $\varphi = s^i$. The transformation $\mathbf{r}(\mathbf{s})$ is of rank n hence the basic tangent vectors $\mathbf{r}_{s^i}$, $i = 1, \dots , n$, at a point P are independent and therefore form the tangent plane at this point (Fig. 4.1 for $n = 2$).

Similarly to the coordinate hypersurface in space (section 2.3) we define a coordinate hypersurface in S^{rn} as an $(n-1)$-dimensional surface lying in S^{rn} along which all of the coordinates $s^1, \dots , s^n$ except one, say s^i, are varied. Thus the s^ith coordinate hypersurface is specified by the parametrization

$$\begin{aligned} &\mathbf{r}[\mathbf{s}_i(s^1, \dots , s^{i-1}, s^{i+1}, \dots , s^n) : S^{n-1} \to R^{n+l} , \\ &\mathbf{s}_i(s^1, \dots , s^{i-1}, s^{i+1}, \dots , s^n) = (s^1, \dots , s^{i-1}, s_0^i, s^{i+1}, \dots , s^n) , \end{aligned} \tag{4.4}$$

where s_0^i fixed, while $\mathbf{r}(\mathbf{s})$ is the function from (4.1). We personify this coordinate hypersurface with the equation $s^i = s_0^i$. Equation (4.4) readily yields that the basic tangent vectors to the hypersurface $s^i = s_0^i$ are the vectors $\mathbf{r}_{s^j}$, $j \neq i$.

A vector lying in the tangent plane to S^{rn} and orthogonal to the coordinate hypersurface $s^i = s^i_0$ in S^{rn} (consequently to its basic tangent vectors $\mathbf{r}_{s^j}$, $j \neq i$) is called a vector orthogonal or a normal vector to this hypersurface in S^{rn}. Let us introduce, analogously to the normal vectors in space, as a basic normal vector to the coordinate hypersurface $s^i = s^i_0$ in S^{rn} the vector designated by ∇s^i for which

$$\nabla s^i \cdot \mathbf{r}_{s^j} = \delta^i_j \,, \quad i, j = 1, \dots, n \,, \quad i \text{ fixed} \,, \tag{4.5}$$

(Fig. 4.1).

The basic normal vectors to the coordinate hypersurface in S^{rn} are described inambiguously in the following section through the basic tangent vectors and elements of the metric tensors of S^{rn}.

4.2 First Groundform

All the properties of an n-dimensional surface S^{rn} which can be described without referring to the surrounding space are called intrinsic properties of the surface and their description constitutes the intrinsic geometry of the surface. Any characteristic of the intrinsic geometry is defined by the surface metric tensor whose elements are specified through the operation of the dot product on the basic tangent vectors.

4.2.1 Covariant Metric Tensor

Definition

The covariant metric tensor of any regular n-dimensional surface S^{rn} represented in the coordinates $s^1, \dots, s^n$ by (4.1) is the matrix (g^{rs}_{ij}), $i, j = 1, \dots, n$, where

$$g^{rs}_{ij} = \mathbf{r}_{s^i} \cdot \mathbf{r}_{s^j} \,, \quad i, j = 1, \dots, n \,. \tag{4.6}$$

In particular, when the surface is a monitor surface i.e. it is identified with the graph of the values of some vector-valued function $\mathbf{u}(\mathbf{s})$ over a domain S^n then this surface is parametrized by the equation $\mathbf{r}(\mathbf{s}) = [\mathbf{s}, \mathbf{r}(\mathbf{s})]$ and consequently

$$g^{rs}_{ij} = \delta^j_i + \frac{\partial \mathbf{u}}{\partial s^i} \cdot \frac{\partial \mathbf{u}}{\partial s^j} \,, \quad i, j = 1, \dots, n \,.$$

Quadratic Form

The differential quadratic form

$$g^{rs}_{ij} \mathrm{d}s^i \mathrm{d}s^j \,, \quad i, j = 1, \dots, n \,,$$

relating to the line elements in space, is called the first groundform or fundamental form of the surface. It represents the value of the square of the length of an elementary displacement $\mathbf{dr}$ (see Fig. 2.4) on the surface. Therefore the length of the curve (4.2) in the surface S^{rn} is computed by the formula

$$l = \int_a^b \sqrt{g_{ij}\frac{\mathrm{d}s^i}{\mathrm{d}t}\frac{\mathrm{d}s^j}{\mathrm{d}t}}\mathrm{d}t\,, \quad i,j = 1,\dots,n\,,$$

which is similar to (2.16).

Let the Jacobian of (g^{rs}_{ij}) be designated by g^{rs}. Then, analogously to the length of the line (4.2), the n-dimensional area of the surface S^{rn} is computed from the formula

$$S = \int_{S^n} \sqrt{g^{rs}}\mathrm{d}\mathbf{s}\,.$$

Basic Parallelepiped

The basic parallelepiped in S^{rn} with respect to the coordinates $s^1,\dots,s^n$ is an n-dimensional parallelepiped whose edges are the basic tangent vectors $\mathbf{r}_{s^i}$, $i = 1,\dots,n$. So the quantity $\sqrt{g^{rs}_{ii}}$ for a fixed index i has the geometrical meaning of the length of the ith edge of the basic parallelepiped (see Fig. 4.2 in the case $n = 2$). Note the uniformly contracted basic parallelepiped represents to a high order of accuracy the cell of the coordinate grid in S^{rn} in the case when the parametric domain S^n is a logical domain (see also Sect. 2.2).

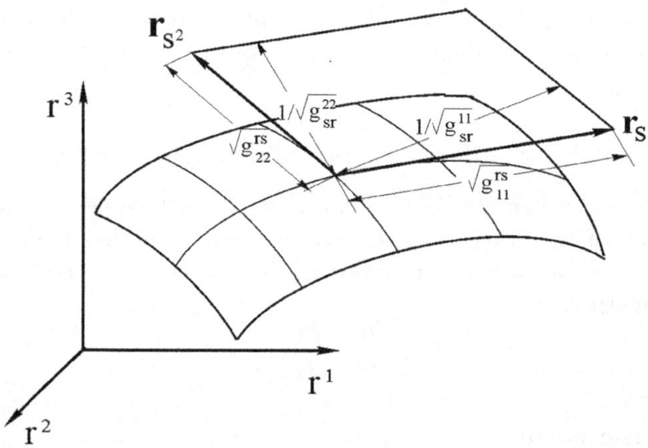

Fig. 4.2. Geometric meaning of the metric elements

4.2.2 Contravariant Metric Tensor

Definition

The contravariant metric tensor of the surface S^{rn} in the coordinates $s^1, \dots, s^n$ is the matrix (g^{ij}_{sr}), $i, j = 1, \dots, n$, inverse to (g^{rs}_{ij}), i.e. the following relations are held

$$g^{rs}_{ij}\, g^{jk}_{sr} = \delta^i_k\,, \quad i, j, k = 1, \dots, n\,. \tag{4.7}$$

Thus analogously to (2.21) we obtain, in the case $n = 2$,

$$\begin{aligned} g^{ij}_{sr} &= (-1)^{i+j} g^{rs}_{3-i\;3-j}/g^{rs}\,, \\ g^{rs}_{ij} &= (-1)^{i+j} g^{rs} g^{3-i\;3-j}_{sr}\,, \quad i, j = 1, 2\,, \end{aligned} \tag{4.8}$$

with fixed indices i and j. Similarly to (2.22) we find, in the case $n = 3$,

$$\begin{aligned} g^{ij}_{sr} &= \frac{1}{g^{rs}}\big(g^{rs}_{i+1\;j+1}\, g^{rs}_{i+2\;j+2} - g^{rs}_{i+1\;j+2}\, g^{rs}_{i+2\;j+1}\big)\,, \\ g^{rs}_{ij} &= g^{rs}\big(g^{i+1\;j+1}_{sr}\, g^{i+2\;j+2}_{sr} - g^{i+1\;j+2}_{sr}\, g^{i+2\;j+1}_{sr}\big)\,, \\ & i, j = 1, 2, 3\,, \quad i, j \text{ fixed}\,, \end{aligned} \tag{4.9}$$

with the convention that any index, say l, is identified with $l \pm 3$.

Computation of Basic Normal Vectors

Using the elements of the contravariant metric tensor we can readily find the expression for the basic normal vector ∇s^i to the coordinate hypersurface $s^i = s^i_0$, satisfying (4.5), through the basic tangent vectors $\mathbf{r}_{s^j}$, $j = 1, \dots, n$. Namely

$$\nabla s^i = g^{ij}_{sr}\mathbf{r}_{s^j}\,, \quad i, j = 1, \dots, n\,. \tag{4.10}$$

Indeed, the condition (4.5) is observed since

$$\nabla s^i \cdot \mathbf{r}_{s^k} = g^{ij}_{sr}\mathbf{r}_{s^j} \cdot \mathbf{r}_{s^k} = g^{ij}_{sr} g^{rs}_{jk} = \delta^i_k\,, \quad i, j, k = 1, \dots, n\,,$$

i.e. the vector ∇s^i is orthogonal to the coordinate hypersurface $s^i = const$ in S^{rn} at the point of consideration. As $\nabla s^i \cdot \mathbf{r}_{s^i} = 1$, i fixed, we conclude that the both basic vectors ∇s^i and $\mathbf{r}_{s^i}$ have the same direction with respect to the hypersurface $s^i = const$ in S^{rn}.

From (4.10) we readily find that

$$\nabla s^i \cdot \nabla s^j = g^{ij}_{sr}\,, \quad i, j = 1, \dots, n\,. \tag{4.11}$$

Note also that for the length of ∇s^i we obtain the following expression

$$|\nabla s^i| = \sqrt{\nabla s^i \cdot \nabla s^i} = \sqrt{g^{ii}_{sr}}\,, \quad i = 1, \dots, n\,, \quad i \text{ fixed}\,.$$

Since the normal vector ∇s^i is orthogonal to the vectors $\mathbf{r}_{s^j}$, $j \neq i$, we found that the distance d_i between the ith $(n-1)$-dimensional faces of the basic parallelepiped formed by the tangent vectors $\mathbf{r}_{s^j}$, $j = 1, \dots, n$, is computed as follows:

$$d_i = \mathbf{r}_{s^i} \cdot \frac{\nabla s^i}{\|\nabla s^i\|} = \frac{1}{\sqrt{g_{sr}^{ii}}} \,, \quad i = 1, \dots, n \,, \quad i \text{ fixed} \,.$$

Thus with respect to the basic parallelepiped the quantity $\sqrt{g_{ii}^{rs}}$ (i fixed) is the length of its ith edge, while the quantity $1/\sqrt{g_{sr}^{ii}}$ (i fixed) is the distance between the parallel $(n-1)$-dimensional faces of the parallelepiped, which are formed by the vectors $\mathbf{r}_{s^j}$, $j \neq i$ (see Fig. 4.2 for $n = 2$).

Normal Vector to a Hypersurface

Formula (4.10) is readily extended to the case of the hypersurface in S^{rn} defined by an equation $\varphi(\mathbf{s}) = 0$. Namely, a normal $\mathbf{n}$ to this hypersurface in S^{rn} is specified by the following formula

$$\mathbf{n} = \varphi_{s^i} g_{sr}^{ik} \mathbf{r}_{s^k} \,, \quad i, k = 1, \dots, n \,, \tag{4.12}$$

that is a generalization of (4.10) obtained for the equation $\varphi(\mathbf{s}) \equiv s^i - const.$ The validity of (4.12) will be proved if we show that

$$\mathbf{n} \cdot \mathbf{t} = 0 \,,$$

where $\mathbf{t}$ is an arbitrary tangent vector to the hypersurface $\varphi(\mathbf{s}) = 0$. Without loss of generality we can assume that $\varphi_{s^n} \neq 0$ at a point $\mathbf{s} \in S^n$ under consideration. Then the equation $\varphi(\mathbf{s}) = 0$ is resolved with respect to s^n, i.e. there exists a function $s^n(s^1, \dots, s^{n-1})$, $(s^1, \dots, s^{n-1}) \in S^{n-1}$ such that

$$\varphi[s^1, \dots, s^{n-1}, s^n(s^1, \dots, s^{n-1})] \equiv 0 \,, \quad (s^1, \dots, s^{n-1}) \in S^{n-1} \,.$$

Therefore, using (4.1), we can locally specify the hypersurface $\varphi(\mathbf{s}) = 0$ in the coordinates $s^1, \dots, s^{n-1}$ by the following parametrization

$$\mathbf{r}^{\varphi}(s^1, \dots, s^{n-1}) = \mathbf{r}[s^1, \dots, s^{n-1}, s^n(s^1, \dots, s^{n-1})] : S^{n-1} \to R^{n+l} \,.$$

Consequently the basic tangent vectors to this hypersurface with respect to the coordinates $s^1, \dots, s^{n-1}$ are computed by

$$\mathbf{r}_{s^i}^{\varphi} = \mathbf{r}_{s^i} + \frac{\partial s^n}{\partial s^i} \mathbf{r}_{s^n} = \mathbf{r}_{s^i} - \frac{\varphi_{s^i}}{\varphi_{s^n}} \mathbf{r}_{s^n} \,, \quad i = 1, \dots, n-1 \,,$$

since $\partial s^n / \partial s^i = -\varphi_{s^i} / \varphi_{s^n}$, $i = 1, \dots, n-1$. Using these relations and (4.12) gives

$$\begin{aligned}\mathbf{n}\cdot\mathbf{r}^{\varphi}_{s^i} &= \varphi_{s^m} g^{mk}_{sr}\mathbf{r}_{s^k}\left(\mathbf{r}_{s^i} - \frac{\varphi_{s^i}}{\varphi_{s^n}}\mathbf{r}_{s^n}\right)\\ &= \varphi_{s^m} g^{mk}_{sr}\left(g^{rs}_{ki} - \frac{\varphi_{s^i}}{\varphi_{s^n}} g^{rs}_{kn}\right)\\ &= \varphi_{s^i} - \frac{\varphi_{s^i}\varphi_{s^n}}{\varphi_{s^n}} = 0\,, \quad i = 1,\dots,n-1\,.\end{aligned}$$

As an arbitrary tangent vector $\mathbf{t}$ to the hypersurface is expanded by $\mathbf{r}^{\varphi}_{s^i}$, $i = 1,\dots,n-1$, we obtain that $\mathbf{n}\cdot\mathbf{t} = 0$, i.e. formula (4.12) gives a real expression for the vector $\mathbf{n}$ normal to the hypersurface $\varphi(\mathbf{s}) = 0$. □

We have, by virtue of (4.12), the following formula for the length of $\mathbf{n}$

$$\begin{aligned}|\mathbf{n}| &= \sqrt{\mathbf{n}\cdot\mathbf{n}} = \sqrt{(\varphi_{s^i} g^{ik}_{sr}\mathbf{r}_{s^k})\cdot(\varphi_{s^m} g^{ml}_{sr}\mathbf{r}_{s^l})}\\ &= \sqrt{\varphi_{s^i}\varphi_{s^m} g^{ik}_{sr} g^{ml}_{sr} g^{rs}_{kl}} \qquad (4.13)\\ &= \sqrt{\varphi_{s^m}\varphi_{s^l} g^{ml}_{sr}}\,, \quad i,k,l,m = 1,\dots,n\,.\end{aligned}$$

4.3 Generalization to Riemannian Manifolds

4.3.1 Definition of the Manifolds

Formula (4.6) readily yields the result that the elements of the covariant metric tensor (g^{rs}_{ij}) and (g^{rv}_{ij}) of the regular surface S^{rn} in arbitrary coordinates $s^1,\dots,s^n$ and $v^1,\dots,v^n$, respectively, are connected by the following relations

$$g^{rs}_{ij} = g^{rv}_{kl}\frac{\partial v^k}{\partial s^i}\frac{\partial v^l}{\partial s^j}\,, \quad i,j,k,l = 1,\dots,n\,. \qquad (4.14)$$

Indeed,

$$\begin{aligned}g^{rs}_{ij} &= \mathbf{r}_{s^i}\cdot\mathbf{r}_{s^j} = \mathbf{r}_{v^k}\frac{\partial v^k}{\partial s^i}\cdot\mathbf{r}_{v^m}\frac{\partial v^m}{\partial s^j}\\ &= \mathbf{r}_{v^k}\cdot\mathbf{r}_{v^m}\frac{\partial v^k}{\partial s^i}\frac{\partial v^m}{\partial s^j} = g^{rv}_{km}\frac{\partial v^k}{\partial s^i}\frac{\partial v^m}{\partial s^j}\,, \quad i,j,k,m = 1,\dots,n\,,\end{aligned}$$

i.e. equations (4.14) are held.

Analogously, the elements of the contravariant metric tensor (g^{ij}_{sr}) and (g^{ij}_{vr}) of the regular surface S^{rn} in the coordinates $s^1,\dots,s^n$ and $v^1,\dots,v^n$, respectively, are connected by

$$g^{ij}_{vr} = g^{kl}_{sr}\frac{\partial v^i}{\partial s^k}\frac{\partial v^j}{\partial s^l}\,, \quad i,j,k,l = 1,\dots,n\,. \qquad (4.15)$$

For showing that the components g^{ij}_{vr} are subject to (4.15), it is sufficient to demonstrate that the matrix

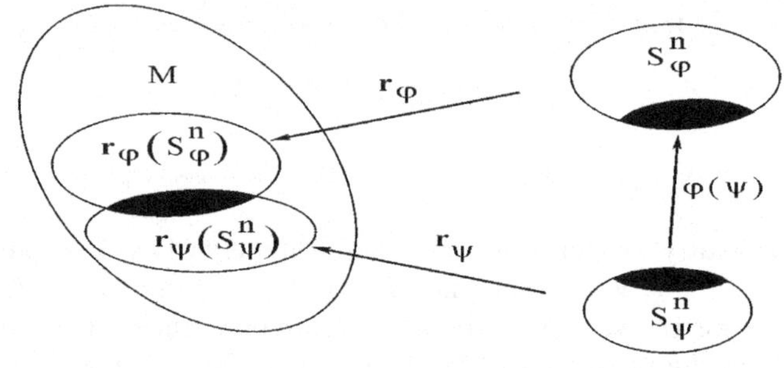

Fig. 4.3. Illustration for Riemannian manifold maps

$$\left(g^{kl}_{sr}\frac{\partial v^i}{\partial s^k}\frac{\partial v^j}{\partial s^l}\right)\,,\qquad i,j,k,l=1,\dots,n\,,$$

coincides with (g^{ij}_{vr}), i.e. it is the inverse of (g^{rv}_{ij}) provided the matrix (g^{ij}_{sr}) is the inverse of (g^{rs}_{ij}). Since (4.14)

$$g^{rv}_{ij}g^{kl}_{sr}\frac{\partial v^j}{\partial s^k}\frac{\partial v^t}{\partial s^l}=g^{rs}_{mp}\frac{\partial s^m}{\partial v^i}\frac{\partial s^p}{\partial v^j}\frac{\partial v^j}{\partial s^k}g^{kl}_{sr}\frac{\partial v^t}{\partial s^l}=\delta^t_i\,,\quad i,j,k,l,m,p,t=1,\dots,n\,,$$

i.e. equations (4.15) are valid.

The relations (4.14) and (4.15) valid for the metrics of arbitrary regular n-dimensional surfaces give rise to the definition of the geometrical objects called Riemannian manifolds which generalize the regular surfaces.

Namely, a point set M^n of R^{n+l}, $l\geq 1$ is called a C^m-differential Riemannian manifold (Fig. 4.3) of dimension n if there is a collection (atlas) Φ of local parametrizations of M (local maps)

$$\begin{aligned}&\mathbf{r}_\varphi(\boldsymbol{\varphi}):S^n_\varphi\to M\,,\quad \varphi\in\Phi\,,\quad \boldsymbol{\varphi}\in S^n_\varphi\,,\\ &\boldsymbol{\varphi}=(\varphi^1,\dots,\varphi^n)\,,\quad \mathbf{r}_\varphi=(r^1_\varphi,\dots,r^{n+l}_\varphi)\,,\end{aligned}\tag{4.16}$$

where $S^n_\varphi\subset R^n$ is an n-dimensional parametric domain, and the same collection of local variable matrices (local covariant metric tensors)

$$(g^{r\varphi}_{ij}(\boldsymbol{\varphi}))\,,\quad i,j=1,\dots,n\,,\quad \boldsymbol{\varphi}\in S^n_\varphi\,,\quad \varphi\in\Phi\,,\tag{4.17}$$

such that

1) $\cup_{\varphi\in\Phi}\mathbf{r}_\varphi(S^n_\varphi)=M^n$, where $\mathbf{r}_\varphi(S^n_\varphi)$ is the image of S^n_φ in M^n built by $\mathbf{r}_\varphi$,
2) for each $\varphi\in\Phi$, $\mathbf{r}_\varphi(S^n_\varphi)$ is an open space of M^n and

$$\mathbf{r}_\varphi:S^n_\varphi\to\mathbf{r}_\varphi(S^n_\varphi)\,,\quad \varphi\in\Phi\,,$$

is a one-to-one continuous mapping,

3) for each $\varphi, \psi \in \Phi$ with $B_{\varphi\psi} \neq \emptyset$, where

$$B_{\varphi\psi} = \mathbf{r}_\varphi(S^n_\varphi) \cap \mathbf{r}_\psi(S^n_\psi) ,$$

the map

$$\boldsymbol{\varphi}(\boldsymbol{\psi}) : S^n_\psi \cap \mathbf{r}^{-1}_\psi(B_{\varphi\psi}) \to R^n ,$$

where $\boldsymbol{\varphi}(\boldsymbol{\psi}) = \mathbf{r}^{-1}_\varphi[\mathbf{r}_\psi(\boldsymbol{\psi})]$, is a C^m – map,

4) for each $\varphi \in \Phi$, $\boldsymbol{\varphi} \in S^n_\varphi$ the matrix $(g^{r\varphi}_{ij}(\boldsymbol{\varphi}))$ is positive, symmetric, and nondegenerate and the functions

$$g^{r\varphi}_{ij}(\boldsymbol{\varphi}) : S^n_\varphi \to R, \quad i,j = 1, \dots, n,$$

are C^m-functions,

5) for each $\varphi, \psi \in \Phi$, $B_{\varphi\psi} \neq \emptyset$,

$$g^{r\psi}_{ij}(\boldsymbol{\psi}) = g^{r\varphi}_{kl}[\boldsymbol{\varphi}(\boldsymbol{\psi})] \frac{\partial \varphi^k}{\partial \psi^i}(\boldsymbol{\psi}) \frac{\partial \varphi^l}{\partial \psi^j}(\boldsymbol{\psi}) , \quad i,j,k,l = 1, \dots, n , \tag{4.18}$$

where

$$\boldsymbol{\varphi} \in S^n_\varphi \cap \mathbf{r}^{-1}_\varphi(B_{\varphi\psi}) , \quad \boldsymbol{\psi} \in S^n_\psi \cap \mathbf{r}^{-1}_\psi(B_{\varphi\psi}) ,$$

$\varphi^i(\boldsymbol{\psi})$, $i = 1, \dots, n$, is the ith component of the function $\boldsymbol{\varphi}(\boldsymbol{\psi}) = \mathbf{r}^{-1}_\varphi(\mathbf{r}_\psi)$,

6) Φ is maximal relative to (2-5) that is if there is one more local parametrization

$$\mathbf{r}_\theta(\boldsymbol{\theta}) : S^n_\theta \to X , \quad S^n_\theta \subset R^n , \quad \boldsymbol{\theta} \in S^n_\theta ,$$

and a nondegenerate, positive, symmetric C^m – matrix function

$$(g^{r\theta}_{ij}(\boldsymbol{\theta})) , \quad \boldsymbol{\theta} \in S^n_\theta ,$$

and their inclusion into local paramotrizations and local matrices, respectively, does not violate the requirements (2-5) – they are in the corresponding collections of Φ.

Here m may be $1, 2, \dots, \infty$. C^m for m finite means all partial derivatives of order less than or equal to m exist and are continuous.

The variables $\varphi^1, \dots, \varphi^n$ of the parametric domain S^n_φ, i.e.

$$\boldsymbol{\varphi} = (\varphi^1, \dots, \varphi^n) \in S^n_\varphi , \quad \varphi \in \Phi ,$$

together with the corresponding transformation

$$\mathbf{r}_\varphi(\boldsymbol{\varphi}) : S^n_\varphi \to M^n$$

are called a local coordinate system or local coordinates of M^n.

Each local covariant metric tensor $(g^{r\varphi}_{ij}(\boldsymbol{\varphi}))$ derives the local contravariant metric tensor $(g^{ij}_{\varphi r}(\boldsymbol{\varphi}))$ as the inverse matrix of $(g^{r\varphi}_{ij}(\boldsymbol{\varphi}))$, i.e.

$$g^{ij}_{\varphi r}(\varphi) = G^{\varphi}_{ij}/\det(g^{r\varphi}_{ij}(\varphi)) \,, \quad i,j = 1,\dots,n \,,$$

where G^{φ}_{ij} is the (ij)th cofactor of $(g^{r\varphi}_{ij}(\varphi))$. Analogously to (4.15) it is readily verified that the equations (4.18) yield

$$g^{ij}_{\psi r}(\psi) = g^{kl}_{\varphi r}[\varphi(\psi)] \frac{\partial \psi^i}{\partial \varphi^k} \frac{\partial \psi^j}{\partial \varphi^l} \,, \quad i,j,k,l = 1,\dots,n \,, \tag{4.19}$$

where $\psi^k(\varphi)$, $k = 1,\dots,n$, is the kth component of the function $\psi(\varphi) = \mathbf{r}^{-1}_{\psi}[\mathbf{r}_{\varphi}(\varphi)]$.

The need of multiple charts is essential for geometries that are not diffeomorphic to an n-dimensional cube. Such geometries appear in various applications, the most complicated and difficult for gridding of them are human organs.

4.3.2 Example of a Riemannian Manifold

In the grid generation theory the Riemannian manifolds appear as tools to control the grids derived by the generalized Laplace equations. Typically these manifolds are formulated as a generalization of monitor surfaces. Namely, let S^{xn} be a regular n-dimensional surface lying in R^{n+k} and represented locally by a mapping

$$\mathbf{x}(\mathbf{s}) : S^n \to R^{n+k} \,, \quad \mathbf{s} = (s^1,\dots,s^n) \,, \quad \mathbf{x} = (x^1,\dots,x^{n+k}) \,,$$

with a covariant metric tensor (g^{xs}_{ij}) in the coordinates $s^1,\dots,s^n$

$$g^{xs}_{ij} = \mathbf{x}_{s^i} \cdot \mathbf{x}_{s^j} \,, \quad i,j = 1,\dots,n \,. \tag{4.20}$$

By a monitor surface over S^{xn} there is meant a regular surface whose points are

$$[\mathbf{x}, \mathbf{f}(\mathbf{x})] \in R^{n+l+k} \,,$$

where $\mathbf{x} \in S^{xn}$, while

$$\mathbf{f}(\mathbf{x}) : S^{xn} \to R^l \,, \quad \mathbf{f} = (f^1,\dots,f^l) \,,$$

is some vector-valued function called a monitor function. The monitor surface is represented in the coordinates $s^1,\dots,s^n$ by the following local parametrization

$$\mathbf{r}(\mathbf{s}) : S^n \to R^{n+k+l} \,, \quad \mathbf{r}(\mathbf{s}) = \{\mathbf{x}(\mathbf{s}), \mathbf{f}[\mathbf{x}(\mathbf{s})]\} \,.$$

An extension of the monitor surface over S^{xn} is produced, for example, by two scalar-valued weight functions $z(\mathbf{s}) > 0$ and $v(\mathbf{s}) \geq 0$ and one vector-valued smooth function $\mathbf{f}(\mathbf{s}) = [f^1(\mathbf{s}),\dots,f^l(\mathbf{s})]$ which form a Riemannian manifold M^n whose points and parametrizations are the same as of S^{xn}, while the elements of the covariant metric tensor in the coordinates $s^1,\dots,s^n$, designated as $g^{\mathbf{s}}_{ij}$, are defined as follows:

$$g^{\mathbf{s}}_{ij} = z(\mathbf{s})g^{xs}_{ij} + v(\mathbf{s})\frac{\partial \mathbf{f}}{\partial s_i}(\mathbf{s}) \cdot \frac{\partial \mathbf{f}}{\partial s_j}(\mathbf{s}) , \quad i,j = 1,\dots,n . \tag{4.21}$$

We shall call the Riemannian manifold with the metric (4.21) imposed by the functions $z(\mathbf{s})$, $v(\mathbf{s})$, and $\mathbf{f}(\mathbf{s})$ as a monitor manifold. The function $\mathbf{f}(\mathbf{s})$ will be referred to as a monitor function, while $z(\mathbf{s})$ and $v(\mathbf{s})$ will be called weight functions.

4.3.3 Christoffel Symbols of Manifolds

Definition of the Symbols

Any covariant metric tensor of a Riemannian manifold M^n, designated by $(g^{\mathbf{s}}_{ij})$ in the local coordinates s^i, $i = 1,\dots,n$, derives the quantities called the Christoffel symbols of the first and second kinds, which are also referred to as the three-index symbols. The symbols of the first kind, designated in the coordinates s^i, $i = 1,\dots,n$, by $[ij,k]^{\mathbf{s}}$, are defined as follows:

$$[ij,k]^{\mathbf{s}} = \frac{1}{2}\Big(\frac{\partial g^{\mathbf{s}}_{jk}}{\partial s^i} + \frac{\partial g^{\mathbf{s}}_{ik}}{\partial s^j} - \frac{\partial g^{\mathbf{s}}_{ij}}{\partial s^k}\Big) , \quad i,j,k = 1,\dots,n . \tag{4.22}$$

Equations (4.22) easily yield the following relations for the first derivatives of the elements of the covariant metric tensor

$$\frac{\partial g^{\mathbf{s}}_{ik}}{\partial s^j} = [ij,k]^{\mathbf{s}} + [kj,i]^{\mathbf{s}} , \quad i,j,k = 1,\dots,n . \tag{4.23}$$

The Christoffel symbols of the second kind, designated in the coordinates s^i, $i = 1,\dots,n$, by ${}^{\mathbf{s}}\Upsilon^l_{ij}$, are defined by the equations

$${}^{\mathbf{s}}\Upsilon^l_{ij} = g^{lm}_{\mathbf{s}}[ij,m]^{\mathbf{s}} , \quad i,j,l,m = 1,\dots,n . \tag{4.24}$$

It is seen at once from (4.22) and (4.24) that the Christoffel symbols are symmetrical in i,j.

Further, when a coordinate system in a formula is fixed we, for simplicity, shall omit the superscript personifying a coordinate system in the Christoffel symbols of the first and second kinds thus designating them merely by $[ij,k]$ and Υ^l_{ij}, respectively.

Analogously to (4.23) we find an expression for the first derivatives of the elements of the contravariant metric tensor through the Christoffel symbols of the second kind

$$\begin{aligned}\frac{\partial g^{ij}_{\mathbf{s}}}{\partial s^k} &= g^{lj}_{\mathbf{s}} g^{\mathbf{s}}_{ml}\frac{\partial g^{im}_{\mathbf{s}}}{\partial s^k} = -g^{lj}_{\mathbf{s}} g^{im}_{\mathbf{s}}\frac{\partial g^{\mathbf{s}}_{ml}}{\partial s^k} \\ &= -g^{lj}_{\mathbf{s}} g^{im}_{\mathbf{s}}([mk,l] + [lk,m]) = -g^{im}_{\mathbf{s}}\Upsilon^j_{km} - g^{lj}_{\mathbf{s}}\Upsilon^i_{lk} , \\ &\qquad i,j,k,l,m = 1,\dots,n .\end{aligned} \tag{4.25}$$

The Christoffel symbols of the second kind also have some relation to the formula of the differentiation of the Jacobian $g^{\mathbf{s}}$ of the metric tensor $(g^{\mathbf{s}}_{ij})$. Indeed the rule of the differentiation of the Jacobian $g^{\mathbf{s}}$ gives

$$\frac{\partial g^{\mathbf{s}}}{\partial s^i} = g^{\mathbf{s}} g^{jm}_{\mathbf{s}} \frac{\partial g^{\mathbf{s}}_{jm}}{\partial s^i} , \quad i,j,m = 1,\dots,n ,$$

and the application of (4.23) to this formula yields

$$\frac{\partial g^{\mathbf{s}}}{\partial s^i} = g^{\mathbf{s}} g^{jm}_{\mathbf{s}}([ji,m]+[mi,j]) = g^{\mathbf{s}}[\Upsilon^j_{ji}+\Upsilon^m_{mi}] = 2g^{\mathbf{s}}\Upsilon^j_{ji} , \quad i,j,m = 1,\dots,n .$$

Remind repeated indices in a single term mean a summation over them so we imply in the above equations

$$\frac{\partial g^{\mathbf{s}}}{\partial s_i} = 2g^{\mathbf{s}}\Upsilon^j_{ji} = 2g^{\mathbf{s}}\sum_{j=1}^{n}\Upsilon^j_{ji} , \quad i,j = 1,\dots,n . \tag{4.26}$$

Symbols for Regular Surfaces

In the case of the regular surface S^{rn} represented by (4.1) whose intrinsic metric is expressed as

$$g^{\mathbf{s}}_{ij} = \mathbf{r}_{s^i} \cdot \mathbf{r}_{s^j} , \quad i,j = 1,\dots,n ,$$

it is readily found, from (4.22), that

$$[ij,k] = \mathbf{r}_{s^i s^j} \cdot \mathbf{r}_{s^k} , \quad i,j,k = 1,\dots,n . \tag{4.27}$$

Using (4.27), (4.24), and (4.10) yields the following formula for the Christoffel symbols of the second kind of the regular surface S^{rn} in the coordinates $s^1,\dots,s^n$:

$$\Upsilon^l_{ij} = g^{lm}_{\mathbf{s}}(\mathbf{r}_{s^i s^j} \cdot \mathbf{r}_{s^m}) = \mathbf{r}_{s^i s^j} \cdot \boldsymbol{\nabla} s^l , \quad i,j,l,m = 1,\dots,n , \tag{4.28}$$

where $\boldsymbol{\nabla} s^l$ is the basic normal vector to the coordinate hypersurface $s^l = c_0$ in S^{rn} (see (4.10)).

Transformation of the Christoffel Symbols

Now we shall establish how the Christoffel symbols of two coordinate systems are related. Let us designate by $g^{\mathbf{s}}_{ij}$ and $g^{\mathbf{v}}_{ij}$ the elements of the covariant metric tensor of a manifold M^n in the coordinates $s^1,\dots,s^n$ and $v^1,\dots,v^n$, respectively. Then, according to (4.18), we have

$$g^{\mathbf{v}}_{ij} = g^{\mathbf{s}}_{kl}\frac{\partial s^k}{\partial v^i}\frac{\partial s^l}{\partial v^j} , \quad i,j,k,l = 1,\dots,n .$$

Differentiating these equations with respect to v^p gives

$$\frac{\partial g^{\mathbf{v}}_{ij}}{\partial v^p} = \frac{\partial g^{\mathbf{s}}_{kl}}{\partial s^m}\frac{\partial s^m}{\partial v^p}\frac{\partial s^k}{\partial v^i}\frac{\partial s^l}{\partial v^j} + 2g^{\mathbf{s}}_{kl}\frac{\partial^2 s^k}{\partial v^i \partial v^p}\frac{\partial s^l}{\partial v^j}, \quad i,j,k,l,m,p = 1,\dots,n,$$

and consequently, applying (4.23),

$$\begin{aligned}\frac{\partial g^{\mathbf{v}}_{ij}}{\partial v^p} &= [ip,j]^{\mathbf{v}} + [jp,i]^{\mathbf{v}} \\ &= ([km,l]^{\mathbf{s}} + [lm,k]^{\mathbf{s}})\frac{\partial s^m}{\partial v^p}\frac{\partial s^k}{\partial v^i}\frac{\partial s^l}{\partial v^j} + 2g^{\mathbf{s}}_{kl}\frac{\partial^2 s^k}{\partial v^i \partial v^p}\frac{\partial s^l}{\partial v^j}, \\ &\qquad i,j,k,l,m,p = 1,\dots,n.\end{aligned}$$

Therefore, after computing by these formulas the following expression

$$-\frac{\partial g^{\mathbf{v}}_{ij}}{\partial v^p} + \frac{\partial g^{\mathbf{v}}_{pi}}{\partial v^j} + \frac{\partial g^{\mathbf{v}}_{pj}}{\partial v^i}, \quad i,j,p = 1,\dots,n,$$

we readily obtain, using first (4.22) and then (4.24),

$$\begin{aligned}[][ij,p]^{\mathbf{v}} &= [km,l]^{\mathbf{s}}\frac{\partial s^l}{\partial v^p}\frac{\partial s^k}{\partial v^i}\frac{\partial s^m}{\partial v^j} + g^{\mathbf{s}}_{kl}\frac{\partial^2 s^k}{\partial v^i \partial v^j}\frac{\partial s^l}{\partial v^p}, \\ {}^{\mathbf{v}}\Upsilon^p_{ij} &= {}^{\mathbf{s}}\Upsilon^m_{kl}\frac{\partial s^k}{\partial v^i}\frac{\partial s^l}{\partial v^j}\frac{\partial v^p}{\partial s^m} + \frac{\partial^2 s^k}{\partial v^i \partial v^j}\frac{\partial v^p}{\partial s^k}, \\ &\qquad i,j,k,l,m,p = 1,\dots,n.\end{aligned} \tag{4.29}$$

Geometric Meanings of the Symbols

Let us consider the Christoffel symbols of a regular n-dimensional surface S^{rn} represented by (4.1) and whose covariant metric tensor is defined by (4.6). We know that the first partial derivatives of the parametrization $\mathbf{r}(\mathbf{s})$ are the tangent vectors forming the tangent plane to the surface and the elements of the covariant metric tensor (g^{rs}_{ij}). It appears that the second partial derivatives of $\mathbf{r}(\mathbf{s})$ are connected with the Christoffel symbols.

Let us designate by $\mathbf{P}$ the operator which projects the vectors from R^{n+l} on the tangent plane to the regular surface $S^{rn} \subset R^{n+l}$ at a point P. Now considering the vector-valued function

$$\mathbf{r}_{s^m s^p} = \frac{\partial^2 \mathbf{r}}{\partial s^m \partial s^p}, \quad m,p = 1,\dots,n,$$

we can expand the vector $\mathbf{P}[\mathbf{r}_{s^m s^p}]$ (lying in the tangent n-dimensional plane) in both the base tangential $\mathbf{r}_{s^i}$, $i = 1,\dots,n$, and normal ∇s^i, $i = 1,\dots,n$, vectors. Applying the formula (2.6) in the case of the tangential vectors, i.e. assuming in (2.6) $\mathbf{a}_i = \mathbf{r}_{s^i}$, $i = 1,\dots,n$, we find

$$\mathbf{P}[\mathbf{r}_{s^m s^p}] = a^{ij}(\mathbf{P}[\mathbf{r}_{s^m s^p}] \cdot \mathbf{r}_{s^j})\mathbf{r}_{s^i} \ , \quad i,j,m,p = 1,\dots,n \ , \tag{4.30}$$

where a^{ij} are the elements of the matrix which is inverse to the matrix (a_{ij})

$$a_{ij} = \mathbf{r}_{s^i} \cdot \mathbf{r}_{s^j} \ , \quad i,j = 1,\dots,n \ .$$

Since $\mathbf{r}_{s^i} \cdot \mathbf{r}_{s^j} = g^{rs}_{ij}$, $i,j = 1,\dots,n$, the functions a^{ij} in (4.30) are the elements of the contravariant metric tensor (g^{ij}_{sr}), i.e.

$$a^{ij} = g^{ij}_{sr} \ , \quad i,j = 1,\dots,n \ .$$

Further, as the operator $\mathbf{P}$ projects the vector $\mathbf{r}_{s^m s^p}$ on the plane formed by the tangent vectors $\mathbf{r}_{s^i}$, $i = 1,\dots,n$, we conclude that

$$\mathbf{P}[\mathbf{r}_{s^m s^p}] \cdot \mathbf{r}_{s^j} = \mathbf{r}_{s^m s^p} \cdot \mathbf{r}_{s^j} \ , \quad j,m,p = 1,\dots,n \ .$$

Therefore equation (4.30) has the following form

$$\mathbf{P}[\mathbf{r}_{s^m s^p}] = g^{ij}_{sr}(\mathbf{r}_{s^m s^p} \cdot \mathbf{r}_{s^j})\mathbf{r}_{s^i} \ , \quad i,j,m,p = 1,\dots,n \ . \tag{4.31}$$

Now, applying (4.28) to this equation, we find

$$\mathbf{P}[\mathbf{r}_{s^m s^p}] = \Upsilon^i_{mp}\mathbf{r}_{s^i} \ , \quad i,m,p = 1,\dots,n \ . \tag{4.32}$$

Thus the Christoffel symbol Υ^i_{mp} of the second kind is the ith component of the vector $\mathbf{P}[\mathbf{r}_{s^m s^p}]$ expanded in the base tangent vectors $\mathbf{r}_{s^i}$ (see Fig. 4.4 where the vector $\mathbf{r}_{s^m s^p}$ is identified with $\mathbf{r}_{mp}$).

Similarly we obtain

$$\mathbf{P}[\mathbf{r}_{s^m s^p}] = [mp, i]\nabla s^i \ , \quad i,m,p = 1,\dots,n \ . \tag{4.33}$$

This formula can also be inferred from (4.32). Indeed multiplying (4.10) by g^{rs}_{ik} gives

$$\mathbf{r}_{s^k} = g^{rs}_{ik}\nabla s^i \ , \quad i,k = 1,\dots,n \ ,$$

and substituting this equation for $\mathbf{r}_{s^i}$ in (4.32), we readily come to (4.33).

So the Christoffel symbols of the first kind represent the components of the vector $\mathbf{P}[\mathbf{r}_{s^m s^p}]$ expanded in the base normal vectors to the coordinate hypersurfaces in S^{rn} (see Fig. 4.4 for $n = 2$).

4.4 Tensors

The theory of multidimensional geometry operates largely with the quantities called tensors. This section gives an introduction to such geometric objects.

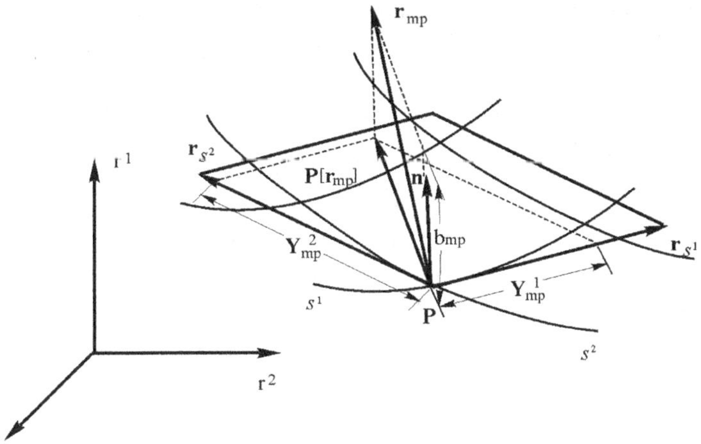

Fig. 4.4. Expension of the vector $\mathbf{r}_{s^m s^p}$ in the base vectors

4.4.1 Definition

Let M^n be some n-dimensional Riemannian manifold, in particular, a regular surface S^{rn}. A tensor of rank $k \geq 0$ at a point P of M^n is a set of values defined for each local coordinate system $s^1, \ldots, s^n$ around this point and indices $\mathbf{i} = (i_1, \ldots, i_k) \in R^k$, $i_j = 1, \ldots, n$, $j = 1, \ldots, k$, such that the values obey certain transformation laws when the coordinates are changed. The number k is called a tensor order or its rank, while the values with the indices are referred to as tensor components.

The tensors are also distinguished by their types. There are two basic tapes for the tensors of order $k > 0$: covariant and contravariant and the third mixed type having the features of the basic types.

Tensors of Order Zero

A quantity which has the same fixed value at the point P in an arbitrary coordinate system is called a scalar, or an invariant, or a tensor of order zero.

Covariant Tensors

In the pure covariant case the components of a tensor $\mathbf{f}$ of order k, whose designation is distinguished by indices being subscripts, for example, by $f^{\mathbf{s}}_{i_1 \ldots i_k}$ for the indices $i_1, \ldots, i_k$ and coordinates $s^1, \ldots, s^n$, are subject to the following relations with respect to arbitrary coordinate systems $s^1, \ldots, s^n$ and $v^1, \ldots, v^n$:

$$f^{\mathbf{s}}_{i_1 \ldots i_k} = f^{\mathbf{v}}_{j_1 \ldots j_k} \frac{\partial v^{j_1}}{\partial s^{i_1}} \cdots \frac{\partial v^{j_k}}{\partial s^{i_k}}\,, \quad i_l\,,\ j_l = 1, \ldots, n\,, \quad l = 1, \ldots, k\,. \tag{4.34}$$

Thus it is sufficient to know the values of the tensor components for some one fixed coordinate system since its values for other systems can be computed by (4.34).

Contravariant Tensors

The components of the pure contravariant tensor **f** of rank $k > 0$, designated with the help of indices being superscripts: by $f_{\mathbf{s}}^{i_1 \dots i_k}$ in the coordinates $s^1, \dots, s^n$, obey the following law with respect to arbitrary coordinate systems $s^1, \dots, s^n$ and $v^1, \dots, v^n$:

$$f_{\mathbf{s}}^{i_1 \dots i_k} = f_{\mathbf{v}}^{j_1 \dots j_k} \frac{\partial s^{i_1}}{\partial v^{j_1}} \cdots \frac{\partial s^{i_k}}{\partial v^{j_k}} , \quad i_l , j_l = 1, \dots , n , \quad l = 1, \dots , k . \tag{4.35}$$

Mixed Tensors

The notion of the pure covariant and contravariant tensors gives rise to the concept of a mixed tensor if it is covariant in some indices and contravariant in the rest of them. Consequently there are used in designations superscripts for contravariant indices and subscripts for covariant indices. Namely, a mixed tensor **f** k times covariant and l times contravariant is a set of values $f_{i_1 \dots i_k}^{j_1 \dots j_l}(\mathbf{s})$ dependent on the coordinate system $s^1, \dots, s^n$ at the point of consideration such that for an arbitrary another coordinate system $v^1, \dots, v^n$

$$\begin{aligned} & f_{i_1 \dots i_k}^{j_1 \dots j_l}(\mathbf{s}) = f_{p_1 \dots p_k}^{m_1 \dots m_l}(\mathbf{v}) \frac{\partial v^{p_1}}{\partial s^{i_1}} \cdots \frac{\partial v^{p_k}}{\partial s^{i_k}} \frac{\partial s^{j_1}}{\partial v^{m_1}} \cdots \frac{\partial s^{j_l}}{\partial v^{m_l}} , \\ & i_a, p_a, j_b, m_b = 1, \dots , n , \quad a = 1, \dots , k , \quad b = 1, \dots , l . \end{aligned} \tag{4.36}$$

4.4.2 Examples of Tensors

Covariant Tensors

A typical covariant tensor of the first order is the vector gradφ where φ is a tensor of order zero. The components of this vector designated as $(\mathrm{grad}\varphi)_i^{\mathbf{s}}$ in the coordinates $s^1, \dots, s^n$ are computed as

$$(\mathrm{grad}\varphi)_i^{\mathbf{s}} = \frac{\partial \varphi(\mathbf{s})}{\partial s^i} = \varphi_{s^i} , \quad i = 1, \dots , n . \tag{4.37}$$

It is obvious that the relations (4.34), for $k = 1$, are held for these values.

Note the second derivatives of φ, i.e. the set

$$\varphi_{ij}^{\mathbf{s}} = \varphi_{s^i s^j}(\mathbf{s}) , \quad i, j = 1, \dots , n ,$$

do not form a tensor of the second rank since

$$\varphi_{s^i s^j}(\mathbf{s}) = \varphi_{v^l v^m} \frac{\partial v^l}{\partial s^i} \frac{\partial v^m}{\partial s^j} + \varphi_{v^l} \frac{\partial^2 v^l}{\partial s^i \partial s^j} , \quad i,j,l,m = 1,\dots,n , \tag{4.38}$$

and the second term in the right-hand part of these equations impedes satisfaction of (4.34) for $k = 2$. However using the Christoffel symbols of the second kind which are not tensors as well, since (4.29), produces the following covariant tensor $\nabla_{ij}(\varphi)$ of the second kind whose (ij)th component designated in the coordinates $s^1,\dots,s^n$ as $\nabla^{\mathbf{s}}_{ij}(\varphi)$ is computed as follows:

$$\nabla^{\mathbf{s}}_{ij}(\varphi) = \varphi_{s^i s^j} - \varphi_{s^k} \Upsilon^k_{ij} , \quad i,j,k = 1,\dots,n . \tag{4.39}$$

Formulas (4.29) and (4.38) readily yield that these components satisfy the condition (4.34) for $k = 2$. Indeed

$$\begin{aligned}
\nabla^{\mathbf{v}}_{ij}(\varphi) &= \varphi_{v^i v^j} - \varphi_{v^k} {}^{\mathbf{v}}\Upsilon^k_{ij} \\
&= \varphi_{s^k s^m} \frac{\partial s^k}{\partial v^i} \frac{\partial s^m}{\partial v^j} + \varphi_{s^k} \frac{\partial^2 s^k}{\partial v^i \partial v^j} \\
&\quad - \varphi_{s^t} \left({}^{\mathbf{s}}\Upsilon^m_{pl} \frac{\partial s^p}{\partial v^i} \frac{\partial s^l}{\partial v^j} \frac{\partial v^k}{\partial s^m} + \frac{\partial^2 s^p}{\partial v^i \partial v^j} \frac{\partial v^k}{\partial s^p} \right) \frac{\partial s^t}{\partial v^k} \\
&= (\varphi_{s^k s^m} - \varphi_{s^p} {}^{\mathbf{s}}\Upsilon^p_{km}) \frac{\partial s^k}{\partial v^i} \frac{\partial s^m}{\partial v^j} = \nabla^{\mathbf{s}}_{km}(\varphi) \frac{\partial s^k}{\partial v^i} \frac{\partial s^m}{\partial v^j} , \\
&\quad i,j,k,l,m,p,t = 1,\dots,n ,
\end{aligned} \tag{4.40}$$

i.e. the quantities $\nabla^{\mathbf{s}}_{km}(\varphi)$ form a covariant tensor of the second kind. This tensor is called a tensor of mixed covariant derivatives of the invariant φ.

Analogous construction over an arbitrary covariant vector $\mathbf{f} = (f^{\mathbf{s}}_i)$ defines a covariant tensor of the second kind called a covariant derivative of this vector. Its (ij)th component, designated by $(\nabla f)^{\mathbf{s}}_{ij}$ in the coordinates $s^1,\dots,s^n$, is computed by the following formula

$$(\nabla f)^{\mathbf{s}}_{ij} = \frac{\partial}{\partial s^j} f^{\mathbf{s}}_i - f^{\mathbf{s}}_k \Upsilon^k_{ij} , \quad i,j,k = 1,\dots,n . \tag{4.41}$$

Tensor relations (4.34), for $n = 2$, are verified for these components similarly as in (4.40).

It is obvious, comparing (4.39) and (4.41), that

$$\nabla^{\mathbf{s}}_{ij}(\varphi) = (\nabla \mathrm{grad}\varphi)^{\mathbf{s}}_{ij} , \quad i,j = 1,\dots,n .$$

In the case of a regular surface $S^{rn} \subset R^{n+l}$ represented by (4.1) another example of a covariant vector is formed through an arbitrary fixed vector $\mathbf{P} \in R^{n+l}$ by the following formula for its components in the coordinates $s^1,\dots,s^n$:

$$f^{\mathbf{s}}_i = \mathbf{P} \cdot \mathbf{r}_{s^i} , \quad i = 1,\dots,n , \tag{4.42}$$

where $\mathbf{r}_{s^i}$ is the ith basic tangent vector of S^{rn} in the coordinates $s^1,\dots,s^n$.

If $\mathbf{P} \in R^{n+l}$ is a vector orthogonal to S^{rn} at a point P then the quantities

$$b^{\mathbf{s}}_{ij} = \mathbf{r}_{s^is^j} \cdot \mathbf{P}\ , \quad i,j = 1,\dots,n\ , \tag{4.43}$$

form a covariant tensor of the second order. Indeed

$$\begin{aligned} b^{\mathbf{v}}_{ij} = \mathbf{r}_{v^iv^j} \cdot \mathbf{P} &= \Big(\frac{\partial s^k}{\partial v^i}\frac{\partial s^m}{\partial v^j}\mathbf{r}_{s^ks^m} + \frac{\partial^2 s^k}{\partial v^i \partial v^j}\mathbf{r}_{s^k}\Big) \cdot \mathbf{P} \\ &= \frac{\partial s^k}{\partial v^i}\frac{\partial s^m}{\partial v^j}\mathbf{r}_{s^ks^m} \cdot \mathbf{P} = b^{\mathbf{s}}_{km}\frac{\partial s^k}{\partial v^i}\frac{\partial s^m}{\partial v^j}\ , \quad i,j,k,m = 1,\dots,n\ . \end{aligned}$$

Since (4.14) or (4.18), a typical example of a symmetric covariant tensor of the second order also gives the metric tensor of any regular m-dimensional surface S^{rn} defined by (4.1) or a manifold M^n.

There is an evident rule for forming a new covariant tensor from two original ones. Namely, let $\mathbf{f}$ and $\mathbf{v}$ be two covariant tensors of the rank k and l, respectively. Then the new tensor $\mathbf{f} \otimes \mathbf{v}$ is the covariant tensor of the rank $k+l$ whose components in the coordinates $s^1,\dots,s^n$ are computed as

$$(\mathbf{f} \otimes \mathbf{v})^{\mathbf{s}}_{i_1\dots i_{k+l}} = f^{\mathbf{s}}_{i_1\dots i_k} v^{\mathbf{s}}_{i_{k+1}\dots i_{k+l}}\ , \quad i_j = 1,\dots,n\ . \tag{4.44}$$

In particular, two smooth functions f and φ specified in the vicinity of a point $P \in S^{rn}$ produce a covariant tensor of the second rank, through the covariant vectors formulated by (4.37):

$$(f \otimes \varphi)^{\mathbf{s}}_{ij} = f_{s^i}\varphi_{s^j}\ , \quad i,j = 1,\dots,n\ . \tag{4.45}$$

Note, generally, this tensor is not symmetric.

Contravariant Tensors

Since (4.15) and (4.19), an example of the contravariant tensor of the second rank is represented by the contravariant metric tensor of S^{rn} and M^n, respectively.

By virtue of (4.15) we can readily conclude that for a fixed vector $\mathbf{P} \in R^{n+l}$ a set of values defined in the coordinates $s^1,\dots,s^n$ as

$$\mathbf{P} \cdot \nabla s^i\ , \quad i = 1,\dots,n\ , \tag{4.46}$$

where ∇s^i is the ith normal vector to the ith coordinate hypersurface in a regular surface S^{rn}, is a contravariant tensor of the first rank. Indeed, by (4.10) and (4.15)

$$\begin{aligned} \mathbf{P} \cdot \nabla s^i &= g^{ij}_{sr}(\mathbf{P} \cdot \mathbf{r}_{s^j}) = g^{km}_{vr}\frac{\partial s^i}{\partial v^k}\frac{\partial s^j}{\partial v^m}\Big(\mathbf{P} \cdot \mathbf{r}_{v^l}\frac{\partial v^l}{\partial s^j}\Big) = g^{km}_{vr}(\mathbf{P} \cdot \mathbf{r}_{v^m})\frac{\partial v^i}{\partial s^k} \\ &= \mathbf{P} \cdot \nabla v^k \frac{\partial s^i}{\partial v^k}\ , \quad i,j,k,l,m = 1,\dots,n\ . \end{aligned}$$

Similarly to the case of the covariant tensors considered above two contravariant tensors **f** and **v** of rank k and l, respectively, form a contravariant tensor $\mathbf{f}\otimes\mathbf{v}$ of the order $k+l$, whose components in the coordinates $s^1, \dots, s^n$ are computed by

$$(\mathbf{f} \otimes \mathbf{v})_{\mathbf{s}}^{i_1 \dots i_{k+l}} = f_{\mathbf{s}}^{i_1 \dots i_k} v_{\mathbf{s}}^{i_{k+1} \dots i_{k+l}}, \quad i_j = 1, \dots, n. \tag{4.47}$$

Mixed Tensors

Examples of the mixed tensors are readily constructed by the product of two tensors one of which is covariant and the order is contravariant. For instance, two tensors of the covariant and contravariant types formed by (4.42) and (4.46), respectively, produce, through a vector $\mathbf{P} \in R^{n+l}$, the following mixed tensor

$$P_j^i(\mathbf{s}) = P_{\mathbf{s}}^i P_j^{\mathbf{s}} = (\mathbf{P} \cdot \nabla s^i)(\mathbf{P} \cdot \mathbf{r}_{s^j}) = g_{sr}^{il}(\mathbf{P} \cdot \mathbf{r}_{s^l})(\mathbf{P} \cdot \mathbf{r}_{s^j}) \qquad i, j, l = 1, \dots, n. \tag{4.48}$$

There also is an operation of covariant differentiation of a contravariant vector $\mathbf{f} = (f_{\mathbf{s}}^i)$ which results in a mixed tensor by the following formula for its components designated by $(\nabla f)_j^i(\mathbf{s})$ in the coordinates $s^1, \dots, s^n$:

$$(\nabla f)_j^i(\mathbf{s}) = \frac{\partial}{\partial s^j} f_{\mathbf{s}}^i + f_{\mathbf{s}}^k \Upsilon_{jk}^i, \quad i, j, k = 1, \dots, n. \tag{4.49}$$

Using the relations (4.29), and (4.35) we easily find that the quantities $(\nabla f)_j^i(\mathbf{s})$ comprise a mixed tensor.

4.4.3 Tensor Operations

The operations over tensors defined at the same point of M^n are addition, multiplication and contraction.

Operation of Addition

The addition operation is carried out over tensors of the same order and type merely by adding the values of their components. A particular case of addition is the operation of subtraction.

Operation of Multiplication

The operation of multiplication is carried out over tensors of arbitrary order and type by multiplying each component of one tensor, say $\mathbf{f}_1$ by every component of another tensor, say $\mathbf{f}_2$, in particular, as in (4.44) and (4.47). As a result the order of the product equals the sum of the orders of the two original tensors $\mathbf{f}_1$ and $\mathbf{f}_2$. The same rule of summation is valid for the type of the product, namely, it is $k_1 + k_2$ times covariant and $l_1 + l_2$ times contravariant if the original tensor $\mathbf{f}_i$, $i = 1, 2$, is k_i times covariant and l_i times contravariant.

The operations of multiplication and addition allow one to formulate a monitor manifold over a physical geometry S^{xn} presented by a parametrization

$$\mathbf{x}(\mathbf{s}) : S^n \to R^{n+k}$$

with the use of covariant vectors $\mathbf{B}^1, \dots, \mathbf{B}^l$. The covariant metric elements $g^{\mathbf{s}}_{ij}$ of the manifold are computed by the following formula

$$g^{\mathbf{s}}_{ij} = \epsilon(\mathbf{s}) g^{xs}_{ij} + B^m_i B^m_j, \quad i, j = 1, \dots, n, \quad m = 1, \dots, l, \tag{4.50}$$

where $\epsilon(\mathbf{s}) > 0$ is an arbitrary function. Analogously a set of contravariant vectors $\mathbf{D}_1, \dots, \mathbf{D}_l$ determines the contravariant metric elements $g^{ij}_{\mathbf{s}}$ of a corresponding monitor manifold over S^{xn}:

$$g^{ij}_{\mathbf{s}} = \epsilon(\mathbf{s}) g^{ij}_{sx} + D^i_m D^j_m, \quad i, j = 1, \dots, n, \quad m = 1, \dots, l.$$

These very metrics prove to be crucial tools in controlling numerical grid properties.

Operation of Contraction

The operation of contraction is carried out over mixed tensors only. Let us take a mixed tensor, say one time covariant and two times contravariant whose components in the coordinates $s^1, \dots, s^n$ are designated, correspondently, as $f^{ij}_k(\mathbf{s})$. Assume the indices j and k the same then summation over them gives quantities designated by $f^{ij}_j(\mathbf{s})$ which are dependent on one index i only. It is readily shown that these quantities form a contravariant tensor of order 1. Indeed

$$f^{ij}_j(\mathbf{s}) = f^{lm}_k(\mathbf{v}) \frac{\partial v^k}{\partial s^j} \frac{\partial s^j}{\partial v^l} \frac{\partial s^i}{\partial v^m} = f^{lm}_l(\mathbf{v}) \frac{\partial s^i}{\partial v^m}, \quad i, k, l, m = 1, \dots, n,$$

i.e. the system $b^i_{\mathbf{s}} = f^{ij}_j(\mathbf{s})$ is a contravariant vector.

Analogously, the operation of contraction is defined for arbitrary mixed tensors by identifying some upper and lower indices in the components and producing summation over them. In particular, if $\mathbf{f}$ is a tensor k times covariant and k times contravariant then the operation of contraction over all indices produces an invariant (tensor of order zero).

For example the operation of contraction over the tensor (4.48) produces the following invariant

$$P^i_i(\mathbf{s}) = g^{ij}_{sr}(\mathbf{P} \cdot \mathbf{r}_{s^i})(\mathbf{P} \cdot \mathbf{r}_{s^j}), \quad i, j = 1, \dots, n.$$

With the operation of contraction one can define an invariant for arbitrary two tensors of the same order k provided one of them is covariant and another contravariant. This invariant is obtained by the composition of two operations: the first is multiplication of the tensors and the second is contraction of the obtained mixed tensor with respect to all indices. For example, the covariant and contravariant metric tensors yield the invariant

$$g_{\mathbf{s}}^{ij} g_{ij}^{\mathbf{s}} = n \,, \quad i,j = 1, \dots, n \,,$$

which equals n at all points of M^n.

4.5 Basic Invariants

This section reviews the most important invariants indispensable in the analysis of grid properties. They are formed by the operation of contraction over the contravariant metric tensor and some covariant tensors of order 2.

4.5.1 Beltrami's Differential Parameters

Mixed Differential Parameters

The successive operations of multiplication and contraction over the covariant tensor (4.45) and the contravariant metric tensor $g_{\mathbf{s}}^{ij}$ of a manifold M^n produce the invariant

$$\nabla(f, \varphi) = f_{s^i} \varphi_{s^j} g_{\mathbf{s}}^{ij} \,, \quad i,j = 1, \dots, n \,, \tag{4.51}$$

called Beltrami's mixed differential parameter of f and φ. In accordance with this designation formula (4.12) is also read as

$$\mathbf{n} = \varphi_{s^j} g_{sr}^{ji} \mathbf{r}_{s^i} = \nabla(\varphi, \mathbf{r}) \,, \quad i,j = 1 \dots, n \,. \tag{4.52}$$

By putting in (4.51) f equal to φ the following invariant is formulated:

$$\nabla(f) = \nabla(f, f) = f_{s^i} f_{s^j} g_{\mathbf{s}}^{ij} \,, \quad i,j = 1, \dots, n \,, \tag{4.53}$$

which is referred to as Beltrami's first differential parameter of f. Thus formula (4.13) with this parameter is read as

$$\|\mathbf{n}\| = \sqrt{\nabla(\varphi)} \,, \tag{4.54}$$

i.e. the length of the normal $\mathbf{n}$ defined by (4.12) is an invariant.

Let $\varphi_1 = const$ and $\varphi_2 = const$ be two hypersurfaces in a regular surface S^{rn}. The angle θ between these hypersurfaces is defined as the angle between the corresponding normals $\mathbf{n}_1$ and $\mathbf{n}_2$ to them. In accordance with (4.52) and (4.54) the cosine of this angle is computed through the Beltrami's differential parameters:

$$\cos\theta = \frac{\nabla(\varphi_1, \mathbf{r}) \cdot \nabla(\varphi_2, \mathbf{r})}{\sqrt{\nabla(\varphi_1)}\sqrt{\nabla(\varphi_2)}} = \frac{\nabla(\varphi_1, \varphi_2)}{\sqrt{\nabla(\varphi_1)}\sqrt{\nabla(\varphi_2)}} \,. \tag{4.55}$$

Second Differential Parameter

Another important Beltrami's differential parameter of a scalar φ is obtained by contracting the mixed tensor formed through the multiplication operation of the tensor of mixed derivatives (4.39) and the contravariant metric tensor. This invariant, designated by $\Delta_B[\varphi]$, is referred to as Beltrami's second differential parameter of φ. Namely

$$\Delta_B[\varphi] = g^{ij}_{\mathbf{s}} \nabla^{\mathbf{s}}_{ij}(\varphi)\,, \quad i,j = 1,\dots,n\,. \tag{4.56}$$

There exists one more important form of the invariant $\Delta_B[\varphi]$ helpful for its computing. To deduce it we note that, from (4.25) and (4.26),

$$\begin{aligned}\frac{1}{\sqrt{g^{\mathbf{s}}}}\frac{\partial}{\partial s^j}(\sqrt{g^{\mathbf{s}}}g^{ij}_{\mathbf{s}}) &= g^{ij}_{\mathbf{s}}\Upsilon^k_{kj} - g^{im}_{\mathbf{s}}\Upsilon^j_{jm} - g^{lj}_{\mathbf{s}}\Upsilon^i_{lj} = \\ &= -g^{lj}_{\mathbf{s}}\Upsilon^i_{lj}\,, \quad i,j,k,l,m = 1,\dots,n\,,\end{aligned} \tag{4.57}$$

and consequently, taking advantage of these relations and (4.39) in (4.56), we obtain the following expression for $\Delta_B[\varphi]$

$$\begin{aligned}\Delta_B[\varphi] &= g^{ij}_{\mathbf{s}}\varphi_{s^i s^j} - \varphi_{s^k} g^{ij}_{\mathbf{s}}\Upsilon^k_{ij} \\ &= g^{ij}_{\mathbf{s}}\varphi_{s^i s^j} + \frac{\varphi_{s^k}}{\sqrt{g^{\mathbf{s}}}}\frac{\partial}{\partial s^j}(\sqrt{g^{\mathbf{s}}}g^{jk}_{\mathbf{s}}) \\ &= \frac{1}{\sqrt{g^{\mathbf{s}}}}\frac{\partial}{\partial s^j}(\sqrt{g^{\mathbf{s}}}g^{jk}_{\mathbf{s}}\varphi_{s^k})\,, \quad i,j,k = 1,\dots,n\,.\end{aligned} \tag{4.58}$$

This very expression is used in Chap. 5 for formulating comprehensive grid generation equations.

4.5.2 Measure of Relative Spacing

In grid technology there is often a need in estimating grid spacing near some hypersurface in S^{rn}. Typically the hypersurface is specified by the equation $\varphi(\mathbf{s}) = 0$. This equation describes one more hypersurface in the parametric domain S^n as well. The parametric mapping $\mathbf{r}(\mathbf{s}) : S^n \to S^{rn}$ transforms a band of the thickness h around the hypersurface in S^n to a band in S^{rn} whose thickness l is computed by the following formula

$$l = \Big(\frac{\partial \mathbf{r}}{\partial \mathbf{n}_1} \cdot \mathbf{n}_2\Big)h + O(h^2)\,, \tag{4.59}$$

where $\mathbf{n}_1$ is a unit normal to the hypersurface in S^n while $\mathbf{n}_2$ is a unit normal to the hypersurface in S^{rn} (Fig. 4.5 for $n = 2$). Since (4.52) and (4.54)

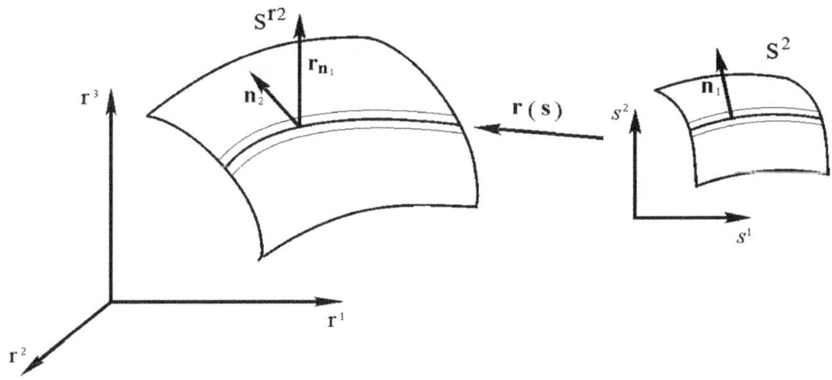

Fig. 4.5. Illustration for the measure of relative spacing

$$\begin{aligned} \mathbf{n}_1 &= (\varphi_{s^1}, \dots, \varphi_{s^n})/\sqrt{\nabla^E(\varphi)} , \\ \mathbf{n}_2 &= \varphi_{s^k} g_{sr}^{ki} \mathbf{r}_{s^i}/\sqrt{\nabla(\varphi)} , \quad i,k = 1,\dots,n , \end{aligned} \tag{4.60}$$

where

$$\nabla^E(\varphi) = \varphi_{s^i}\varphi_{s^j}\delta^i_j = (\varphi_{s^1})^2 + \dots + (\varphi_{s^n})^2 , \quad i,j = 1,\dots,n ,$$

is Beltrami's first differential parameter of $\varphi(\mathbf{s})$ in the Euclidean metric δ^i_j of S^n. Substituting (4.60) and the relation

$$\frac{\partial \mathbf{r}}{\partial \mathbf{n}_1} = \frac{1}{\sqrt{\nabla^E(\varphi)}} \varphi_{s^j}\mathbf{r}_{s^j} , \quad j = 1,\dots,n ,$$

in (4.59) yields

$$\begin{aligned} l &= \frac{\varphi_{s^j}\mathbf{r}_{s^j} \cdot (\varphi_{s^k} g_{sr}^{ki} \mathbf{r}_{s^i})}{\sqrt{\nabla^E(\varphi)}\sqrt{\nabla(\varphi)}} h + O(h^2) \\ &= \sqrt{\frac{\nabla^E(\varphi)}{\nabla(\varphi)}} h + O(h^2) , \quad i,j,k = 1,\dots,n . \end{aligned}$$

So the invariant

$$s(\varphi) = \sqrt{\nabla^E(\varphi)}/\sqrt{\nabla(\varphi)} \tag{4.61}$$

defined through Beltrami's first differential parameters of φ can be considered as a measure of relative spacing produced by the parametric transformation $\mathbf{r}(\mathbf{s})$ near the hypersurface $\varphi(\mathbf{s}) = 0$ in S^{rn}.

When a logical domain Ξ^n introduced for generating grids is considered as one more parametric domain then the quantity (4.61) with the identification

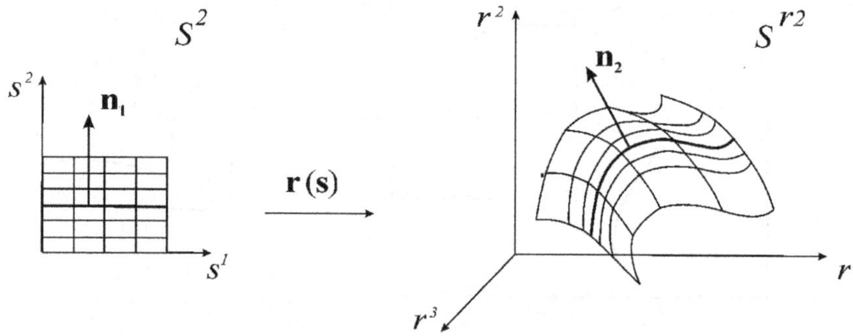

Fig. 4.6. Illustration for the measure of relative clustering

$S^n = \Xi^n$ (i.e. $\nabla^E(\varphi) = (\varphi_{\xi^1})^2 + \cdots + (\varphi_{\xi^n})^2$) is referred to as a measure of relative grid spacing near the hypersurface $\varphi(\mathbf{s}) = 0$ in S^{rn}. In particular, let the hypersurface $\varphi(\mathbf{s}) = 0$ be the grid hypersurface $\xi^i(\mathbf{s}) - c = 0$. As $\nabla^E(\xi^i) = 1$, $\nabla(\xi^i) = g^{ii}_{\xi r}$ for i fixed, hence (4.61) has the following form

$$s(\xi^i) = 1/\sqrt{g^{ii}_{\xi r}}\,, \quad i \text{ fixed}\,. \tag{4.62}$$

4.5.3 Measure of Relative Clustering

The rate of change of relative spacing in the direction $\mathbf{n}_2$ normal to the hypersurface $\varphi(\mathbf{s}) = 0$ in S^{rn} is called a measure of relative clustering near this hypersurface (see Fig. 4.6). Designating this measure by v we find, since (4.60) and (4.61),

$$\begin{aligned} v(\varphi) &= \frac{\mathrm{d}}{\mathrm{d}\mathbf{n}_2} s(\varphi) = \frac{1}{\sqrt{\nabla(\varphi)}} \varphi_{s^j} g^{ji}_{sr} \frac{\partial}{\partial s^i} s(\varphi) \\ &= \frac{1}{\sqrt{\nabla(\varphi)}} \nabla(\varphi, s(\varphi))\,, \quad i,j,k = 1,\ldots,n\,. \end{aligned} \tag{4.63}$$

i.e. it is defined through the Beltrami's first and mixed differential parameters.

In particular, if $\varphi(\mathbf{s}) \equiv \xi^i(\mathbf{s}) - c$, where ξ^i is the ith grid coordinate, using (4.62) and (4.63) yields

$$v(\xi^i) = \frac{1}{\sqrt{g^{ii}_{\xi r}}} \nabla\Big(\xi^i, \frac{1}{\sqrt{g^{ii}_{\xi r}}}\Big)\,, \quad i \text{ fixed}\,. \tag{4.64}$$

It will be shown in Chap. 6 that the measure (4.64) can also be expressed through Beltrami's second differential parameters and the so called mean curvature of the grid hypersurface $\xi^i = const.$

4.5.4 Mean Curvature

One more invariant of a regular surface S^{rn}, important in grid technology, is obtained from the covariant tensor (4.43) and contravariant metric tensor of S^{rn}

$$\sigma = g^{ij}_{sr}\mathbf{r}_{s^i s^j} \cdot \mathbf{P} , \quad i,j = 1, \ldots , n . \tag{4.65}$$

When the surface S^{rn} lies in a surface $S^{r(n+1)}$ and the vector $\mathbf{P}$ being orthogonal to S^{rn} belongs also to the tangent plane to $S^{r(n+1)}$ then the invariant σ from (4.65), scaled by the factor $1/(n\|\mathbf{P}\|)$ and designated as K_m, i.e.

$$K_m = \frac{1}{n\|\mathbf{P}\|} g^{ij}_{sr}\mathbf{r}_{s^i s^j} \cdot \mathbf{P} , \quad i,j = 1, \ldots , n , \tag{4.66}$$

is called the mean curvature of S^{rn} in $S^{r(n+1)}$ with the respect to the normal $\mathbf{P}$.

In particular for $n = 1$ (S^{r1} is a curve) the invariant K_m is referred to as the geodesic curvature of the curve S^{r1} in the surface S^{r2}.

In the following section a formula for this invariant will be established for an arbitrary hypersurface lying in some regular surface. If this hypersurface is found from the equation $\varphi(\mathbf{s}) = 0$ then the mean curvature is defined by Beltrami's first and second differential parameters of φ.

An amusing role played by the mean curvature in grid technology is demonstrated in Chap. 6.

4.6 Geometry of Hypersurfaces

We consider in this section the geometric characteristics which appear when a regular n-dimensional hypersurface S^{xn} represented by a set of local parametrizations

$$\mathbf{x}(\mathbf{s}) : S^n \to R^{n+l} , \quad S^n \subset R^n ,$$

is a subset of an $(n + 1)$-dimensional surface $S^{r(n+1)}$ specified by local parametrizations

$$\mathbf{r}(\mathbf{s}) : S^{n+1} \to R^{n+l} , \quad S^{n+1} \subset R^{n+1} .$$

This situation occurs in the grid technology when a scalar-valued monitor function is considered or grid hypersurfaces are analyzed.

4.6.1 Normal Vector to a Hypersurface

A unit vector $\mathbf{n}$ at a point $\mathbf{x} \in S^{xn}$ which lies in the tangent plane to $S^{r(n+1)}$ at this point and orthogonal to the tangent plane to S^{xn} at the same point is called the unit normal vector to the surface S^{xn} in S^{rn}.

General Case

A normal vector is readily computed by the formula analogous to (2.25). Namely, let P be a point of the surface S^{xn} and let vectors $\mathbf{x}_1, \ldots, \mathbf{x}_n$ and $\mathbf{r}_1, \ldots, \mathbf{r}_{n+1}$ be the basic tangent vectors at this point of S^{xn} and S^{rn} respectively. Now we form, analogously to (2.25), the following $(n+1)\times(n+1)$ matrix

$$\mathbf{A} = \begin{pmatrix} \mathbf{r}_1 & \ldots & \mathbf{r}_{n+1} \\ \mathbf{x}_1 \cdot \mathbf{r}_1 & \ldots & \mathbf{x}_1 \cdot \mathbf{r}_{n+1} \\ \ldots & \ldots & \ldots \\ \mathbf{x}_n \cdot \mathbf{r}_1 & \ldots & \mathbf{x}_n \cdot \mathbf{r}_{n+1} \end{pmatrix}, \tag{4.67}$$

which derives the vector

$$\mathbf{b} = \det(\mathbf{A}) . \tag{4.68}$$

The rank of the $n \times (n+1)$ matrix obtained from $\mathbf{A}$ by eliminating its top row equals n hence $\mathbf{b} \neq \mathbf{0}$. As

$$\mathbf{b} \cdot \mathbf{x}_i = \begin{pmatrix} \mathbf{x}_i \cdot \mathbf{r}_1 & \ldots & \mathbf{x}_i \cdot \mathbf{r}_{n+1} \\ \mathbf{x}_1 \cdot \mathbf{r}_1 & \ldots & \mathbf{x}_1 \cdot \mathbf{r}_{n+1} \\ \ldots & \ldots & \ldots \\ \mathbf{x}_n \cdot \mathbf{r}_1 & \ldots & \mathbf{x}_n \cdot \mathbf{r}_{n+1} \end{pmatrix} = 0 ,$$

The vector $\mathbf{b}$ is orthogonal to $\mathbf{x}_1, \ldots, \mathbf{x}_n$ and consequently to the surface S^{xn} at the point of consideration. Thus for the unit normal vector $\mathbf{n}$ to the surface S^{xn} in $S^{r(n+1)}$ we find

$$\mathbf{n} = \mathbf{b}/|\mathbf{b}| . \tag{4.69}$$

In a one-dimensional case, i.e. when $n = 1$, we find from (4.67) and (4.68)

$$\mathbf{b} = (\mathbf{x}_1 \cdot \mathbf{r}_2)\mathbf{r}_1 - (\mathbf{x}_1 \cdot \mathbf{r}_1)\mathbf{r}_2$$

(Fig. 4.7), while for $n = 2$ we obtain, similar to (2.26),

$$\mathbf{b} = [(\mathbf{x}_1 \cdot \mathbf{r}_{k+1})(\mathbf{x}_2 \cdot \mathbf{r}_{k+2}) - (\mathbf{x}_1 \cdot \mathbf{r}_{k+2})(\mathbf{x}_2 \cdot \mathbf{r}_{k+1})]\mathbf{r}_k , \quad k = 1, 2, 3 ,$$

where any index, say l, is identified with $l \pm 3$. Thus the vector $\mathbf{b}$ obtained by the formula (4.68) can be considered as the tensor product of the vectors $\mathbf{x}_1, \ldots, \mathbf{x}_n$.

When the surface S^{xn} lies in R^{n+1} and is determined by the equation $\xi(x^1, \ldots, x^{n+1}) = c$ then there also is the well-known formula for the unit normal $\mathbf{n}$:

$$\mathbf{n} = \mathrm{grad}\xi/|\mathrm{grad}\xi| = (\xi_{x^1}, \ldots, \xi_{x^{n+1}})/\sqrt{(\xi_{x^1})^2 + \ldots + (\xi_{x^{n+1}})^2} . \tag{4.70}$$

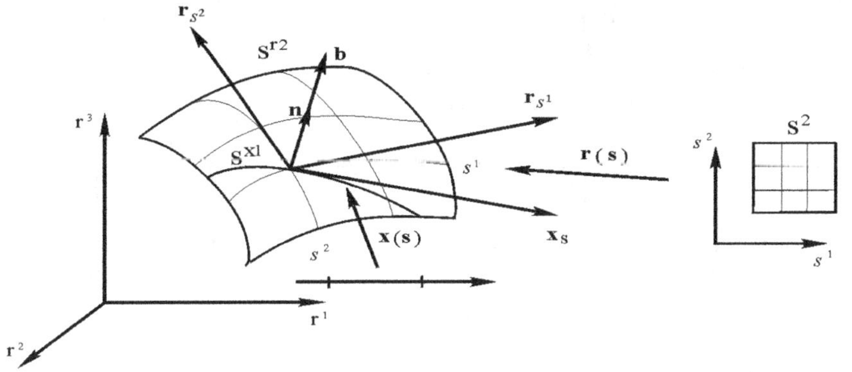

Fig. 4.7. Normal vector to the curve on the surface

Note if the surface S^{xn} is a coordinate hypersurface of $S^{r(n+1)}$ in some coordinates $s^1, \dots, s^{n+1}$, for example, it is defined by the equation $s^1 = s_0^1$ then S^{xn} is represented in the coordinates $s^2, \dots, s^{n+1}$ by

$$\mathbf{x}(s^2, \dots, s^{n+1}) = \mathbf{r}(s_0^1, s^2, \dots, s^{n+1}) .$$

Therefore

$$\mathbf{x}_i = \mathbf{r}_{s^i}(s_0^1, s^2, \dots, s^{n+1}) , \quad i = 2, \dots, n+1 ,$$

and consequently (4.67) and (4.68) result in

$$\mathbf{A} = \begin{pmatrix} \mathbf{r}_{s^1} & \dots & \mathbf{r}_{s^{n+1}} \\ g_{21}^{rs} & \dots & g_{2n+1}^{rs} \\ \dots & \dots & \dots \\ g_{n+11}^{rs} & \dots & g_{n+1n+1}^{rs} \end{pmatrix} ,$$

$$\mathbf{b}^1 = G^{1i}\mathbf{r}_{s^i} , \quad i = 1, \dots, n+1 ,$$

where G^{1i} is the $(1i)$th cofactor of the matrix (g_{ij}^{rs}), $i, j = 1, \dots, n+1$. Since

$$G^{1i} = g^{rs} g_{sr}^{1i} , \quad i = 1, \dots, n+1 ,$$

we obtain that

$$\mathbf{b}^1 = g^{rs} g_{sr}^{1i} \mathbf{r}_{s^i} , \quad i = 1, \dots, n+1 ,$$

and comparing this expression with (4.10) gives, in this case,

$$\mathbf{b}^1 = g^{rs} \nabla s^1 .$$

Analogously one readily shows that the vector $\mathbf{b}^i$ expressed as follows:

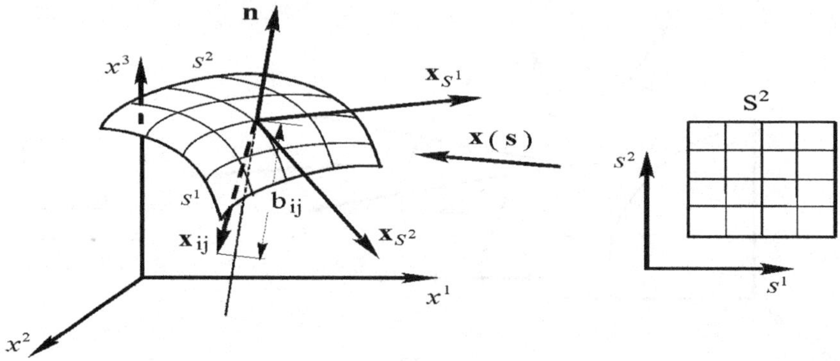

Fig. 4.8. Base vectors of the two-dimensional surface in R^3

$$\mathbf{b}^i = g^{rs} g^{ij}_{sr} \mathbf{r}_{s^j} \ , \quad i,j = 1,\dots,n+1 \ , \tag{4.71}$$

or using (4.10)

$$\mathbf{b}^i = g^{rs} \nabla s^i \ , \quad i = 1,\dots,n+1 \ , \tag{4.72}$$

is orthogonal to the coordinate hypersurface $s^i = s^i_0$.

Note by the vectors $\mathbf{x}_1,\dots,\mathbf{x}_n$ and $\mathbf{r}_1,\dots,\mathbf{r}_{n+1}$ in (4.67) for defining the unit normal vector $\mathbf{n}$ to S^{xn} in $S^{r(n+1)}$ by (4.69) one can use arbitrary sets of vectors if only they are independent and lie in the corresponding tangent planes.

Hypersurface in a Domain

Let $S^{r(n+1)} = D^{n+1} \subset R^{n+1}$ be an $(n+1)$-dimensional domain containing the n-dimensional surface S^{xn} represented by

$$\mathbf{x}(s) : S^n \to R^{n+1} \ .$$

Since a unit normal vector $\mathbf{n}$ to S^{xn} in D^{n+1} is orthogonal to the tangential vectors $\mathbf{x}_{s^i}$, $i = 1,\dots,n$, the vectors

$$\mathbf{x}_{s^1},\dots,\mathbf{x}_{s^n}, \mathbf{n} \ ,$$

constitute the basis of R^{n+1}. These vectors are called the base vectors of S^{xn} in R^{n+1} (Fig. 4.8 for $n = 2$). The vector $\mathbf{x}_{s^m s^p}$, $m,p = 1,\dots,n$, at a point $P \in S^{xn}$ is expanded in these vectors as

$$\mathbf{x}_{s^m s^p} = \mathbf{P}[\mathbf{x}_{s^m s^p}] + (\mathbf{x}_{s^m s^p} \cdot \mathbf{n})\mathbf{n} \ , \quad m,p = 1,\dots,n \ , \tag{4.73}$$

Where $\mathbf{P}$ is the operator which projects vectors in R^{n+1} on the tangent plane to S^{xn}. Taking advantage of (4.32) yields

$$\mathbf{x}_{s^m s^p} = \Upsilon^i_{mp}\mathbf{x}_{s^i} + (\mathbf{x}_{s^m s^p} \cdot \mathbf{n})\mathbf{n}\,, \quad i,m,p = 1,\dots,n\,, \tag{4.74}$$

(see Fig 4.4 for $n = 2$ with the identification $\mathbf{x} = \mathbf{r}$, $\mathbf{x}_{s^m s^p} = \mathbf{r}_{mp}$, $\mathbf{x}_{s^m s^p} \cdot \mathbf{n} = b_{mp}$). Similarly, using (4.33) in (4.73),

$$\mathbf{x}_{s^m s^p} = [mp,i]\boldsymbol{\nabla} s^i + (\mathbf{x}_{s^m s^p} \cdot \mathbf{n})\mathbf{n}\,, \quad i,m,p = 1,\dots,n\,, \tag{4.75}$$

where

$$\boldsymbol{\nabla} s^i = g^{ij}_{sx}\mathbf{x}_{s^j}\,, \quad i,j = 1,\dots,n\,.$$

For example, let S^{xn} be a monitor surface over $S^n \subset R^n$ with a scalar-valued monitor function $f(\mathbf{s})$, i.e. S^{xn} is represented in the parametric coordinates $s^1,\dots,s^n$ by

$$\mathbf{x}(\mathbf{s}) : S^n \to R^{n+1}\,, \quad \mathbf{x}(\mathbf{s}) = [\mathbf{s}, f(\mathbf{s})]\,,$$

It is readily verified that for the elements of the covariant and contravariant metric tensors of S^{xn} in the coordinates $s^1,\dots,s^n$ we have

$$\begin{aligned} g^{xs}_{ij} &= \mathbf{x}_{s^i} \cdot \mathbf{x}_{s^j} = \delta^i_j + f_{s_i} f_{s_j}\,, \quad i,j = 1,\dots,n\,, \\ g^{ij}_{sx} &= \delta^i_j - \frac{1}{g^{xs}}\frac{\partial f}{\partial s_i}\frac{\partial f}{\partial s_j}\,, \quad i,j = 1,\dots,n\,, \end{aligned} \tag{4.76}$$

where

$$g^{xs} = \det(g^{xs}_{ij}) = 1 + (f_{s^1})^2 + \ldots + (f_{s^n})^2 = 1 + \nabla^E(f)\,.$$

Since

$$\mathbf{x}_{s^i} = (\underbrace{0,\dots,0}_{i-1},1,\underbrace{0,\dots,0}_{n-i}, f_{s^i})\,, \quad i = 1,\dots,n\,,$$

it is obvious that one unit normal vector $\mathbf{n}$ to S^{xn} in D^{n+1} can be computed as follows:

$$\mathbf{n} = \frac{1}{\sqrt{g^{xs}}}(f_{s^1},\dots,f_{s^n},-1)\,. \tag{4.77}$$

Therefore in this case the expansion (4.74) has the form

$$\mathbf{x}_{s^m s^p} = \Upsilon^i_{mp}\mathbf{x}_{s^i} - \frac{1}{\sqrt{g^{xs}}} f_{s^m s^p}\mathbf{n}\,, \quad i,m,p = 1,\dots,n\,. \tag{4.78}$$

Since (4.76)

$$g^{il}_{mp} f_{s^l} = \left(\delta^i_l - \frac{1}{g^{xs}} f_{s^i} f_{s^l}\right) f_{s^l} = \frac{1}{g^{xs}} f_{s^i}\,, \quad i,l,m,p = 1,\dots,n\,,$$

therefore from (4.24) and (4.27) we find

$$\Upsilon^i_{mp} = g^{li}_{sx}[mp,l] = g^{li}_{sx} f_{s^m s^p} f_{s^l} = \frac{1}{g^{xs}} f_{s^m s^p} f_{s^i}\,, \quad i,l,m,p = 1,\dots,n\,.$$

So (4.78) results in

$$\mathbf{x}_{s^m s^p} = \frac{1}{g^{xs}} f_{s^m s^p}(f_{s^i}\mathbf{x}_{s^i} - \sqrt{g^{xs}}\mathbf{n})\,, \quad i,m,p = 1,\dots,n\,.$$

4.6.2 Second Fundamental Form

Assuming in the coordinates $s^1, \ldots, s^n$, similarly to (4.43),

$$b_{ij} = \mathbf{x}_{s^i s^j} \cdot \mathbf{n} \, , \quad i,j = 1, \ldots, n \, , \tag{4.79}$$

where $\mathbf{n}$ is the unit normal to the surface S^{xn} in $S^{r(n+1)}$ (Figs. 4.7 and 4.8 for $n = 2$) we define the so called second fundamental form of the surface S^{xn} in $S^{r(n+1)}$ by

$$b_{ij} \mathrm{d}s^i \mathrm{d}s^j \, , \quad i,j = 1, \ldots, n \, .$$

The covariant tensor (b_{ij}) reflects the local warping of the surface S^{xn} in $S^{r(n+1)}$.

4.6.3 Surface Curvatures

Multidimensional Case

The covariant tensor (b_{ij}) and the contravariant tensor (g^{ij}_{sx}) of the surface S^{xn} in $S^{r(n+1)}$ define the mixed tensor (K^i_j), where

$$K^i_j = g^{ik}_{sx} b_{kj} \, , \quad i,k,j = 1, \ldots, n \, . \tag{4.80}$$

An nth part of the trace of (K^i_j), namely, the quantity

$$K_m = \frac{1}{n} \mathrm{tr}(K^i_j) = \frac{1}{n} g^{ij}_{sx} b_{ij} \, , \quad i,j = 1, \ldots, n \, , \tag{4.81}$$

is called the mean curvature of the surface S^{xn} in $S^{r(n+1)}$. From (4.79 – 4.81) we readily conclude that the quantity K_m is invariant of parametrizations of S^{xn} and $S^{r(n+1)}$.

Two-Dimensional Case

When $n = 2$ then the determinant of (K^i_j), i.e.

$$K_G = \det(K^i_j) \, , \quad i,j = 1,2 \, ,$$

is called the Gaussian curvature of S^{x2} in S^{r3}. It is obvious that

$$K_G = \det(b_{ij}) \det(g^{ij}_{sx}) = \frac{1}{g^{xs}} \det(b_{ij}) \, , \tag{4.82}$$

where $g^{xs} = \det(g^{xs}_{ij}) = \det(\mathbf{x}_{s^i} \cdot \mathbf{x}_{s^j})$.

4.6.4 Formulas of the Mean Curvature

Formula for the Mean Curvature of a Monitor Surface over a Domain

Let S^{xn} be the monitor surface over the domain S^n parametrized by

$$\mathbf{x}(\mathbf{s}) : S^n \to R^{n+1}, \quad \mathbf{x}(\mathbf{s}) = [\mathbf{s}, f(\mathbf{s})] .$$

Then, taking into account (4.77), we find

$$b_{ij} = \mathbf{x}_{s^i s^j} \cdot \mathbf{n} = -\frac{1}{\sqrt{g^{\mathbf{s}}}} f_{s^i s^j} , \quad i,j = 1,\dots,n .$$

where $g^{\mathbf{s}} = \det(\mathbf{x}_{s^i} \cdot \mathbf{x}_{s^j})$. Using (4.76), (4.77), and (4.81) gives

$$K_m = \frac{1}{n\sqrt{g^{\mathbf{s}}}} \left[\frac{1}{g^{\mathbf{s}}} f_{s^i} f_{s^j} f_{s^i s^j} - \nabla^2[f] \right] , \quad i,j = 1,\dots,n . \tag{4.83}$$

Formula for the Mean Curvature of the Hypersurface Specified by the Equation $\varphi(\mathbf{s}) = 0$

Multidimensional Case

Let the surface $S^{r(n+1)}$ be represented by a parametrization

$$\mathbf{r}(\mathbf{s}) : S^{n+1} \to R^{n+l} , \quad \mathbf{s} = (s^1,\dots,s^{n+1}) , \quad l \geq 1 ,$$

while the n-dimensional surface S^{xn} in $S^{r(n+1)}$ is the image through $\mathbf{r}(\mathbf{s})$ of the n-dimensional surface in S^{n+1} defined from the equation $\varphi(\mathbf{s}) = 0$, i.e. the hypersurface S^{xn} consists of the points $\mathbf{r}(\mathbf{s})$, $\mathbf{s} \in S^{n+1}$, such that $\varphi(\mathbf{s}) = 0$. In this paragraph we will find a formula for the mean curvature of this hypersurface S^{xn} in $S^{r(n+1)}$.

Let the equation $\varphi(\mathbf{s}) = 0$ be locally resolved with respect to s^{n+1}, i.e. there is a function $s^{n+1}(s^1,\dots,s^n)$ such that

$$\varphi[s^1,\dots,s^n,s^{n+1}(s^1,\dots,s^n)] \equiv 0 , \quad (s^1,\dots,s^n) \in S^n ,$$

for some n-dimensional domain S^n. This is possible if $\varphi_{s^{n+1}} \neq 0$ at some point $\mathbf{s} \in S^{n+1}$. To make things definite we assume that $\varphi_{s^{n+1}} < 0$. Then the hypersurface S^{xn} is represented locally by the coordinates $s^1,\dots,s^n$ from the domain S^n as

$$\mathbf{x}(s^1,\dots,s^n) : S^n \to R^{n+l} , \tag{4.84}$$

where

$$\mathbf{x}(s^1,\dots,s^n) = \mathbf{r}(s^1,\dots,s^n,s^{n+1}(s^1,\dots,s^n)) .$$

Since

$$\frac{\partial s^{n+1}}{\partial s^i} = -\frac{\varphi_{s^i}}{\varphi_{s^{n+1}}} , \qquad i = 1, \dots , n ,$$

the basic tangent vectors to S^{xn} in the coordinates $s^1, \dots , s^n$ are subject to the equations

$$\mathbf{x}_{s^i} = \mathbf{r}_{s^i} + \frac{\partial s^{n+1}}{\partial s^i}\mathbf{r}_{s^{n+1}} = \frac{1}{\varphi_{s^{n+1}}}(\varphi_{s^{n+1}}\mathbf{r}_{s^i} - \varphi_{s^i}\mathbf{r}_{s^{n+1}}) , \qquad i = 1, \dots , n , \tag{4.85}$$

therefore the values for the elements g^{xs}_{ij}, $i, j = 1, \dots , n$, of the covariant metric tensor of S^{xn} in the coordinates $s^1, \dots , s^n$ can be computed by the following formula

$$\begin{aligned} g^{xs}_{ij} &= \mathbf{x}_{s^i} \cdot \mathbf{x}_{s^j} \\ &= \frac{1}{(\varphi_{s^{n+1}})^2}[(\varphi_{s^{n+1}})^2 g^{rs}_{ij} - \varphi_{s^{n+1}}(\varphi_{s^j} g^{rs}_{n+1 i} + \varphi_{s^i} g^{rs}_{n+1 j}) \\ &\quad + \varphi_{s^i}\varphi_{s^j} g^{rs}_{n+1 n+1}] , \quad i, j = 1, \dots , n , \end{aligned} \tag{4.86}$$

where the quantities g^{rs}_{kl}, $k, l = 1, \dots , n+1$, are the elements of the covariant metric tensor of $S^{r(n+1)}$ in the coordinates $s^1, \dots , s^{n+1}$ computed by the formula

$$g^{rs}_{kl} = \mathbf{r}_{s^k} \cdot \mathbf{r}_{s^l} , \quad k, l = 1, \dots , n+1 .$$

For the unit normal vector $\mathbf{n}$ to S^{xn} in $S^{r(n+1)}$ we have from (4.12)

$$\mathbf{n} = \mathbf{b}/|\mathbf{b}| , \tag{4.87}$$

where

$$\begin{aligned} \mathbf{b} &= \varphi_{s^l} g^{lk}_{sr}\mathbf{r}_{s^k} , & l, k = 1, \dots , n+1 , \\ |\mathbf{b}| &= \sqrt{\varphi_{s^l}\varphi_{s^j} g^{lj}_{sr}} = \sqrt{\nabla(\varphi)} , & j, l = 1, \dots , n+1 . \end{aligned} \tag{4.88}$$

Since our assumption that $\varphi_{s^{n+1}} < 0$, equations (4.87) and (4.88) result in

$$\mathbf{n} \cdot \mathbf{r}_{s^{n+1}} = \frac{1}{\sqrt{\nabla\varphi}}\varphi_{s^l} g^{lk}_{sr}\mathbf{r}_{s^k} \cdot \mathbf{r}_{s^{n+1}} = \frac{1}{\sqrt{\nabla\varphi}}\varphi_{s^{n+1}} < 0 ,$$

i.e. the vectors $\mathbf{n}$ and $\mathbf{r}_{s^{n+1}}$ have opposite directions with respect to the tangent plane to S^{xn} in $S^{r(n+1)}$.

Now we compute the elements g^{ij}_{sx} of the contravariant metric tensor of the hypersurface S^{xn} in the coordinates $s^1, \dots , s^n$.

First note that the following $n+1$ vectors

$$\mathbf{x}_1, \mathbf{x}_2, \dots , \mathbf{x}_n, \mathbf{n} ,$$

where $\mathbf{n}$ is defined by (4.87), comprise a basis of R^{n+1}. As $\mathbf{n}$ is orthogonal to $\mathbf{x}_i$, $i = 1, \dots, n$, so, in accordance with (2.6), any vector $\mathbf{v} \in R^{n+1}$ is represented through the vectors of this basis by

$$\mathbf{v} = g^{ij}_{sx}(\mathbf{v} \cdot \mathbf{x}_j)\mathbf{x}_i + (\mathbf{v} \cdot \mathbf{n})\mathbf{n} \,, \quad i, j = 1, \dots, n \,. \tag{4.89}$$

Note every basic normal vector ∇s^i, $i = 1, \dots, n$, to the coordinate surface $s^i = c$ in $S^{r(n+1)}$ defined, in accordance with (4.10), by

$$\nabla s^i = g^{il}_{sr}\mathbf{r}_{s^l} \,, \quad l = 1, \dots, n+1 \,, \tag{4.90}$$

is orthogonal to $\mathbf{x}_k$, $k = 1, \dots, n$, and $k \neq i$. Indeed, from (4.85) and (4.89),

$$\begin{aligned} \mathbf{x}_k \cdot \nabla s^i &= \frac{1}{\varphi_{s^{n+1}}}(\varphi_{s^{n+1}}\mathbf{r}_{s^k} - \varphi_{s^k}\mathbf{r}_{s^{n+1}}) \cdot g^{il}_{\mathbf{s}}\mathbf{r}_{s^l} \\ &= \delta^k_i - \frac{\varphi_{s^k}}{\varphi_{s^{n+1}}}\delta^{n+1}_i = 0 \,, \\ & l = 1, \dots, n+1 \,, \quad i, k = 1, \dots, n \,, \quad i \neq k \,. \end{aligned} \tag{4.91}$$

Therefore, from (4.87 – 4.91), we obtain

$$\begin{aligned} \nabla s^i &= g^{ij}_{sx}(\nabla s^i \cdot \mathbf{x}_{s^i})\mathbf{x}_{s^j} + (\nabla s^i \cdot \mathbf{n})\mathbf{n} \\ &= g^{ij}_{sx}\mathbf{x}_{s^j} + \frac{1}{\sqrt{\nabla(\varphi)}}\varphi_{s^l}g^{il}_{sr}\mathbf{n} \,, \\ & l = 1, \dots, n+1 \,, \quad i, j = 1, \dots, n \,, \ i \text{ fixed} \,. \end{aligned} \tag{4.92}$$

Thus by expanding the basic tangent vectors $\mathbf{r}_{s^l}$, $l = 1, \dots, n$, through $\mathbf{x}_1, \dots, \mathbf{x}_n$, and $\mathbf{n}$ and availing us of this expansion in (4.90) we can compute the contravariant elements g^{ij}_{sx}, $i, j = 1, \dots, n$, of S^{xn} in the coordinates $s^1, \dots, s^n$.

In order to find this expansion we first note that, in accordance with (4.85),

$$\mathbf{r}_{s^k} = \frac{1}{\varphi_{s^{n+1}}}(\varphi_{s^{n+1}}\mathbf{x}_{s^k} + \varphi_{s^k}\mathbf{r}_{s^{n+1}}) \,, \quad k = 1, \dots, n \,. \tag{4.93}$$

Multiplying this equation by

$$\varphi_{s^l}g^{lk}_{sr}/\sqrt{\nabla(\varphi)} \,, \quad l = 1, \dots, n+1 \,, \quad k = 1, \dots, n \,,$$

and using (4.87) and (4.88) we find

$$\begin{aligned} \mathbf{n} - \frac{\varphi_{s^l}g^{ln+1}_{sr}}{\sqrt{\nabla(\varphi)}}\mathbf{r}_{s^{n+1}} &= \frac{\varphi_{s^l}g^{lk}_{sr}}{\sqrt{\nabla(\varphi)}}\mathbf{x}_{s^k} + \frac{\nabla(\varphi) - \varphi_{s^l}\varphi_{s^{n+1}}g^{ln+1}_{sr}}{\sqrt{\nabla(\varphi)}\varphi_{s^{n+1}}}\mathbf{r}_{s^{n+1}} \,, \\ & l = 1, \dots, n+1 \,, \quad k = 1, \dots, n \,. \end{aligned}$$

Therefore

$$\mathbf{r}_{s^{n+1}} = \frac{\varphi_{s^{n+1}}}{\sqrt{\nabla(\varphi)}}\left(\mathbf{n} - \frac{\varphi_{s^l} g^{lk}_{sr}}{\sqrt{\nabla(\varphi)}}\mathbf{x}_{s^k}\right), \qquad l = 1,\dots,n+1, \quad k = 1,\dots,n, \tag{4.94}$$

and using this equation in (4.93) gives

$$\mathbf{r}_{s^j} = \mathbf{x}_{s^j} + \frac{1}{\sqrt{\nabla(\varphi)}}\left(\varphi_{s^j}\mathbf{n} - \frac{\varphi_{s^j}\varphi_{s^l} g^{lk}_{sr}}{\sqrt{\nabla(\varphi)}}\mathbf{x}_{s^k}\right), \qquad l = 1,\dots,n+1, \quad j,k = 1,\dots,n. \tag{4.95}$$

Substituting (4.94) and (4.95) in (4.90) we obtain

$$\begin{aligned}
\boldsymbol{\nabla} s^i &= g^{ij}_{sr}\mathbf{r}_{s^j} + g^{in+1}_{sr}\mathbf{r}_{s^{n+1}} = g^{ij}_{sr}\mathbf{x}_{s^j} + \frac{1}{\sqrt{\nabla(\varphi)}}\varphi_{s^j} g^{ji}_{sr}\mathbf{n} \\
&\quad - \frac{1}{\nabla(\varphi)}\varphi_{s^j}\varphi_{s^l} g^{ij}_{sr} g^{lk}_{sr}\mathbf{x}_{s^k} + \frac{1}{\sqrt{\nabla(\varphi)}}\varphi_{s^{n+1}} g^{in+1}_{sr}\mathbf{n} \\
&\quad - \frac{1}{\nabla(\varphi)}\varphi_{s^{n+1}}\varphi_{s^l} g^{lk}_{sr} g^{in+1}_{sr}\mathbf{x}_{s^k} \\
&= \left(g^{ij}_{sr} - \frac{1}{\nabla(\varphi)}\varphi_{s^p}\varphi_{s^l} g^{ip}_{sr} g^{lj}_{sr}\right)\mathbf{x}_{s^j} + \frac{1}{\sqrt{\nabla(\varphi)}}\varphi_{s^l} g^{il}_{sr}\mathbf{n}, \\
&\quad i,j,k = 1,\dots,n, \quad l,p = 1,\dots,n+1.
\end{aligned}$$

Comparing this expansion with (4.92) we find

$$g^{ij}_{sx} = g^{ij}_{sr} - \frac{1}{\nabla(\varphi)}\varphi_{s^p}\varphi_{s^l} g^{ip}_{sr} g^{jl}_{sr}, \qquad i,j = 1,\dots,n, \quad l,p = 1,\dots,n+1. \tag{4.96}$$

This very matrix is the contravariant metric tensor of the subsurface S^{xn} in the coordinates $s^1,\dots,s^n$. In order to check this statement we note first that for arbitrary symmetric matrices (a_{kl}) and (b^{kl}), $k,l = 1,\dots,n+1$, the following relations are held:

$$\begin{aligned}
b^{ij} a_{jk} &= b^{il} a_{lk} - b^{in+1} a_{n+1k}, \\
b^{ij}\varphi_{s^j} a_{kn+1} &= b^{il}\varphi_{s^l} a_{kn+1} - b^{in+1}\varphi_{s^{n+1}} a_{kn+1}, \\
b^{ij}\varphi_{s^k} a_{n+1j} &= b^{il}\varphi_{s^k} a_{n+1l} - b^{in+1}\varphi_{s^k} a_{n+1k+1}, \\
&\quad i,j,k = 1,\dots,n, \quad l = 1,\dots,n+1.
\end{aligned} \tag{4.97}$$

Availing us of these relations we obtain

$$
\begin{aligned}
g^{ij}_{s x} g^{x s}_{j k} &= \Big(g^{ij}_{s r} - \frac{1}{\nabla(\varphi)} \varphi_{s^p} \varphi_{s^l} g^{ip}_{s r} g^{jl}_{s r} \Big) \\
&\times \Big[g^{r s}_{j k} - \frac{1}{\varphi_{s^{n+1}}} (\varphi_{s^j} g^{r s}_{n+1 k} + \varphi_{s^k} g^{r s}_{n+1 j}) + \frac{\varphi_{s^j} \varphi_{s^k}}{\varphi_{s^{n+1}}} g^{r s}_{n+1 n+1} \Big] \\
&= g^{il}_{s r} g^{r s}_{l k} - g^{i n+1}_{s r} g^{r s}_{n+1 k} - \frac{1}{\varphi_{s^{n+1}}} g^{r s}_{n+1 k} (g^{il}_{s r} \varphi_{s^l} - g^{i n+1}_{s r} \varphi_{s^{n+1}}) \\
&\quad - \frac{\varphi_{s^k}}{\varphi_{s^{n+1}}} (g^{il}_{s r} g^{r s}_{l n+1} - g^{i n+1}_{s r} g^{r s}_{n+1 n+1}) \\
&\quad + \frac{\varphi_{s^k}}{(\varphi_{s^{n+1}})^2} g^{r s}_{n+1 n+1} (g^{il}_{s r} \varphi_{s^l} - g^{i n+1}_{s r} \varphi_{s^{n+1}}) \\
&\quad - \frac{\varphi_{s^p} \varphi_{s^m}}{\nabla(\varphi)} g^{ip}_{s r} \Big[g^{ml}_{s r} g^{r s}_{l k} - g^{m n+1}_{s r} g^{r s}_{n+1 k} \\
&\quad - \frac{1}{\varphi_{s^{n+1}}} g^{r s}_{k n+1} (g^{ml}_{s r} \varphi_{s^l} - g^{m n+1}_{s r} \varphi_{s^{n+1}}) \\
&\quad - \frac{\varphi_{s^k}}{\varphi_{s^{n+1}}} (g^{ml}_{s r} g^{r s}_{l n+1} - g^{m n+1}_{s r} g^{r s}_{n+1 n+1}) \\
&\quad - \frac{\varphi_{s^k}}{(\varphi_{s^{n+1}})^2} g^{r s}_{n+1 n+1}) (g^{ml}_{s r} \varphi_{s^l} - g^{m n+1}_{s r} \varphi_{s^{n+1}}) \Big] \\
&= \delta^i_k - \frac{\varphi_{s^l}}{\varphi_{s^{n+1}}} g^{il}_{s r} g^{r s}_{k n+1} + \frac{\varphi_{s^k} \varphi_{s^l}}{(\varphi_{s^{n+1}})^2} g^{il}_{s r} g^{r s}_{n+1 n+1} \\
&\quad + \frac{\varphi_{s^p}}{\varphi_{s^{n+1}}} g^{ip}_{s r} g^{r s}_{k n+1} - \frac{\varphi_{s^p} \varphi_{s^k}}{(\varphi_{s^{n+1}})^2} g^{ip}_{s r} g^{r s}_{n+1 n+1} = \delta^i_k \ , \\
&\quad i, j, k = 1, \dots, n \ , \quad m, l, p = 1, \dots, n+1 \ ,
\end{aligned}
$$

i.e. the matrix $(g^{ij}_{s x})$, $i, j = 1, \dots, n$, whose elements are defined by (4.96) is the inverse to the covariant metric tensor $(g^{x s}_{ij})$, $i, j = 1, \dots, n$, of $S^{x n}$, expressed by (4.86). Consequently the elements $g^{ij}_{s x}$ specified by (4.96) comprise the contravariant metric tensor of $S^{x n}$ in the coordinates $s^1, \dots, s^n$.

Now we proceed to the computation of the quantity $\mathbf{x}_{s^i s^j} \cdot \mathbf{n}$. From (4.85) we find

$$
\mathbf{x}_{s^i s^j} = \frac{1}{(\varphi_{s^{n+1}})^2} L_{ij}[\mathbf{r}] - \frac{1}{(\varphi_{s^{n+1}})^3} L_{ij}[\varphi] \mathbf{r}_{s^{n+1}} \ , \quad i, j = 1, \dots, n \ , \tag{4.98}
$$

where L_{ij} is the operator defined at a function $v(s^1, \dots, s^{n+1})$ as

$$
\begin{aligned}
L_{ij}[v] &= (\varphi_{s^{n+1}})^2 v_{s^i s^j} - \varphi_{s^{n+1}} (\varphi_{s^j} v_{s^i s^{n+1}} + \varphi_{s^i} v_{s^j s^{n+1}}) \\
&\quad + \varphi_{s^i} \varphi_{s^j} v_{s^{n+1} s^{n+1}} \ , \quad i, j = 1, \dots, n \ .
\end{aligned} \tag{4.99}
$$

Therefore, using (4.87) and (4.88),

$$L_{ij}[\mathbf{r}]\cdot\mathbf{n} = \frac{1}{\sqrt{\nabla(\varphi)}}[(\varphi_{s^{n+1}})^2\varphi_{s^l}g^{lk}_{sr}\mathbf{r}_{s^is^j}\cdot\mathbf{r}_{s^k}$$
$$-\varphi_{s^{n+1}}\varphi_{s^l}g^{lk}_{sr}(\varphi_{s^j}\mathbf{r}_{s^is^{n+1}}\cdot\mathbf{r}_{s^k}+\varphi_{s^i}\mathbf{r}_{s^js^{n+1}}\cdot\mathbf{r}_{s^k})$$
$$+\varphi_{s^i}\varphi_{s^j}\varphi_{s^l}g^{lr}_{sr}\mathbf{r}_{s^{n+1}s^{n+1}}\cdot\mathbf{r}_{s^k}] \tag{4.100}$$
$$= \frac{\varphi_{s^l}}{\sqrt{\nabla(\varphi)}}[(\varphi_{s^{n+1}})^2\Upsilon^l_{ij}-\varphi_{s^{n+1}}(\varphi_{s^j}\Upsilon^l_{in+1}+\varphi_{s^i}\Upsilon^l_{jn+1})$$
$$+\varphi_{s^i}\varphi_{s^j}\Upsilon^l_{n+1n+1}]\,,\quad i,j=1,\dots,n\,,\quad l,k=1,\dots,n+1\,.$$

where Υ^l_{ij}, $k,l,p=1,\dots,n+1$, are the Christoffel symbols of the second kind of $S^{r(n+1)}$ in the coordinates $s^1,\dots,s^{n+1}$. Analogously

$$L_{ij}[\varphi]\mathbf{r}_{s^{n+1}}\cdot\mathbf{n} = \frac{\varphi_{s^{n+1}}}{\sqrt{\nabla(\varphi)}}L_{ij}[\varphi]\,,\quad i,j=1,\dots,n\,. \tag{4.101}$$

Thus for the mean curvature of the hypersurface S^{xn} in $S^{r(n+1)}$ we have, from (4.81) and (4.96),

$$K_m = \frac{1}{n}g^{ij}_{sx}\mathbf{x}_{s^is^j}\cdot\mathbf{n}$$
$$= \frac{1}{2(\varphi_{s^{n+1}})^2\sqrt{\nabla(\varphi)}}\left(g^{ij}_{sr}-\frac{1}{\nabla(\varphi)}\varphi_{s^p}\varphi_{s^l}g^{ip}_{sr}g^{jl}_{sr}\right)$$
$$\times\{(\varphi_{s^{n+1}})^2(\varphi_{s^l}\Upsilon^l_{ij}-\varphi_{s^is^j})-\varphi_{s^{n+1}}\varphi_{s^j}(\varphi_{s^l}\Upsilon^l_{in+1}-\varphi_{s^is^{n+1}})$$
$$-\varphi_{s^{n+1}}\varphi_{s^i}(\varphi_{s^l}\Upsilon^l_{jn+1}-\varphi_{s^js^{n+1}}) \tag{4.102}$$
$$+\varphi_{s^i}\varphi_{s^j}(\varphi_{s^l}\Upsilon^l_{n+1n+1}-\varphi_{s^{n+1}s^{n+1}})\}$$
$$= -\frac{1}{n(\varphi_{s^{n+1}})^2\sqrt{\nabla(\varphi)}}\left(g^{ij}_{sr}-\frac{1}{\nabla(\varphi)}\varphi_{s^p}\varphi_{s^l}g^{ip}_{sr}g^{jl}_{sr}\right)d_{ij}\,,$$
$$i,j=1,\dots,n\,,\quad l,p=1,\dots,n+1\,,$$

where

$$d_{kl} = (\varphi_{s^{n+1}})^2\nabla_{kl}(\varphi)-\varphi_{s^{n+1}}\varphi_{s^k}\nabla_{n+1l}(\varphi)-\varphi_{s^{n+1}}\varphi_{s^l}\nabla_{n+1k}(\varphi)$$
$$+\varphi_{s^k}\varphi_{s^l}\nabla_{n+1n+1}(\varphi)\,,\quad k,l,=1,n+1\,,$$

while

$$\nabla_{kl}(\varphi) = \varphi_{s^ks^l}-\varphi_{s^p}\Upsilon^p_{kl}\,,\quad k,l,=1,n+1\,,$$

is the mixed covariant derivative of φ with respect to s^k and s^l in the metric of $S^{r(n+1)}$.

In order to compute (4.102) we use an analog of the formulas (4.97) which states that the following combinations of the same matrices (a_{kl}) and (b^{kl}), as in (4.97), are subject to the relations

$$
\begin{aligned}
&b^{ij}a_{ij} = b^{kl}a_{kl} - 2b^{kn+1}a_{kn+1} + b^{n+1n+1}a_{n+1n+1} \ , \\
&b^{ij}\varphi_{s^i}a_{jn+1} = b^{ij}\varphi_{s^j}a_{in+1} = a_{ln+1}(b^{kl}\varphi_{s^k} - b^{n+1l}\varphi_{s^{n+1}}) \\
&-a_{n+1n+1}(b^{ln+1}\varphi_{s^l} - b^{n+1n+1}\varphi_{s^{n+1}}) \ , \\
&i,j = 1,\dots,n \ , \quad k,l = 1,\dots,n+1 \ .
\end{aligned}
\tag{4.103}
$$

Therefore

$$
\begin{aligned}
&b^{ij}(a_{ij} - \frac{1}{\varphi_{s^{n+1}}}(\varphi_{s^i}a_{n+1j} + \varphi_{s^j}a_{n+1i}) - \frac{\varphi_{s^i}\varphi_{s^j}}{(\varphi_{s^n})^2}a_{n+1n+1}) \\
&= b^{kl}a_{kl} - 2b^{kn+1}a_{kn+1} + b^{n+1n+1}a_{n+1n+1} \\
&-\frac{2}{\varphi_{s^{n+1}}}[a_{ln+1}(b^{kl}\varphi_{s^k} - b^{n+1l}\varphi_{s^{n+1}}) \\
&-a_{n+1n+1}(b^{ln+1}\varphi_{s^l} - b^{n+1n+1}\varphi_{s^{n+1}})] \\
&+\frac{a_{n+1n+1}}{(\varphi_{s^{n+1}})^2}(b^{kl}\varphi_{s^k}\varphi_{s^l} - 2b^{kn+1}\varphi_{s^k}\varphi_{s^{n+1}} + b^{n+1n+1}\varphi_{s^{n+1}}\varphi_{s^{n+1}}) \\
&= b^{kl}a_{kl} - \frac{2}{\varphi_{s^{n+1}}}b^{kl}\varphi_{s^k}a_{ln+1} + \frac{a_{n+1n+1}}{(\varphi_{s^{n+1}})^2}b^{kl}\varphi_{s^k}\varphi_{s^l} \ , \\
&i,j = 1,\dots,n \ , \quad k,l = 1,\dots,n+1 \ .
\end{aligned}
\tag{4.104}
$$

Assuming now

$$
\begin{aligned}
b^{kl} &= g^{kl}_{sr} - \frac{1}{\nabla(\varphi)}\varphi_{s^p}\varphi_{s^t}g^{pk}_{sr}g^{ll} \ , \\
a_{kl} &= \nabla_{kl}(\varphi) \ , \quad k,l,p,t = 1,\dots,n+1 \ ,
\end{aligned}
$$

we obtain from (4.102) and (4.104)

$$
\begin{aligned}
K_m &= \frac{1}{n}g^{ij}_{sx}\mathbf{x}_{s^is^j}\cdot\mathbf{n} \\
&= -\frac{1}{2\sqrt{\nabla(\varphi)}}[(g^{kl}_{sr} - \frac{1}{\nabla(\varphi)}\varphi_{s^p}\varphi_{s^t}g^{pk}_{sr}g^{tl})\nabla_{kl}(\varphi) \\
&+\frac{2}{\varphi_{s^{n+1}}}\left(g^{kl}_{sr}\varphi_{s^k} - \frac{1}{\nabla(\varphi)}\varphi_{s^p}\varphi_{s^k}g^{pk}_{sr}\varphi_{s^t}g^{lt}\right)\nabla_{ln+1}(\varphi) \\
&-\frac{1}{(\varphi_{s^{n+1}})^2}\left(g^{kl}_{sr}\varphi_{s^k}\varphi_{s^l} - \frac{1}{\nabla(\varphi)}\varphi_{s^p}\varphi_{s^k}g^{pk}_{sr}\varphi_{s^l}\varphi_{s^t}g^{lt}\right)\nabla_{n+1n+1}(\varphi) \\
&= \frac{1}{n(\nabla(\varphi))^{3/2}}[\varphi_{s^p}\varphi_{s^t}g^{pk}_{sr}g^{tl}_{sr} - \nabla(\varphi)g^{kl}_{sr}]\nabla_{kl}(\varphi) \ , \\
&i,j = 1,\dots,n \ , \quad k,l,p,t = 1,\dots,n+1 \ .
\end{aligned}
\tag{4.105}
$$

It is obvious that the same formula for K_m is obtained if $\varphi_{s^i}(\mathbf{s}) \neq 0$ for some i, $1 \le i \le n$, and $\mathbf{s} \in S^{n+1}$. □

Hypersurface in a Domain

In the case $S^{r(n+1)}$ is an $(n+1)$-dimensional domain with the Euclidean metric $g^{rs}_{kl} = \delta^k_l$ the equation (4.105) results in

$$K_m = \frac{1}{n|\text{grad}\ \varphi|^3}[\varphi_{s^k}\varphi_{s^l}\varphi_{s^k s^l} - |\text{grad}\ \varphi|^2\varphi_{s^p s^p}]\ , \qquad k,l,p = 1,\dots,n+1\ . \tag{4.106}$$

In particular, let S^{xn} be an n-dimensional sphere of a radius ρ, i.e.

$$\varphi(\mathbf{s}) \equiv \rho^2 - \sum_{i=1}^{n+1}(s^i)^2.$$

Then, at the points of the sphere,

$$\begin{aligned} &|\text{grad}\varphi|^2 = 4\rho^2\ , \\ &\varphi_{s^k s^k} = -2(n+1)\ , \quad k = 1,\dots,n+1\ , \\ &\varphi_{s^k}\varphi_{s^l}\varphi_{s^k s^l} = -8\rho^2\ , k,l = 1,\dots,n+1\ . \end{aligned}$$

Thus, from (4.106),

$$K_m = \frac{1}{8n\rho^3}[-8\rho^2 + 8(n+1)\rho^2] = \frac{1}{\rho}\ .$$

If S^{xn} is a monitor surface over a domain S^n defined by the values of a scalar-valued function $f(\mathbf{s})$, then this surface in R^{n+1} is also specified by the equation

$$\varphi(s^1,\dots,s^{n+1}) \equiv f(s^1,\dots,s^n) - s^{n+1} = 0\ .$$

Consequently

$$|\text{grad}\varphi| = \sqrt{g^{\mathbf{s}}} = \sqrt{1 + f_{s^i}f_{s^i}}\ , \quad i = 1,\dots,n\ ,$$

$$\varphi_{s^k s^l} = \begin{cases} 0\ , & k = n+1 \text{ or } l = n+1\ , \\ f_{s^k s^l}\ , & k \neq n+1 \text{ and } l \neq n+1\ , \end{cases}$$

therefore, from (4.106), we have the following formula for K_m

$$K_m = \frac{1}{n(g^{\mathbf{s}})^{3/2}}(f_{s^i}f_{s^j}f_{s^i s^j} - g^{\mathbf{s}}f_{s^i s^i})\ , \quad i = 1,\dots,n\ ,$$

which coincides with (4.83).

Expression Through Beltrami's Differential Parameters

Note that similarly to the formula (4.25)

$$\frac{\partial}{\partial s^k} g^{ij}_{sr} = -g^{im}_{sr} \Upsilon^j_{mk} - g^{lj}_{sr} \Upsilon^i_{lk} \,, \quad i,j,l,m = 1,\dots,n+1 \,.$$

Taking advantage of these relations yields

$$\begin{aligned} \frac{\partial}{\partial s^i} \nabla(\varphi) &= \frac{\partial}{\partial s^i} (\varphi_{s^k} \varphi_{s^l} g^{kl}_{sr}) = 2 g^{kl}_{sr} \varphi_{s^l} (\varphi_{s^k s^i} - \varphi_{s^m} \Upsilon^m_{ki}) \\ &= 2 g^{kl}_{sr} \varphi_{s^l} \nabla_{ki}(\varphi) \,, \quad i,k,l,m = 1,\dots,n+1 \,, \end{aligned} \tag{4.107}$$

and consequently

$$\begin{aligned} &\varphi_{s^p} \varphi_{s^t} g^{pk}_{sr} g^{tl}_{sr} \nabla_{kl}(\varphi) = \frac{1}{2} \varphi_{s^p} g^{pk}_{sr} \frac{\partial}{\partial s^k} \nabla(\varphi) \\ &= \frac{1}{2} \nabla(\varphi, \nabla(\varphi)) \,, \quad k,l,p,t = 1,\dots,n+1 \,. \end{aligned} \tag{4.108}$$

Further, in accordance with (4.56),

$$g^{kl}_{sr} \nabla_{kl}(\varphi) = \Delta_B[\varphi] \,, \quad k,l = 1,\dots,n+1 \,. \tag{4.109}$$

Therefore (4.105) is transformed, with the help of (4.108) and (4.109), to

$$K_m = -\frac{1}{n} \Big[\frac{\Delta_B[\varphi]}{\sqrt{\nabla(\varphi)}} + \nabla\Big(\varphi, \frac{1}{\sqrt{\nabla(\varphi)}}\Big) \Big] \,. \tag{4.110}$$

This formula does not require the knowledge of a normal to the hypersurface $\varphi(\mathbf{s}) = 0$ in S^{rn} so it is used for determining the mean curvature of such a hypersurface in an arbitrary Riemannian manifold.

Another Form

One more formula for the mean curvature is found from the following relation

$$\begin{aligned} \Delta_B[\varphi] + \sqrt{\nabla(\varphi)} \nabla\Big(\varphi, \frac{1}{\sqrt{\nabla(\varphi)}}\Big) &= \frac{1}{\sqrt{g^{rs}}} \frac{\partial}{\partial s^j} (\sqrt{g^{rs}} g^{ij}_{sr} \varphi_{s^i}) \\ + \sqrt{\nabla(\varphi)} g^{ij}_{sr} \varphi_{s^i} \frac{\partial}{\partial s^j} \frac{1}{\sqrt{\nabla(\varphi)}} &= \sqrt{\frac{\nabla(\varphi)}{g^{rs}}} \frac{\partial}{\partial s^j} \Big(\sqrt{\frac{g^{rs}}{\nabla(\varphi)}} g^{ij}_{sr} \varphi_{s^i} \Big) \,, \\ & \qquad i,j = 1,\dots,n+1 \,. \end{aligned}$$

Hence equation (4.110) also becomes

$$K_m = -\frac{1}{n\sqrt{g^{rs}}} \frac{\partial}{\partial s^j} \Big(\sqrt{\frac{g^{rs}}{\nabla(\varphi)}} g^{ij}_{sr} \varphi_{s^i} \Big) \,, \quad i,j = 1,\dots,n+1 \,. \tag{4.111}$$

In particular, for the n-dimensional coordinate hypersurface $\varphi \equiv s^i - c_0 = 0$, we have

$$\nabla(\varphi) = g^{ii}_{sr} , \quad i \text{ fixed} ,$$

therefore, in accordance with (4.111), the mean curvature of the coordinate hypersurface $s^i = c_0$ is expressed as follows:

$$K_m = -\frac{1}{n\sqrt{g^{rs}}} \frac{\partial}{\partial s^j} \left(\sqrt{\frac{g^{rs}}{g^{ii}_{sr}}} g^{ij}_{sr} \right) , \quad i,j = 1, \dots , n+1 , \quad i \text{ fixed} . \quad (4.112)$$

One-Dimensional Case

When $n = 1$, i.e. S^{xn} is a curve, while $S^{r(n+1)}$ is a two-dimensional surface then (g^{rs}_{ij}) is a 2×2 matrix and its elements satisfy the relations

$$g^{rs}_{ij} = (-1)^{i+j} g^{rs} g^{3-i3-j}_{sr} , \quad i,j = 1,2 ,$$

where $g^{rs} = \det(g^{rs}_{ij})$. Substituting these relations in (4.86) for $n = 1$ we find the following expression for the metric element g^{xs}_{11} of the curve S^{x1} in S^{r2} represented by the equation $\varphi(\mathbf{s}) = 0$, $\mathbf{s} = (s^1, s^2)$,

$$g^{xs}_{11} = \frac{1}{(\varphi_{s^2})^2} g^{rs} \varphi_{s^l} \varphi_{s^k} g^{lk}_{sr} = \frac{1}{(\varphi_{s^2})^2} g^{rs} \nabla(\varphi) , \quad k,l = 1,2 .$$

Therefore the contravariant metric element of S^{x1} is expressed as

$$g^{11}_{sx} = (\varphi_{s^2})^2 / [g^{rs} \nabla(\varphi)] .$$

Further note that the formula (4.99) with $n = 1$ has the following form

$$L_{11}[v] = |\mathrm{grad}\varphi|^2 v_{s^l s^l} - \varphi_{s^k} \varphi_{s^l} v_{s^k s^l} , \quad k,l = 1,2 ,$$

hence equation (4.105) for $n = 1$ is transformed to

$$\begin{aligned} \sigma &= \frac{1}{g^{rs}(\nabla(\varphi))^{3/2}} [|\mathrm{grad}\varphi|^2 (\varphi_{s^l} \Upsilon^l_{kk} - \varphi_{s^k s^k}) \\ &- \varphi_{s^k} \varphi_{s^p} (\varphi_{s^l} \Upsilon^l_{kp} - \varphi_{s^k s^p})] \\ &= \frac{1}{g^{rs}(\nabla(\varphi))^{3/2}} [\varphi_{s^k} \varphi_{s^l} \nabla_{kl}(\varphi) - |\mathrm{grad}\varphi|^2 \nabla_{kk}(\varphi)] \\ &- (-1)^{k+l} \frac{1}{g^{rs}(\nabla(\varphi))^{3/2}} \varphi_{s^{3-k}} \varphi_{s^{3-l}} \nabla_{kl}(\varphi) , \; k,l,p = 1,2 , \end{aligned} \quad (4.113)$$

where $\sigma = K_m$ is the geodesic curvature of S^{x1} in S^{r2}.

The geodesic curvature of the coordinate line $s^i = c_0$ is also computed from (4.112) by the formula

$$\sigma_i = -(-1)^{i+j} \frac{1}{\sqrt{g^{rs}}} \frac{\partial}{\partial s^j} \left[\frac{1}{\sqrt{g^{rs}_{3-i3-i}}} g^{rs}_{3-i3-j} \right] , \qquad (4.114)$$
$$i, j = 1, 2 , \quad i \text{ fixed} ,$$

i.e.

$$\sigma_1 = -\frac{1}{\sqrt{g^{rs}}} \left[\frac{\partial}{\partial s^1} \left(\frac{1}{\sqrt{g^{rs}_{22}}} g^{rs}_{22} \right) - \frac{\partial}{\partial s^2} \left(\frac{1}{\sqrt{g^{rs}_{22}}} g^{rs}_{12} \right) \right] ,$$
$$\sigma_2 = -\frac{1}{\sqrt{g^{rs}}} \left[\frac{\partial}{\partial s^2} \left(\frac{1}{\sqrt{g^{rs}_{11}}} g^{rs}_{11} \right) - \frac{\partial}{\partial s^1} \left(\frac{1}{\sqrt{g^{rs}_{11}}} g^{rs}_{12} \right) \right] ,$$

where σ_i, $i = 1, 2$, is the geodesic curvature of the coordinate curve $s^i = c_0$.

Computation of the Mean Curvature in the Case of a Parametric Representation

Multidimensional Case

Let the coordinates $s^1, \dots, s^{n+1}$ of the points of the n-dimensional surface S^{xn} in $S^{r(n+1)}$ be represented locally by the following transformation

$$\mathbf{s}(\mathbf{t}) : T^n \to S^{n+1} , \quad \mathbf{t} = (t^1, \dots, t^n) , \qquad (4.115)$$

where $T^n \subset R^n$ is some n-dimensional domain of R^n. Then S^{xn} is parametrized in the coordinates $t^1, \dots, t^n$ by

$$\mathbf{x}(\mathbf{t}) = \mathbf{r}[\mathbf{s}(\mathbf{t})] : T^n \to R^{n+l} , \qquad (4.116)$$

where

$$\mathbf{r}(\mathbf{s}) : S^{n+1} \to R^{n+l} , \quad l \geq 1 ,$$

is the parametrization of $S^{r(n+1)}$. So the basic tangent vectors $\mathbf{x}_i$, $i = 1, \dots, n$, to the hypersurface S^{xn} in $S^{r(n+1)}$ are expressed as follows:

$$\mathbf{x}_i = \mathbf{r}_{s^k} \frac{ds^k}{dt^i} , \quad i = 1, \dots, n , \quad k = 1, \dots, n+1 . \qquad (4.117)$$

For finding a normal to S^{xn} in $S^{r(n+1)}$ we use formulas (4.68) and (4.69) which require the computation of the determinant of the matrix (4.67) which has, in accordance with (4.117), the following form

$$\mathbf{A} = \begin{pmatrix} \mathbf{r}_{s^1} & \cdots & \mathbf{r}_{s^{n+1}} \\ g^{rs}_{1l} \frac{ds^l}{dt^1} & \cdots & g^{rs}_{n+1l} \frac{ds^l}{dt^1} \\ \cdot & \cdots & \cdot \\ g^{rs}_{1l} \frac{ds^l}{dt^n} & \cdots & g^{rs}_{n+1l} \frac{ds^l}{dt^n} \end{pmatrix} , \quad l = 1, \dots, n+1 . \qquad (4.118)$$

We readily see that the matrix **A** is composed as the product of the covariant metric tensor and the following matrix **B**

$$\mathbf{B} = \begin{pmatrix} g_{sr}^{j1}\mathbf{r}_{s^j} & \cdots & g_{sr}^{jn+1}\mathbf{r}_{s^j} \\ \dfrac{ds^1}{dt^1} & \cdots & \dfrac{ds^{n+1}}{dt^1} \\ \cdot & \cdots & \cdot \\ \dfrac{ds^1}{dt^n} & \cdots & \dfrac{ds^{n+1}}{dt^n} \end{pmatrix} , \quad j = 1, \dots , n+1 , \qquad (4.119)$$

namely,

$$\mathbf{A} = \mathbf{B}(g_{ij}^{rs}) .$$

Therefore, analogously to formula (4.68), a normal vector **b** to S^{xn} in $S^{r(n+1)}$ is also computed by

$$\mathbf{b} = \det \mathbf{B} .$$

In accordance with the rule of the computation of the determinant of a matrix we find from (4.119)

$$\mathbf{b} = -(-1)^i g_{sr}^{ji} (\det \mathbf{D}_i) \mathbf{r}_{s^j} , \quad i, j = 1, \dots , n+1 , \qquad (4.120)$$

where $\mathbf{D}_i$ is the $n \times n$ matrix obtained by deleting the ith column of the $n \times (n+1)$ matrix (ds^i/dt^j), $i = 1, \dots , n+1$, $j = 1, \dots , n$, i.e.

$$\mathbf{D}_i = \begin{pmatrix} \dfrac{ds^1}{dt^1} & \cdots & \dfrac{ds^{i-1}}{dt^1} & \dfrac{ds^{i+1}}{dt^1} & \cdots & \dfrac{ds^{n+1}}{dt^1} \\ \cdot & \cdots & \cdot & \cdot & \cdots & \cdot \\ \dfrac{ds^1}{dt^n} & \cdots & \dfrac{ds^{i-1}}{dt^n} & \dfrac{ds^{i+1}}{dt^n} & \cdots & \dfrac{ds^{n+1}}{dt^n} \end{pmatrix} .$$

Hence

$$|\mathbf{b}|^2 = \mathbf{b} \cdot \mathbf{b} = (-1)^{i+k} g_{sr}^{ki} \det \mathbf{D}_i \det \mathbf{D}_k , \quad i, k = 1, \dots , n+1 . \qquad (4.121)$$

As for

$$\mathbf{x}_{t^i t^j} = \frac{\partial^2 \mathbf{x}(\mathbf{t})}{\partial t^i \partial t^j} , \quad i, j = 1, \dots , n ,$$

we have from (4.116)

$$\mathbf{x}_{t^i t^j} = \mathbf{r}_{s^m s^l} \frac{ds^m}{dt^i} \frac{ds^l}{dt^j} + \mathbf{r}_{s^m} \frac{d^2 s^m}{dt^i dt^j} , \quad i, j = 1, \dots , n , \quad l, m = 1, \dots , n+1 ,$$

therefore

$$b_{ij} = \mathbf{x}_{t^i t^j} \cdot \mathbf{n} = \mathbf{x}_{t^i t^j} \cdot \frac{\mathbf{b}}{|\mathbf{b}|}$$

$$= -(-1)^p \frac{1}{|\mathbf{b}|} g^{pk}_{sr} \det \mathbf{D}_p \Big(\frac{ds^m}{dt^i}\frac{ds^l}{dt^j} \mathbf{r}_{s^m s^l} \cdot \mathbf{r}_{s^k} + \frac{d^2 s^m}{dt^i dt^j} \mathbf{r}_{s^m} \cdot \mathbf{r}_{s^k} \Big)$$

$$= -(-1)^p \frac{1}{|\mathbf{b}|} \det \mathbf{D}_p \Big(\Upsilon^p_{ml} \frac{ds^m}{dt^i}\frac{ds^l}{dt^j} + \frac{d^2 s^p}{dt^i dt^j} \Big) ,$$

$$i,j = 1,\dots,n , \quad k,l,m,p = 1,\dots,n+1 ,$$

and consequently

$$K_m = \frac{1}{2} g^{ij}_{tx} b_{ij} = \frac{1}{2|\mathbf{b}|} \det \mathbf{C} , \quad i,j = 1,\dots,n , \tag{4.122}$$

where g^{ij}_{tx} is the (ij)th element of the contravariant metric tensor of S^{xn} in the coordinates $t^1,\dots,t^n$, while

$$\mathbf{C} = \begin{pmatrix} g^{ij}_{tx}\Big(\Upsilon^1_{ml} \frac{\partial s^m}{\partial t^i}\frac{\partial s^l}{\partial t^j} + \frac{\partial^2 s^1}{\partial t^i \partial t^j}\Big) & \cdots & g^{ij}_{tx}\Big(\Upsilon^{n+1}_{ml} \frac{\partial s^m}{\partial t^i}\frac{\partial s^l}{\partial t^j} + \frac{\partial^2 s^{n+1}}{\partial t^i \partial t^j}\Big) \\ \frac{\partial s^1}{\partial t^1} & \cdots & \frac{\partial s^{n+1}}{\partial t^1} \\ \cdot & \cdots & \cdot \\ \frac{\partial s^1}{\partial t^n} & \cdots & \frac{\partial s^{n+1}}{\partial t^n} \end{pmatrix} .$$

Now we establish a relation between the elements of the contravariant metric tensor (g^{ij}_{tx}), $i,j = 1,\dots,n$, of S^{xn} in the coordinates $t^1,\dots,t^n$ and the metric elements of $S^{r(n+1)}$ in the coordinates $s^1,\dots,s^{n+1}$. For this purpose we notice that the vectors

$$\mathbf{x}_1, \dots, \mathbf{x}_n, \mathbf{n} ,$$

where $\mathbf{n} = \mathbf{b}/|\mathbf{b}|$, constitute a basis for the tangent plane to $S^{r(n+1)}$. The vectors $\mathbf{r}_{s^k}$, $k = 1,\dots,n+1$, are expanded in these basis by the formula (2.6), namely,

$$\mathbf{r}_{s^k} = g^{ij}_{tx}(\mathbf{r}_{s^k}\mathbf{x}_i) \cdot \mathbf{x}_j + (\mathbf{r}_{s^k} \cdot \mathbf{n})\mathbf{n} , \quad i,j = 1,\dots,n , \quad k = 1,\dots,n+1 ,$$

therefore, using (4.117) and (4.120),

$$g^{rs}_{kl} = \mathbf{r}_{s^k} \cdot \mathbf{r}_{s^l} = [g^{ij}_{tx}(\mathbf{r}_{s^k} \cdot \mathbf{x}_i)\mathbf{x}_j + (\mathbf{r}_{s^k} \cdot \mathbf{n})\mathbf{n}] \cdot [g^{kp}_{tx}(\mathbf{r}_{s^l} \cdot \mathbf{x}_p)\mathbf{x}_k + (\mathbf{r}_{s^l} \cdot \mathbf{n})\mathbf{n}]$$

$$= g^{ij}_{tx} g^{kp}_{tx} g^{xt}_{kj} (\mathbf{r}_{s^k} \cdot \mathbf{x}_i)(\mathbf{r}_{s^l} \cdot \mathbf{x}_p) + (\mathbf{r}_{s^k} \cdot \mathbf{n})(\mathbf{r}_{s^l} \cdot \mathbf{n})$$

$$= g^{ij}_{tx}(\mathbf{r}_{s^k} \cdot \mathbf{x}_i)(\mathbf{r}_{s^l} \cdot \mathbf{x}_j) + \frac{1}{|\mathbf{b}|^2}(\mathbf{r}_{s^k} \cdot \mathbf{b})(\mathbf{r}_{s^l} \cdot \mathbf{b})$$

$$= g^{ij}_{tx} g^{rs}_{km} \frac{\partial s^m}{\partial t^i} g^{rs}_{lp} \frac{\partial s^p}{\partial t^j} + (-1)^{k+l} \frac{(g^{rs})^2}{|\mathbf{b}|^2} \det \mathbf{D}_k \det \mathbf{D}_l ,$$

$$i,j = 1,\dots,n , \quad k,l = 1,\dots,n+1 , \quad k,l \text{ fixed} .$$

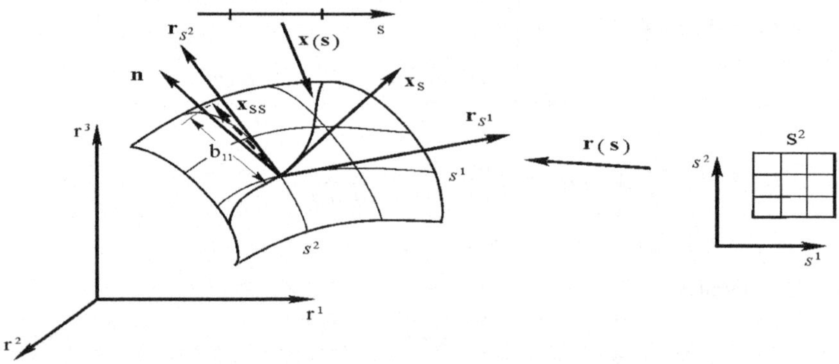

Fig. 4.9. Illustration for a curve on the surface

Multiplying these relations by g^{lh}_{sr} yields

$$\delta^k_h = g^{ij}_{tx} g^{rs}_{km} \frac{\partial s^m}{\partial t^i} \frac{\partial s^h}{\partial t^j} + (-1)^{k+l} \frac{(g^{rs})^2}{|\mathbf{b}|^2} \det \mathbf{D}_k \det \mathbf{D}_l g^{lh}_{sr} \, , \qquad (4.123)$$
$$i,j = 1,\dots,n \, , \quad k,l,h = 1,\dots,n+1 \, , \quad k, \text{ fixed} \, .$$

Further multiplication of (4.123) by g^{kp}_{sr} gives

$$g^{hp}_{sr} = g^{ij}_{tx} \frac{\partial s^p}{\partial t^i} \frac{\partial s^h}{\partial t^j} + (-1)^{m+l} \frac{(g^{rs})^2}{|\mathbf{b}|^2} g^{hl}_{sr} g^{mp}_{sr} \det \mathbf{D}_m \det \mathbf{D}_l \, , \qquad (4.124)$$
$$i,j = 1,\dots,n \, , \quad h,l,m,p = 1,\dots,n+1 \, .$$

Therefore

$$g^{ij}_{tx} \Upsilon^p_{ml} \frac{\partial s^m}{\partial t^i} \frac{\partial s^l}{\partial t^j} = g^{ml}_{sr} \Upsilon^p_{ml} - d^m d^l \Upsilon^p_{ml} \, , \; i,j = 1,\dots,n \, , \; l,m,p = 1,\dots,n+1 \, ,$$

where

$$d^k = (-1)^m \frac{g^{rs}}{|\mathbf{b}|} g^{mk}_{sr} \det \mathbf{D}_m \, , \quad k,m = 1,\dots,n+1 \, .$$

One-Dimensional Subsurface

In the case $n = 1$ (Fig. 4.8), when the surface S^{x1} is a curvilinear line in S^{r2}, the twice mean curvature is called the geodesic curvature of the curve S^{x1} in S^{r2}. Typically the geodesic curvature is designated by σ. It is obvious that in this case $K_G = \sigma$. Besides this it is evident that the geodesic curvature of S^{x1} in S^{r2} is its curvature determined by (3.8) if S^{r2} is a two-dimensional domain.

The geodesic curvature of S^{x1} in S^{r2} can be computed through the elements of the first groundforms of S^{x1} and S^{r2}. In order to prove this we assume that the coordinates s^1 and s^2 of S^{x1} are specified by the function

$$\mathbf{s}(t) : [a, b] \to S^2 ,$$

i.e. S^{x1} is parametrized as

$$\mathbf{r}[\mathbf{s}(t)] : [a, b] \to R^{l+2} ,$$

where

$$\mathbf{r}(\mathbf{s}) : S^2 \to R^{2+l} , \quad l \geq 0$$

is the parametrization of S^{r2}. The geodesic curvature of S^{x1} in S^{r2} is defined through the scalar product of

$$\mathbf{r}_{tt} = \frac{\mathrm{d}^2\mathbf{r}[\mathbf{s}(t)]}{\mathrm{d}t^2} = \mathbf{r}_{s^i s^j} \frac{\mathrm{d}s^i}{\mathrm{d}t} \frac{\mathrm{d}s^j}{\mathrm{d}t} + \mathbf{r}_{s^i} \frac{\mathrm{d}^2 s^i}{\mathrm{d}t^2} , \quad i, j = 1, 2 ,$$

and a normal to this curve in S^{r2}. Namely,

$$\sigma = \frac{1}{g^{rt}} \mathbf{r}_{tt} \cdot \mathbf{n} ,$$

where

$$g^{rt} = \frac{\mathrm{d}\mathbf{r}[\mathbf{s}(t)]}{\mathrm{d}t} \cdot \frac{\mathrm{d}\mathbf{r}[\mathbf{s}(t)]}{\mathrm{d}t} = g_{ij}^{rs} \frac{\mathrm{d}s^i}{\mathrm{d}t} \frac{\mathrm{d}s^j}{\mathrm{d}t} , \quad i, j = 1, 2 ,$$

while the normal $\mathbf{n}$ may be computed by (4.69). This formula yields

$$\mathbf{n} = \mathbf{b}/|\mathbf{b}|$$

where

$$\mathbf{b} = \left(\frac{\mathrm{d}s^i}{\mathrm{d}t} \mathbf{r}_{s^i} \cdot \mathbf{r}_{s^2}\right) \mathbf{r}_{s^1} - \left(\frac{\mathrm{d}s^i}{\mathrm{d}t} \mathbf{r}_{s^i} \cdot \mathbf{r}_{s^1}\right) \mathbf{r}_{s^2}$$
$$= (g_{2i}^{rs} \mathbf{r}_{s^1} - g_{1i}^{rs} \mathbf{r}_{s^2}) \frac{\mathrm{d}s^i}{\mathrm{d}t} = -(-1)^i g^{rs} g_{sr}^{ik} \frac{\mathrm{d}s^{3-i}}{\mathrm{d}t} \mathbf{r}_{s^k} , \quad i, k = 1, 2 .$$

From this equation we readily find

$$|\mathbf{b}|^2 = \mathbf{b} \cdot \mathbf{b} = (-1)^{i+j} (g^{rs})^2 g_{sr}^{ik} g_{kl}^{rs} g_{sr}^{jl} \frac{\mathrm{d}s^{3-i}}{\mathrm{d}t} \frac{\mathrm{d}s^{3-j}}{\mathrm{d}t}$$
$$= (-1)^{i+j} (g^{rs})^2 g_{sr}^{ij} \frac{\mathrm{d}s^{3-i}}{\mathrm{d}t} \frac{\mathrm{d}s^{3-j}}{\mathrm{d}t} = g^{rs} g^{rt} , \quad i, j, k, l = 1, 2 ,$$

and consequently

$$\mathbf{n} = -(-1)^i \sqrt{\frac{g^{rs}}{g^{rt}}} g_{sr}^{ik} \frac{\mathrm{d}s^{3-i}}{\mathrm{d}t} \mathbf{r}_{s^k} , \quad i, k = 1, 2 .$$

Thus using the above formulas gives the following expression for the geodesic curvature of S^{x1} in S^{r2}

$$\begin{aligned}\sigma &= -(-1)^k \frac{\sqrt{g^{rs}}}{(g^{rt})^{3/2}} \left(\mathbf{r}_{s^i s^j}\frac{\mathrm{d}s^i}{\mathrm{d}t}\frac{\mathrm{d}s^j}{\mathrm{d}t} + \mathbf{r}_{s^i}\frac{\mathrm{d}^2 s^i}{\mathrm{d}t^2}\right)\cdot\left(g^{lk}_{sr}\frac{\mathrm{d}s^{3-k}}{\mathrm{d}t}\mathbf{r}_{s^l}\right) \\ &= -(-1)^k \frac{\sqrt{g^{rs}}}{(g^{rt})^{3/2}}\frac{\mathrm{d}s^{3-k}}{\mathrm{d}t}\left(\frac{\mathrm{d}^2 s^k}{\mathrm{d}t^2} + \Upsilon^k_{ij}\frac{\mathrm{d}s^i}{\mathrm{d}t}\frac{\mathrm{d}s^j}{\mathrm{d}t}\right), \quad i,j,k,l = 1,2\,,\end{aligned}$$

or using the matrix notation

$$\sigma = \frac{\sqrt{g^{rs}}}{(g^{rt})^{3/2}}\det(A) \tag{4.125}$$

where

$$A = \begin{pmatrix} \dfrac{\mathrm{d}^2 s^1}{\mathrm{d}t^2} + \Upsilon^1_{ij}\dfrac{\mathrm{d}s^i}{\mathrm{d}t}\dfrac{\mathrm{d}s^j}{\mathrm{d}t} & \dfrac{\mathrm{d}^2 s^2}{\mathrm{d}t^2} + \Upsilon^2_{ij}\dfrac{\mathrm{d}s^i}{\mathrm{d}t}\dfrac{\mathrm{d}s^j}{\mathrm{d}t} \\ \dfrac{\mathrm{d}s^1}{\mathrm{d}t} & \dfrac{\mathrm{d}s^2}{\mathrm{d}t} \end{pmatrix}.$$

Note formula (4.125) is determined through the elements of the first groundform only, therefore it is used to formulated the geodesic curvature of curves in arbitrary two-dimensional Riemannian manifolds too.

4.7 Relations to the Principal Curvatures of Two-Dimensional Surfaces

4.7.1 Second Fundamental Form

The coefficients of the second fundamental form

$$b_{ij}\mathrm{d}s^i\mathrm{d}s^j\,, \quad i,j = 1,2\,,$$

of the surface S^{x2} in R^3 represented by

$$\mathbf{x}(\mathbf{s}) : S^2 \to R^3\,, \quad \mathbf{x} = (x^1, x^2, x^3)\,, \quad \mathbf{s} = (s^1, s^2)\,, \tag{4.126}$$

are defined by the dot products of the second derivatives of the vector function $\mathbf{x}(\mathbf{s})$ and the unit normal vector $\mathbf{n}$ to the surface at the point $\mathbf{s}$ under consideration:

$$b_{ij} = \mathbf{x}_{s^i s^j}\cdot\mathbf{n}\,, \quad i,j = 1,2\,.$$

Thus, from (2.26) we obtain for b_{ij}, $i,j = 1,2$,

$$b_{ij} = \frac{1}{\sqrt{g^{rs}}}\left[\frac{\partial^2 x^l}{\partial s^i \partial s^j}\left(\frac{\partial x^{l+1}}{\partial s^1}\frac{\partial x^{l+2}}{\partial s^2} - \frac{\partial x^{l+2}}{\partial s^1}\frac{\partial x^{l+1}}{\partial s^2}\right)\right], \quad l = 1,2,3\,, \tag{4.127}$$

with the identification convention for the superscripts that k is equivalent to $k\pm3$. Correspondingly, for the monitor surface with the scalar-valued monitor function $f(\mathbf{s})$, we obtain

$$b_{ij} = \frac{1}{\sqrt{1+(f_{s^1})^2+(f_{s^2})^2}} f_{s^i s^j} , \quad i,j = 1,2 . \tag{4.128}$$

The tensor (b_{ij}) reflects the local warping of the surface, namely its deviation from the tangent plane at the point under consideration. In particular, if $(b_{ij}) \equiv 0$ at all points of S^2 then the surface is a plane.

4.7.2 Principal Curvatures

Let a curve on the surface $S^{x2} \subset R^3$ be defined by the intersection of a plane containing the normal $\mathbf{n}$ with the surface. Taking into account (3.8), we obtain for the curvature of this curve

$$k = \frac{b_{ij}\mathrm{d}s^i\mathrm{d}s^j}{g^{rs}_{ij}\mathrm{d}s^i\mathrm{d}s^j} , \quad i,j = 1,2 . \tag{4.129}$$

Here $(\mathrm{d}s^1,\ \mathrm{d}s^2)$ is the direction of the curve, i.e. $\mathrm{d}s^i = c(\mathrm{d}s^i/\mathrm{d}\varphi)$, where $\mathbf{s}(\varphi)$ is a curve parametrization. The two extreme quantities K_{I} and K_{II} of the values of k are called the principal curvatures of the surface at the point under consideration. In order to compute the principal curvatures, we consider the following relation for the value of the curvature:

$$(b_{ij} - kg^{rs}_{ij})\mathrm{d}s^i\mathrm{d}s^j = 0 , \quad i,j = 1,2 , \tag{4.130}$$

which follows from (4.129). In order to find the maximum and minimum values of k, the usual method of equating to zero the derivative with respect to $\mathrm{d}s^i$ is applied. Thus the components of the $(\mathrm{d}s^1,\ \mathrm{d}s^2)$ direction giving an extreme value of k are subject to the restriction

$$(b_{ij} - kg^{rs}_{ij})\mathrm{d}s^j = 0 , \quad i,j = 1,2 ,$$

which, in fact, is the eigenvalue problem for curvature. One finds the eigenvalues k by setting the determinant of this equation equal to zero, obtaining thereby the secular equation for k:

$$\det(b_{ij} - kg^{rs}_{ij}) = 0 , \quad i,j = 1,2 .$$

This equation, written out in full, is a quadratic equation

$$k^2 - g^{ij}_{sr}b_{ij}k + [b_{11}b_{22} - (b_{12})^2]/g^{rs} = 0 ,$$

with two roots, which are the maximum and minimum values K_{I} and K_{II} of the curvature k:

$$K_{\mathrm{I,II}} = \frac{1}{2}g^{ij}_{sr}b_{ij} \pm \sqrt{\frac{1}{4}(g^{ij}_{sr}b_{ij})^2 - \frac{1}{g^{rs}}[b_{11}b_{22} - (b_{12})^2]} . \tag{4.131}$$

Mean Curvature

One half of the sum of the principal curvatures is, in fact, the mean surface curvature:

$$K_{\mathrm{m}} = \frac{1}{2} g_{sr}^{ij} b_{ij} = \frac{1}{2}(K_{\mathrm{I}} + K_{\mathrm{II}}) \,, \quad i,j = 1,2 \,. \tag{4.132}$$

In the case of the monitor surface represented by the scalar-valued function $f(s^1, s^2)$, we obtain using (4.128)

$$K_{\mathrm{m}} = \frac{f_{s^1s^1}[1 + (f_{s^2})^2] + f_{s^2s^2}[1 + (f_{s^1})^2] - 2f_{s^1} f_{s^2} f_{s^1s^2}}{2[1 + (f_{s^1})^2 + (f_{s^2})^2]^{3/2}} \,.$$

A surface whose mean curvature is zero, i.e. $K_{\mathrm{I}} = -K_{\mathrm{II}}$, possesses the following unique property. Namely, if a surface bounded by a specified contour has a minimum area then its mean curvature is zero. Conversely, of all the surfaces bounded by a curve whose length is sufficiently small, the minimum area is possessed by the surface whose mean curvature is zero. So the surface whose mean curvature is equal zero at all its points is referred to as a minimal surface.

Gaussian Curvature

Taking into account (4.82), we readily see that the Gaussian curvature is the product of the two principal curvatures K_{I} and K_{II}, i.e.

$$K_{\mathrm{G}} = K_{\mathrm{I}} K_{\mathrm{II}} = \frac{1}{g^{rs}}[b_{11}b_{22} - (b_{12})^2] \,. \tag{4.133}$$

In terms of the derivatives of a scalar-valued function $f(\mathbf{s})$ representing the monitor surface S^{r2} we have, from (4.128) and (4.133),

$$K_{\mathrm{G}} = \frac{f_{s^1s^1} f_{s^2s^2} - (f_{s^1s^2})^2}{[1 + (f_{s^1})^2 + (f_{s^2})^2]^2} \,. \tag{4.134}$$

Formula Dependent on Christoffel Symbols

In the case of a general two-dimensional surface represented by (4.126) an expression for the quantity $b_{11}b_{22} - (b_{12})^2$ can be obtained through derivatives of the elements of the metric tensor and the coefficients of the second fundamental form of the surface. This is accomplished by using the expansion (4.74) which yields the following relation:

$$\mathbf{x}_{s^1s^1} \cdot \mathbf{x}_{s^2s^2} - \mathbf{x}_{s^1s^2} \cdot \mathbf{x}_{s^1s^2} = g_{ij}^{rs}(\Upsilon_{11}^i \Upsilon_{22}^j - \Upsilon_{12}^i \Upsilon_{12}^j) + b_{11}b_{22} - (b_{12})^2 \,,$$

$$i,j = 1,2 \,. \tag{4.135}$$

The left-hand part of (4.135) equals

$$\mathbf{x}_{s^1 s^1} \cdot \mathbf{x}_{s^2 s^2} - \mathbf{x}_{s^1 s^2} \cdot \mathbf{x}_{s^1 s^2} = \frac{\partial}{\partial s^1}(\mathbf{x}_{s^2 s^2} \cdot \mathbf{x}_{s^1}) - \frac{\partial}{\partial s^2}(\mathbf{x}_{s^1 s^2} \cdot \mathbf{x}_{s^1}) \,. \quad (4.136)$$

Since

$$\mathbf{x}_{s^2 s^2} \cdot \mathbf{x}_{s^1} = [22,1] = \frac{1}{2}\Big(2\frac{\partial g_{12}^{rs}}{\partial s^2} - \frac{\partial g_{22}^{rs}}{\partial s^1}\Big) \,,$$

$$\mathbf{x}_{s^1 s^2} \cdot \mathbf{x}_{s^1} = [12,1] = \frac{1}{2}\frac{\partial g_{11}^{rs}}{\partial s^2} \,,$$

we obtain from (4.136)

$$\mathbf{x}_{s^1 s^1} \cdot \mathbf{x}_{s^2 s^2} - \mathbf{x}_{s^1 s^2} \cdot \mathbf{x}_{s^1 s^2} = \frac{1}{2}\Big(2\frac{\partial^2 g_{12}^{rs}}{\partial s^1 \partial s^2} - \frac{\partial^2 g_{11}^{rs}}{\partial s^2 \partial s^2} - \frac{\partial^2 g_{22}^{rs}}{\partial s^1 \partial s^1}\Big) \,.$$

Therefore (4.133) results in

$$K_{\mathrm{G}} = \frac{1}{g^{rs}}\Big[\frac{1}{2}\Big(2\frac{\partial^2 g_{12}^{rs}}{\partial s^1 \partial s^2} - \frac{\partial^2 g_{11}^{rs}}{\partial s^2 \partial s^2} - \frac{\partial^2 g_{22}^{rs}}{\partial s^1 \partial s^1}\Big) - g_{ij}^{rs}(\Upsilon_{11}^i \Upsilon_{22}^j - \Upsilon_{12}^i \Upsilon_{12}^j)\Big] \,, \quad i,j = 1,2 \,.$$

Applying (4.8) transforms this equation to

$$K_{\mathrm{G}} = -\frac{1}{g^{rs}}\Big[\frac{1}{2}\frac{\partial^2}{\partial s^i \partial s^j}(g^{rs} g_{sr}^{ij}) + g_{ij}^{rs}(\Upsilon_{11}^i \Upsilon_{22}^j - \Upsilon_{12}^i \Upsilon_{12}^j)\Big] \,, \quad i,j = 1,2 \,. \quad (4.137)$$

□

This equation depends on the elements of the first surface groundform only. Therefore it can be applied to compute the Gauss curvature of two-dimensional Riemannian manifolds with arbitrary metric tensors.

Since (4.24)

$$g_{ij}^{rs}(\Upsilon_{kl}^i \Upsilon_{mp}^j) = g_{sr}^{ij}[kl,i][mp,j] \,, \quad i,j,k,l,m,p = 1,2 \,,$$

therefore (4.137) has also the following form

$$K_G = -\frac{1}{g^{rs}}\Big\{\frac{1}{2}\frac{\partial^2}{\partial s^i \partial s^j}(g^{rs} g_{sr}^{ij}) + g_{ij}^{rs}([11,i][22,j] - [12,i][12,j])\Big\} \,, \quad i,j = 1,2 \,. \quad (4.138)$$

In particular, in the case of the spherical metric

$$g_{ij}^{\mathbf{s}} = v(\mathbf{s})\delta_j^i \,, \quad i,j = 1,2 \,,$$

we readily find from (4.138)

$$\begin{aligned} K_{\mathrm{G}} &= -\frac{1}{[v(\mathbf{s})]^2}\Big\{\frac{1}{2}\nabla^2[v] + \frac{1}{v(\mathbf{s})}([11,i][22,i] - [12,i][12,i])\Big\} \\ &= -\frac{1}{2[v(\mathbf{s})]^2}\Big\{\nabla^2[v] - \frac{1}{v(\mathbf{s})}[(v_{s^1})^2 + (v_{s^2})^2]\Big\} \\ &= -\frac{1}{2v(\mathbf{s})}\nabla^2[\ln v] \,, \quad i = 1,2 \,. \end{aligned} \quad (4.139)$$

Formula Dependent on Metric Elements

Now we establish one more important expression for K_G. For this purpose we compute the quantity

$$d = \det A$$

where

$$A = \begin{pmatrix} g^{rs}_{11} & g^{rs}_{12} & g^{rs}_{22} \\ \frac{\partial g^{rs}_{11}}{\partial s^1} & \frac{\partial g^{rs}_{12}}{\partial s^1} & \frac{\partial g^{rs}_{22}}{\partial s^1} \\ \frac{\partial g^{rs}_{11}}{\partial s^2} & \frac{\partial g^{rs}_{12}}{\partial s^2} & \frac{\partial g^{rs}_{22}}{\partial s^2} \end{pmatrix} .$$

Using (4.22 – 4.24) gives

$$\begin{aligned} d = {} & g^{rs}_{11}\Big(\frac{\partial g^{rs}_{12}}{\partial s^1}\frac{\partial g^{rs}_{22}}{\partial s^2} - \frac{\partial g^{rs}_{12}}{\partial s^2}\frac{\partial g^{rs}_{22}}{\partial s^1}\Big) - g^{rs}_{12}\Big(\frac{\partial g^{rs}_{11}}{\partial s^1}\frac{\partial g^{rs}_{22}}{\partial s^2} - \frac{\partial g^{rs}_{11}}{\partial s^2}\frac{\partial g^{rs}_{22}}{\partial s^1}\Big) \\ & + g^{rs}_{22}\Big(\frac{\partial g^{rs}_{11}}{\partial s^1}\frac{\partial g^{rs}_{12}}{\partial s^2} - \frac{\partial g^{rs}_{11}}{\partial s^2}\frac{\partial g^{rs}_{12}}{\partial s^1}\Big) \\ & = 2g^{rs}_{11}\{([11,2]+[21,1])[22,2] - ([12,2]+[22,1])[21,2]\} \\ & -4g_{12}([11,1][22,2] - [12,1][21,2]) \\ & +2g^{rs}_{22}\{[11,1]([12,2]+[22,1]) - [12,1]([11,2]+[21,1])\} \\ & = 2g^{rs}_{11}[(\Upsilon^k_{11}g^{rs}_{k2} + \Upsilon^k_{21}g^{rs}_{k1})\Upsilon^l_{22}g^{rs}_{l2} - (\Upsilon^k_{12}g^{rs}_{k2} + \Upsilon^k_{22}g^{rs}_{k1})\Upsilon^l_{21}g^{rs}_{l2}] \\ & -4g^{rs}_{12}(\Upsilon^k_{11}g^{rs}_{k1}\Upsilon^l_{22}g^{rs}_{l2} - \Upsilon^k_{12}g^{rs}_{k1}\Upsilon^l_{21}g^{rs}_{l2}) \\ & +2g^{rs}_{22}[\Upsilon^k_{11}g^{rs}_{k1}(\Upsilon^l_{12}g^{rs}_{l2} + \Upsilon^l_{22}g^{rs}_{l1}) - \Upsilon^k_{12}g^{rs}_{k1}(\Upsilon^l_{11}g^{rs}_{l2} + \Upsilon^l_{21}g^{rs}_{l1})] \\ & = 2(\Upsilon^k_{11}\Upsilon^l_{22} - \Upsilon^k_{12}\Upsilon^l_{12})(g^{rs}_{11}g^{rs}_{k2}g^{rs}_{l2} - 2g^{rs}_{12}g^{rs}_{k1}g^{rs}_{l2} + g^{rs}_{22}g^{rs}_{k1}g^{rs}_{l1}) \\ & +2(g^{rs}_{11}\Upsilon^k_{21}\Upsilon^l_{22} + g^{rs}_{22}\Upsilon^k_{11}\Upsilon^l_{12})(g^{rs}_{k1}g^{rs}_{l2} - g^{rs}_{l1}g^{rs}_{k2}) \\ & = 2g^{rs}[g^{rs}_{11}(\Upsilon^1_{11}\Upsilon^1_{22} - \Upsilon^1_{12}\Upsilon^1_{12}) + g^{rs}_{22}(\Upsilon^2_{11}\Upsilon^2_{22} - \Upsilon^2_{12}\Upsilon^2_{12}) \\ & +2g^{rs}_{12}(\Upsilon^2_{11}\Upsilon^1_{22} - \Upsilon^2_{12}\Upsilon^1_{12}) + g^{rs}_{11}(\Upsilon^1_{21}\Upsilon^2_{22} - \Upsilon^2_{21}\Upsilon^1_{22}) \\ & +g^{rs}_{22}(\Upsilon^1_{11}\Upsilon^2_{12} - \Upsilon^2_{11}\Upsilon^1_{12})] \\ & = 2g^{rs}[g^{rs}_{ij}(\Upsilon^i_{11}\Upsilon^j_{22} - \Upsilon^i_{12}\Upsilon^j_{12}) + g^{rs}_{12}(\Upsilon^2_{11}\Upsilon^1_{22} - \Upsilon^1_{11}\Upsilon^2_{22}) \\ & +g^{rs}_{11}(\Upsilon^1_{21}\Upsilon^2_{22} - \Upsilon^2_{21}\Upsilon^1_{22}) + g^{rs}_{22}(\Upsilon^1_{11}\Upsilon^2_{12} - \Upsilon^2_{11}\Upsilon^1_{12})] , \quad k,l = 1,2 . \end{aligned} \tag{4.140}$$

On the other hand

$$
\begin{aligned}
&\Upsilon^k_{ki}\frac{\partial}{\partial s^j}(g^{rs}g^{ij}_{sr}) = (-1)^{i+1}\Upsilon^k_{ki}\left(\frac{\partial}{\partial s^1}g^{rs}_{23-i} - \frac{\partial}{\partial s^2}g^{rs}_{13-i}\right) \\
&= \Upsilon^k_{k1}([21,2]-[22,1]) + \Upsilon^k_{k2}([12,1]-[11,2]) \\
&= \Upsilon^k_{k1}(\Upsilon^l_{21}g^{rs}_{l2} - \Upsilon^l_{22}g^{rs}_{l1}) + \Upsilon^k_{k2}(\Upsilon^l_{12}g^{rs}_{l1} - \Upsilon^l_{11}g^{rs}_{l2}) \\
&= g^{rs}_{l2}(\Upsilon^k_{k1}\Upsilon^l_{21} - \Upsilon^k_{k2}\Upsilon^l_{11}) + g^{rs}_{l1}(\Upsilon^k_{k2}\Upsilon^l_{12} - \Upsilon^k_{k1}\Upsilon^l_{22}) \\
&= g^{rs}_{12}[\Upsilon^1_{21}(\Upsilon^1_{11}+\Upsilon^2_{21}) - \Upsilon^1_{11}(\Upsilon^1_{12}+\Upsilon^2_{22})] \\
&+g^{rs}_{21}[\Upsilon^2_{12}(\Upsilon^1_{12}+\Upsilon^2_{22}) - \Upsilon^2_{22}(\Upsilon^1_{11}+\Upsilon^2_{21})] \\
&= 2g^{rs}_{12}(\Upsilon^1_{21}\Upsilon^2_{21} - \Upsilon^1_{11}\Upsilon^2_{22}) + g^{rs}_{22}[\Upsilon^2_{21}(\Upsilon^1_{11}+\Upsilon^2_{21}) - \Upsilon^2_{11}(\Upsilon^1_{12}+\Upsilon^2_{22})] \\
&+g^{rs}_{11}[\Upsilon^1_{12}(\Upsilon^1_{12}+\Upsilon^2_{22}) - \Upsilon^1_{22}(\Upsilon^1_{11}+\Upsilon^2_{21})] \\
&= g^{rs}_{ij}(\Upsilon^i_{12}\Upsilon^j_{12} - \Upsilon^i_{11}\Upsilon^j_{22}) + g^{rs}_{11}(\Upsilon^1_{12}\Upsilon^2_{22} - \Upsilon^1_{22}\Upsilon^2_{21}) \\
&+g^{rs}_{22}(\Upsilon^2_{21}\Upsilon^1_{11} - \Upsilon^2_{11}\Upsilon^1_{12}) + g^{rs}_{12}(\Upsilon^2_{11}\Upsilon^1_{22} - \Upsilon^1_{11}\Upsilon^2_{22}) \, .
\end{aligned}
\tag{4.141}
$$

Availing us of (4.140) and (4.141) we find

$$
d - 2g^{rs}\Upsilon^k_{ki}\frac{\partial}{\partial s^j}(g^{rs}g^{ij}_{sr}) = 4g^{rs}g^{rs}_{ij}(\Upsilon^i_{11}\Upsilon^j_{22} - \Upsilon^i_{12}\Upsilon^j_{12}) \, , \quad i,j=1,2 \, . \tag{4.142}
$$

Since (4.25)

$$
\Upsilon^k_{ki} = -\sqrt{g^{rs}}\frac{\partial}{\partial s^i}\left(\frac{1}{\sqrt{g^{rs}}}\right) \, , \quad i,k=1,2 \, ,
$$

therefore, using (4.137), the equation (4.142) becomes

$$
\begin{aligned}
K_G &= -\frac{1}{g^{rs}}\left[\frac{1}{2}\frac{\partial^2}{\partial s^i \partial s^j}(g^{rs}g^{ij}_{sr}) + \frac{d}{4g^{rs}}\right. \\
&\left.+\frac{\sqrt{g^{rs}}}{2}\frac{\partial}{\partial s^i}\left(\frac{1}{\sqrt{g^{rs}}}\right)\frac{\partial}{\partial s^j}(g^{rs}g^{ij}_{sr})\right] \\
&= -\frac{d}{4(g^{rs})^2} - \frac{1}{2\sqrt{g^{rs}}}\frac{\partial}{\partial s^i}\left[\frac{1}{\sqrt{g^{rs}}}\frac{\partial}{\partial s^j}(g^{rs}g^{ij}_{sr})\right] \, , \quad i,j=1,2 \, . \, \Box
\end{aligned}
\tag{4.143}
$$

In particular, when the coordinate system s^1, s^2 is orthogonal, then $d = 0$, $g^{12}_{sr} = 0$ and (4.143) yields in this case

$$
K_G = -\frac{1}{2\sqrt{g^{rs}}}\left[\frac{\partial}{\partial s^1}\left(\frac{1}{\sqrt{g^{rs}}}\frac{\partial}{\partial s^1}g^{rs}_{22} + \frac{\partial}{\partial s^2}\left(\frac{1}{\sqrt{g^{rs}}}\frac{\partial}{\partial s^2}g^{rs}_{11}\right)\right] \, . \tag{4.144}
$$

Classification of Points of Two-Dimensional Surfaces

A surface point is called elliptic if $K_G > 0$, i.e. both K_I and K_{II} are both negative or both positive at the point of consideration. A saddle or hyperbolic point has principal curvatures of opposite sign, and therefore has negative Gaussian curvature. A parabolic point has one principal curvature vanishing and, consequently, a vanishing Gaussian curvature. This classification of points is prompted by the form of the curve which is obtained by the intersection of the surface with a slightly offset tangent plane. For an elliptic point the curve is an ellipse; for a saddle point it is a hyperbola. It is a pair of lines (degenerate conic) at a parabolic point, and it vanishes at a planar point, where both principal curvatures are zero.

Part II

Algorithms and Applications of Advanced Grid Technology

The current part of the monograph including Chaps. 5–7 gives an account of the geometrization of the popular comprehensive grid methods and presents an important extension of the methods, related to the application of the theory of Riemannian manifolds to the formulation of grid equations and grid functionals.

As comprehensive tools for generating grids there are chosen the operators of Beltrami and diffusion in an arbitrary metric. These operators give rise to the formulation of Beltrami and diffusion equations, respectivelly. It is established in Chap. 5 that an arbitrary one-to-one, smooth multidimensional transformation deriving a numerical grid in a domain or on a surface by the mapping approach is realized by the solution of the system of Beltrami equations in a suitable monitor metric specified in the physical geometry. The system can be interpreted as the multidimensional equidistribution principle in which the monitor metric tensor is an extension of a scalar-valued weight function. While the required grid properties are realized through the specification of suitable monitor metrics that can be formulated as a combination of metric tensors with weights, each of which is responsible for providing one individual grid property.

Taking advantage of the tensor relations, the Beltrami and diffusion equations are converted in Chap. 6 into compact forms convenient for the numerical treatment by the available algorithms. Some of these forms of equations rule out division by the Jacobian of intermediate transformations. So solutions of these equations allow one to use singular intermediate transformations as demonstrated in Chap. 7. Relations of the mean curvature of the monitor surfaces to the equations for grid generation are also exhibited in Chap. 6. In this chapter the technique of the multidimensional differential geometry is also applied to the analysis of the qualitative behavior of the grids generated by the mapping approach. For this purpose a new characteristic of grid clustering is formulated. A relation between this grid measure and some geometric characteristics of grid hypersurfaces and the monitor functions forming the monitor metric is established. The well-known results for grids generated by the inverted Laplace equations about node clustering near concave boundary segments of domains and node rarefaction near convex ones are extended, using the relation, to arbitrary boundary segments and to more general elliptic equations formulated for generating grids. On the basis of the formula the monitor functions are readily estimated in the diffusion and Beltrami elliptic models of grid equations to provide grid clustering or, if it is reasonable, grid rarefaction near arbitrary segments of physical geometries.

Chap. 7 gives a description of numerical codes for generating grids with the help of the elliptic grid equations and energy functionals obtained in Chaps. 5 and 6 by changing mutually dependent and independent variables in the basic grid models, formulated in Chap. 5. The application of the grid technology advocated in the book to some gas-dynamics problems is also discussed in Chap. 7.

5 Comprehensive Grid Models

This chapter formulates differential and variational models for generating grids in domains and on surfaces of arbitrary dimensions. The formulations are based on the operators of Beltrami and diffusion with respect to monitor metrics.

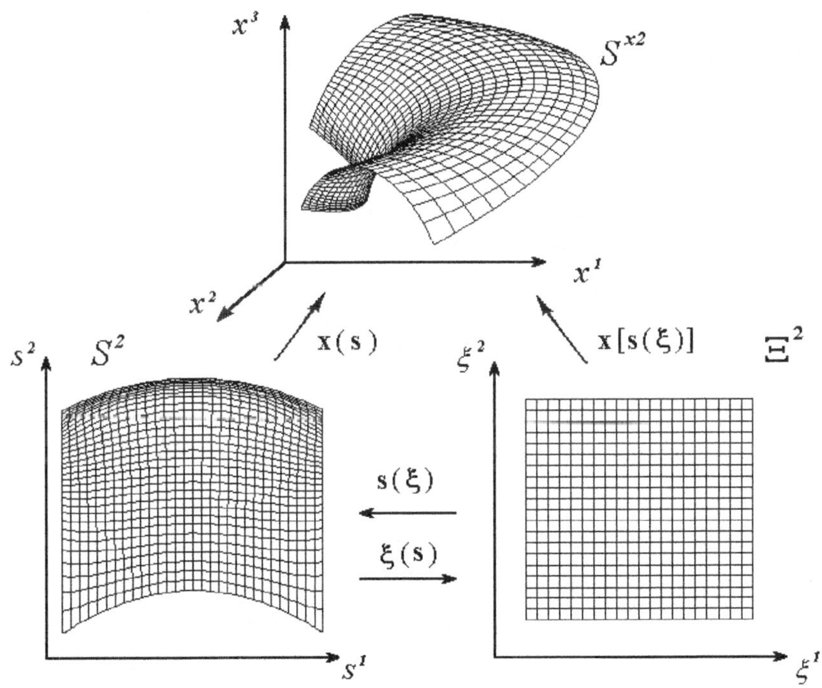

Fig. 5.1. Scheme of the Mapping Approach for Grid Generation

For the purpose of unification of the formulation of grid models for generating meshes in arbitrary physical geometries a physical domain, surface, or curve is considered in a unified manner as a single geometric object, referred to as an n-dimensional regular surface, locally represented by a parametrization

$$\mathbf{x}(\mathbf{s}) : S^n \to \mathbf{R}^{n+k}, \ \mathbf{x} = (x^1, \dots, x^{n+k}), \ \mathbf{s} = (s^1, \dots, s^n), \ n \geq 1 , \qquad (5.1)$$

where S^n is an n-dimensional parametric domain (an interval if $n = 1$), while $\mathbf{x}(\mathbf{s})$ is a smooth vector-valued function of rank n at all points $\mathbf{s} \in S^n$. We shall designate by S^{xn} the regular surface parametrized by (5.1). Note, when $k = 0$ then S^{xn} is a domain $X^n \subset \mathbf{R}^n$ which itself can naturally be considered as a parametric domain S^n for X^n with $\mathbf{x}(\mathbf{s})$ being the identical, i.e., $\mathbf{x}(\mathbf{s}) \equiv \mathbf{s}$ (Fig. 5.2), or another parametrization.

Our further findings are related to the mapping approach of grid generation (see Sect. 1.3.1). By this approach the process of numerical grid generation on S^{xn} is turned to finding an intermediate smooth nondegenerate transformation

$$\mathbf{s}(\boldsymbol{\xi}) : \Xi^n \to S^n , \quad \boldsymbol{\xi} = (\xi^1, \dots, \xi^n) \qquad (5.2)$$

between S^n and a suitable computational (logical) domain Ξ^n and consequently the mesh points on S^{xn} are the images through

$$\mathbf{x}[\mathbf{s}(\boldsymbol{\xi})] : \Xi^n \to \mathbf{R}^{n+k}$$

of the nodes of a reference grid in Ξ^n (see Figs. 5.1 and 5.2).

The computational domain Ξ^n and the cells of the reference mesh can be either rectangular (Fig. 5.1) or of a different shape (for example the logical domain may be a trapezoid while its cells may be triangles as in Fig. 5.2) depending on the shape of the physical geometry and/or the numerical algorithm (finite differences, finite volumes, finite elements, spectral elements, etc.) applied to solve a particular physical problem.

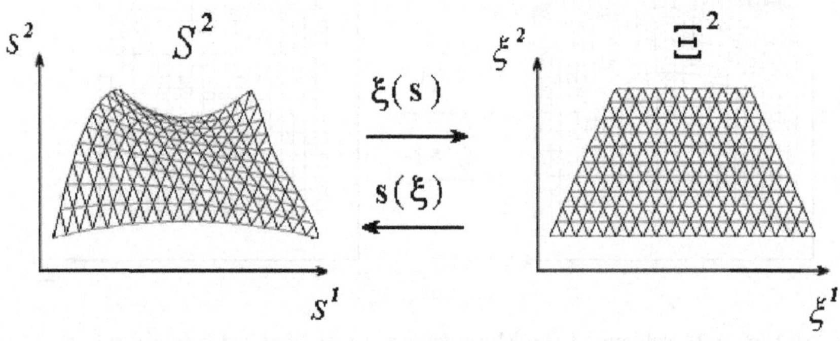

Fig. 5.2. Scheme for Generating Triangular Grids

The coordinates $\xi^1, \dots, \xi^n$ of the surface S^{xn} specified by composition of the parametrizations (5.1) and (5.2) are called logical or grid coordinates, while the coordinates $s^1, \dots, s^n$ in (5.1) are referred to as parametric coordinates. We assume further that the parametric coordinates $s^1, \dots, s^n$ in S^n are the Cartesian coordinates.

For the purpose of providing efficient control of grid generation in the geometry S^{xn} we introduce a notion of a monitor manifold M^n over S^{xn}. The points of M^n are the points of S^{xn} while its metric may differ from the metric of S^{xn}, in particular, in the coordinates $s^1, \dots, s^n$ it can be defined in the form (4.21) or (4.50).

5.1 Formulation of Differential Grid Generators

This section reviews mathematical models for robust grid generators based on the differential operators of Beltrami and diffusion.

5.1.1 Beltramian Operator

The operator is formulated on a set of twice differentiable functions $\mathbf{v}(\mathbf{x})$ defined on an arbitrary Riemannian manifold M^n with a covariant metric tensor $(g_{ij}^{\mathbf{x}})$ in local coordinates x^i, $i = 1, \dots, n$, as the following elliptic operator

$$\Delta_B[\mathbf{v}](\mathbf{x}) = \frac{1}{\sqrt{g^{\mathbf{x}}}} \frac{\partial}{\partial x^j} \left(\sqrt{g^{\mathbf{x}}} g_{\mathbf{x}}^{jk} \frac{\partial \mathbf{v}(\mathbf{x})}{\partial x^k} \right), \quad j, k = 1, \dots, n, \tag{5.3}$$

where $\mathbf{x} = (x^1, \dots, x^n)$, $(g_{\mathbf{x}}^{jk})$, $j, k = 1, \dots, n$, is the contravariant metric tensor of the manifold in the coordinates x^i, $i = 1, \dots, n$, and $g^{\mathbf{x}} = \det(g_{ij}^{\mathbf{x}})$.

As it was mentioned we observe a convention that the summation is carried out over repeated indices unless otherwise noted thus (5.3), in fact, is

$$\Delta_B[\mathbf{v}](\mathbf{x}) = \frac{1}{\sqrt{g^{\mathbf{x}}}} \sum_{j=1}^{n} \frac{\partial}{\partial x^j} \left(\sqrt{g^{\mathbf{x}}} \sum_{k=1}^{n} g_{\mathbf{x}}^{jk} \frac{\partial \mathbf{v}(\mathbf{x})}{\partial x^k} \right).$$

Notice the matrix $(g_{\mathbf{x}}^{ij})$ is the inverse of the matrix $(g_{ij}^{\mathbf{x}})$ and vice versa, hence the elements of the covariant and contravariant metric tensors are connected by the relations

$$g_{\mathbf{x}}^{ij} g_{jk}^{\mathbf{x}} = g_{ij}^{\mathbf{x}} g_{\mathbf{x}}^{jk} = \delta_i^k, \quad i, j, k = 1, \dots, n,$$

where $\delta_i^k = 0, 1$ if $k \neq i$, $k = i$.

It was shown in Sect. 4.5.1 (see equations (4.56) and (4.58)) that the formula (5.3) is an invariant of parametrizations of the manifold M^n. We present here one more inference of this fact by virtue of the identity (2.48). Let s^i, $i = 1, \dots, n$, be another local coordinate system in M^n and $(g_{ij}^{\mathbf{s}})$

and $(g_{\mathbf{s}}^{ij})$, $i,j = 1,\dots,n$, be the covariant and contravariant metric tensors, respectively, of M^n in these coordinates. Using the metric relations

$$g_{\mathbf{x}}^{ij} = g_{\mathbf{s}}^{kl}\frac{\partial x^i}{\partial s^k}\frac{\partial x^j}{\partial s^l}\,, \quad i,j,k,l = 1,\dots,n\,,$$

$$g^{\mathbf{x}}J^{-2} = g^{\mathbf{s}}\,, \quad g^{\mathbf{x}} = g^{\mathbf{s}}(J)^2\,,$$

where

$$J = \det\Big(\frac{\partial s^i}{\partial x^j}\Big)\,, \quad g^{\mathbf{s}} = \det(g_{ij}^{\mathbf{s}})\,, \quad i,j = 1,\dots,n\,,$$

and the fundamental identity

$$\frac{\partial}{\partial x^j}\Big(J\frac{\partial x^j}{\partial s^i}\Big) = 0\,, \quad i,j = 1,\dots,n\,,$$

(see 2.48) gives

$$\begin{aligned}
\Delta_B[\mathbf{v}](\mathbf{x}(\mathbf{s})) &= \frac{1}{\sqrt{g^{\mathbf{x}}}}\frac{\partial}{\partial x^j}\Big(\sqrt{g^{\mathbf{x}}}J^{-1}Jg_{\mathbf{s}}^{mk}\frac{\partial x^j}{\partial s^m}\frac{\partial x^i}{\partial s^k}\frac{\partial \mathbf{v}[\mathbf{x}(\mathbf{s})]}{\partial s^p}\frac{\partial s^p}{\partial x^i}\Big)\\
&= \frac{1}{\sqrt{g^{\mathbf{x}}}}\frac{\partial}{\partial x^j}\Big(\sqrt{g^{\mathbf{s}}}J\frac{\partial x^j}{\partial s^m}g_{\mathbf{s}}^{mk}\frac{\partial \mathbf{v}[\mathbf{x}(\mathbf{s})]}{\partial s^k}\Big)\\
&= \frac{J}{\sqrt{g^{\mathbf{x}}}}\frac{\partial}{\partial x^j}\Big(\sqrt{g^{\mathbf{s}}}g_{\mathbf{s}}^{mk}\frac{\partial \mathbf{v}[\mathbf{x}(\mathbf{s})]}{\partial s^k}\Big)\frac{\partial x^j}{\partial s^m}\\
&= \frac{1}{\sqrt{g^{\mathbf{s}}}}\frac{\partial}{\partial s^m}\Big(\sqrt{g^{\mathbf{s}}}g_{\mathbf{s}}^{mp}\frac{\partial \mathbf{v}[\mathbf{x}(\mathbf{s})]}{\partial s^p}\Big) = \Delta_B[\mathbf{v}(\mathbf{x})](\mathbf{s})\,,\\
&\quad i,j,k,m,p = 1,\dots,n\,. \ \square
\end{aligned}$$

So the value of the Beltramian operator at a function $\mathbf{v}(\mathbf{s})$ is an invariant of the choice of a parametrization of the manifold M^n. This invariant is called Beltrami's second differential parameter of the function $\mathbf{v}(\mathbf{s})$.

5.1.2 Boundary Value Problem for Grid Equations

The Beltramian operator allows one to formulate a mathematical model for generating grids in arbitrary smooth Riemannian manifolds. Though for the practical purpose it is sufficient to consider the manifolds as the monitor manifolds over a physical geometry S^{xn}.

Let M^n be an n-dimensional manifold with the metric tensor $(g_{ij}^{\mathbf{s}})$ in the local coordinates s^i, $i = 1,\dots,n$, whose values lie in some parametric domain $S^n \subset R^n$. Thus there is a local map $\mathbf{x}(\mathbf{s}) : S^n \to M^n$. Analogously to the case of the physical geometry S^{xn}, a local grid in M^n is found by mapping a reference grid in a standard logical domain Ξ^n into M^n by the composition of $\mathbf{x}(\mathbf{s})$ and some one-to-one intermediate smooth transformation

$$\mathbf{s}(\boldsymbol{\xi}) : \Xi^n \to S^n , \quad \boldsymbol{\xi} = (\xi^1, \ldots , \xi^n) , \quad \mathbf{s} = (s^1, \ldots , s^n) ,$$

i.e. by $\mathbf{x}(\mathbf{s}(\boldsymbol{\xi})) : \Xi^n \to M^n$ (see Fig. 5.3).

Note the parametrization $\mathbf{x}(\mathbf{s})$ also generates a grid in M^n by mapping some grid in S^n. However, this grid may be unsatisfactory and as a rule it is not independent of parametrizations. Besides this, if the geometry of S^n is complex, the grid generation in S^n may require serious efforts. The role of the intermediate transformation $\mathbf{s}(\boldsymbol{\xi})$ is to make the grid on M^n satisfy the necessary properties, in particular, the property of independence of the choice of a parametrization. While the role of the logical domain Ξ^n is to replace the parametrization domain S^n with a standard parametric domain (n-dimensional cube, simplex, prism, etc) having a simpler shape.

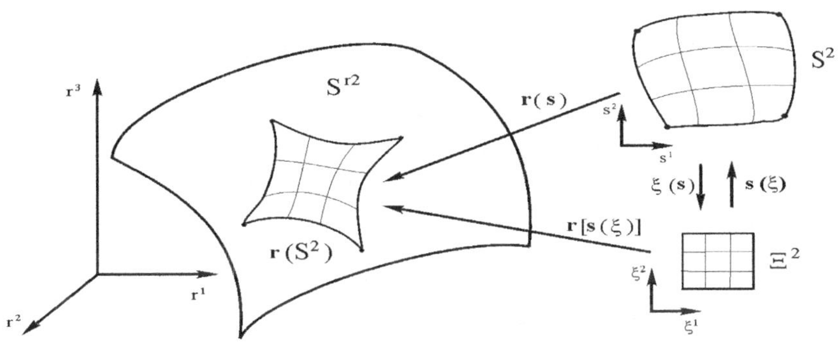

Fig. 5.3. Illustration for Grid Generation

The logical domain Ξ^n, its reference grid, and the parametrization $\mathbf{x}(\mathbf{s})$ are chosen by the user. Therefore the local grid on M^n with the required properties is defined when a suitable intermediate transformation $\mathbf{s}(\boldsymbol{\xi})$ is found. One of the ways to find this transformation is to use the operator of Beltrami in the metric of M^n. Namely, $\mathbf{s}(\boldsymbol{\xi})$ may be determined as the inverse of the transformation

$$\boldsymbol{\xi}(\mathbf{s}) : S^n \to \Xi^n , \quad \boldsymbol{\xi}(\mathbf{s}) = [\xi^1(\mathbf{s}), \ldots , \xi^n(\mathbf{s})]$$

which is a solution of the following Dirichlet boundary value problem

$$\Delta_B[\boldsymbol{\xi}](\mathbf{s}) \equiv \frac{1}{\sqrt{g^{\mathrm{s}}}} \frac{\partial}{\partial s^j} \left(\sqrt{g^{\mathrm{s}}} g_{\mathrm{s}}^{jk} \frac{\partial \boldsymbol{\xi}}{\partial s^k} \right) = 0 , \quad j, k = 1, \ldots , n ,$$

$$\Gamma[\boldsymbol{\xi}](\mathbf{s}) \quad \equiv \boldsymbol{\xi}|_{\partial S^n} = \varphi(\mathbf{s}) : \partial S^n \to \partial \Xi^n , \quad \varphi(\mathbf{s}) = [\varphi^1(\mathbf{s}), \ldots , \varphi^n(\mathbf{s})] ,$$

or in an equivalent form for the components $\xi^i(\mathbf{s})$ (since $g^{\mathrm{s}} > 0$)

$$\frac{\partial}{\partial s^j}\left(\sqrt{g^{\mathbf{s}}}g^{jk}_{\mathbf{s}}\frac{\partial \xi^i}{\partial s^k}\right) = 0\,, \quad i,j,k = 1,\dots,n\,,$$
$$\xi^i|_{\partial S^n} = \varphi^i(\mathbf{s})\,, \quad i = 1,\dots,n\,, \tag{5.4}$$

where $g^{jk}_{\mathbf{s}}$ are the (jk)th element of the contravariant metric tensor of M^n in the parametric coordinates $s^1,\dots,s^n$, $g^{\mathbf{s}} = \det(g^{\mathbf{s}}_{ij})$, ∂S^n and $\partial \Xi^n$ are the boundaries of S^n and Ξ^n, respectively, while $\varphi(\mathbf{s})$ is a one-to-one continuous map between the boundaries of S^n and Ξ^n. So we see that the transformation $\boldsymbol{\xi}(\mathbf{s})$, satisfying (5.4) is the function whose Beltrami's second differential parameter is equal to the zero vector $\mathbf{0}$. In the theory of Riemannian manifolds the equations (5.4) are called generalized Laplace equations or Beltrami equations of the second order. We shall call them further as Beltrami equations.

The functions $\xi^1,\dots,\xi^n$ satisfying (5.4) form a curvilinear coordinate system in S^n, S^{xn}, and M^n. These curvilinear coordinates are further referred to as grid coordinates.

Since the value of the operator $\Delta_B[\boldsymbol{\xi}]$ is independent of parametrizations of M^n we obtain by the solution of (5.4) the same grid in M^n regardless of the original coordinate system in M^n provided the boundary conditions in (5.4) for different coordinates are consistent. Assuming $s^i = \xi^i$, $i = 1,\dots,n$, in (5.4) we also find that this system is equivalent to the following system

$$\frac{\partial}{\partial \xi^j}(\sqrt{g^{\boldsymbol{\xi}}}g^{ji}_{\boldsymbol{\xi}}) = 0\,, \quad i,j = 1,\dots,n\,, \tag{5.5}$$

where $(g^{ji}_{\boldsymbol{\xi}})$, $i,j = 1,\dots,n$, is the contravariant metric tensor of M^n in the grid coordinates $\xi^1,\dots,\xi^n$,

$$g^{\boldsymbol{\xi}} = J^2 g^{\mathbf{s}}\,, \quad J = \det\left(\frac{\partial s^i}{\partial \xi^j}\right) = 1/\det\left(\frac{\partial \xi^i}{\partial s^j}\right)\,.$$

Note, in order to find grid nodes through the intermediate transformation $\mathbf{s}(\boldsymbol{\xi})$ the inverse of which is a solution of the problem (5.4) there is no necessity to compute the transformation $\mathbf{s}(\boldsymbol{\xi})$ at all points $\boldsymbol{\xi} \in \Xi^n$. It is sufficient to solve numerically an equivalent inverted boundary value problem obtained by interchanging in (5.4) dependent and independent variables, i.e. assuming that variables $\xi^1,\dots,\xi^n$ are independent while the quantities $s^1,\dots,s^n$ are dependent . The transformed problem with respect to the components $s^i(\boldsymbol{\xi})$, $i = 1,\dots,n$, of the intermediate transformation $\mathbf{s}(\boldsymbol{\xi})$ should be solved on the reference grid in Ξ^n. The values of this very numerical solution

$$\mathbf{s}(\boldsymbol{\xi}) = [s^1(\boldsymbol{\xi}),\dots,s^n(\boldsymbol{\xi})]$$

at the points of the reference grid determine grid nodes in S^n and consequently on M^n by mapping them through $\mathbf{x}(\mathbf{s})$.

The linear elliptic system in (5.4) is of divergence form hence its solution is subject to the maximum principle. Therefore the grid nodes produced through (5.4) will be inside of S^{xn} if the target domain Ξ^n is convex. Moreover, for $n = 2$ there is valid a general theorem of Rado from which a particular corollary follows that the transformation $\boldsymbol{\xi}(\mathbf{s})$ obtained by the solution of the Dirichlet boundary value problem (5.4) with an arbitrary metric is nondegenerate if Ξ^2 is convex and the boundary mapping determined by the Dirichlet condition is a one-to-one continuous map between the boundaries of S^{x2} and Ξ^2. Note, in the case $n > 2$, this property may be breached (see Farrel and Jones (1996)).

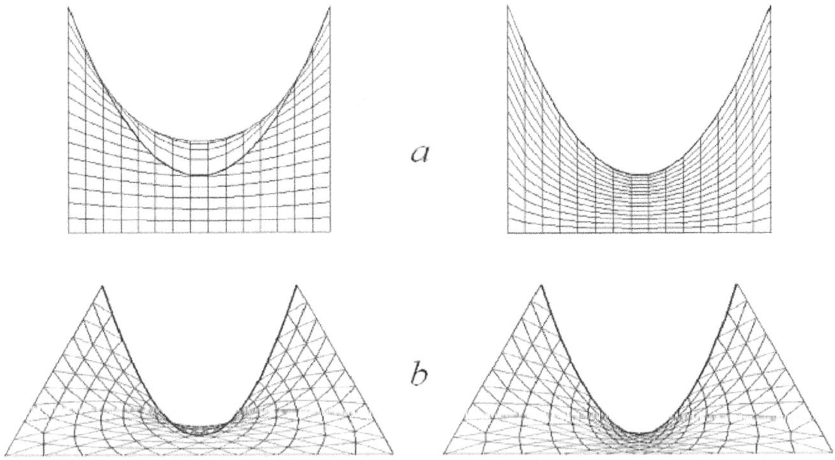

Fig. 5.4. Quadrilateral (a) and triangular (b) grids in concave domains generated by the solution of equations (5.6) (left-hand) and by the solution of equations (5.4) (right-hand); both with respect to the Euclidean metric

Another formulation of a grid model through the generalized Laplace equations with respect to the intermediate transformation $\mathbf{s}(\boldsymbol{\xi})$:

$$\frac{\partial}{\partial \xi^j}\left(\sqrt{g} g^{jk} \frac{\partial s^i}{\partial \xi^k}\right) = 0 , \quad i, j, k = 1, \dots , n , \tag{5.6}$$

where g^{jk} are contravariant metric components of a monitor manifold over Ξ^n, $g = det(g_{jk}) = 1/det(g^{jk})$, was proposed by Godunov and Prokopov (1967), and Ryskin and Leal (1983). These equations seem to be more natural than the equations in (5.4) for the implementation into numerical codes as they are linear and of divergent form with respect to the intermediate transformation $\mathbf{s}(\boldsymbol{\xi})$. However such divergent model, owing to the maximum principle, does not guarantee that all grid points will be inside of the physical geometry S^{xn} when the parametric domain S^n is not convex (Fig. 5.4, left), let alone the grid cells may be folded. Providing grid nondegeneracy through the solution of equations (5.6) depends on devising a suitable monitor metric in Ξ^n, what hasn't been made so far in a general form.

The considerations mentioned are the major reasons why the formulation of grid systems through the Beltrami operator is reasonable to make with respect to the function $\boldsymbol{\xi}(\mathbf{s}) : S^n \to \Xi^n$ that is the inverse of the intermediate transformation $\mathbf{s}(\boldsymbol{\xi}) : \Xi^n \to S^n$.

5.1.3 Interpretation as a Multidimensional Equidistribution Principle

In a one-dimensional case the equation in (5.4) is equivalent to:

$$\frac{\mathrm{d}}{\mathrm{d}s}\Big(\sqrt{g^s} g_s^{11} \frac{\mathrm{d}\xi}{\mathrm{d}s}\Big) \equiv \frac{\mathrm{d}}{\mathrm{d}s}\Big(\sqrt{g_s^{11}} \frac{\mathrm{d}\xi}{\mathrm{d}s}\Big) = 0 \,,$$

since $g^s = g_{11}^s = 1/g_s^{11}$. This equation presents the well known differential formulation of the equidistribution principle with the weight function $w(s) = \sqrt{g_s^{11}}$ defined through the monitor metric g_{11}^s. The multidimensional equation in (5.4) may also be interpreted as a formulation of the multidimensional equidistribution principle with the matrix-valued weight function

$$\mathbf{w}(\mathbf{s}) = \{\sqrt{g^s} g_{\mathbf{s}}^{jk}\} \,, \quad j,k = 1,\dots,n \,,$$

in the coordinates s^i, $i = 1,\dots,n$, defined through the metric $g_{ij}^{\mathbf{s}}$, $i,j = 1,\cdots,n$. By this interpretation the problem of the grid generation through the solution of (5.4) is naturally resolved by the choice of the matrix-valued weight function or, what is the same, with a specification of the metric, called a monitor metric, in M^n.

Relation to Beltramian Equations of the First Order

The Beltramian system of the first order for generating quasiconformal mappings $\boldsymbol{\xi}(\mathbf{s}) : S^n \to \Xi^n$ between the parametric domain S^n of a Riemannian manifold M^n with the metric $g_{ij}^{\mathbf{s}}$ and Ξ^n with the Euclidean metric has the following form

$$\sqrt{g^{\xi s}} \frac{\partial \mathbf{s}}{\partial \xi^i} = \sqrt{g^{\mathbf{s}}} \boldsymbol{\nabla}(\xi^i);, \quad i = 1,\dots,n \,, \tag{5.7}$$

here

$$\boldsymbol{\nabla}(\xi^i) = (g_{\mathbf{s}}^{1m} \frac{\partial \xi^i}{\partial s^m}, \dots, g_{\mathbf{s}}^{nm} \frac{\partial \xi^i}{\partial s^m}) \,, \quad i,m = 1,\dots,n \,,$$

$$g^{\xi s} = \det\Big(\frac{\partial \boldsymbol{\xi}}{\partial s^i} \cdot \frac{\partial \boldsymbol{\xi}}{\partial s^j}\Big) \,.$$

Note if $n = 2$, $g^{\mathbf{s}}_{ij} = \delta^i_j$, then (5.7) is the Caouchy–Riemann system

$$\frac{\partial \xi^1}{\partial s^1} = \frac{\partial \xi^2}{\partial s^2}\,, \qquad \frac{\partial \xi^1}{\partial s^2} = -\frac{\partial \xi^2}{\partial s^1}\,,$$

i.e. $\boldsymbol{\xi}(\mathbf{s})$, in this case, is a conformal mapping. So the relations (5.7) are generally considered as a condition of conformality of the mapping $\boldsymbol{\xi}(\mathbf{s}) : S^n \to \Xi^n$ with respect to the Riemannian metric $g^{\mathbf{s}}_{ij}$ introduced in S^n.

From the identity (2.48) we have

$$\mathrm{div}(\sqrt{g^{\xi s}}\frac{\partial \mathbf{s}}{\partial \xi^i}) \equiv \frac{\partial}{\partial s^j}\left(\sqrt{g^{\xi s}}\frac{\partial s^j}{\partial \xi^i}\right) \equiv 0\,, \quad i,j = 1,\dots,n\,,$$

therefore applying the operator *div* to the system (5.7) yields the system of the second order presented in (5.4) and equivalent to the Beltrami equations.

Note that the contrary assertion is not valid, i. e. the system in (5.4) does not lead, in general, to the system (5.7).

5.1.4 Realization of Specified Grids

Grids in Domains

Let S^{xn} be an n-dimensional domain $X^n \subset R^n$ which has a local grid obtained with the aid of a nondegenerate smooth transformation

$$\mathbf{x}(\boldsymbol{\xi}) : \Xi^n \to X^n\,, \quad \mathbf{x} = (x^1,\dots,x^n)\,, \quad \boldsymbol{\xi} = (\xi^1,\dots,\xi^n)\,, \tag{5.8}$$

i.e. the local grid in X^n is the image of a reference grid in Ξ^n through the coordinate transformation $\mathbf{x}(\boldsymbol{\xi})$. Note the reference grid in Ξ^n can be indiscriminately structured or unstructured.

Let S^n be the image of the domain Ξ^n in X^n through $\mathbf{x}(\boldsymbol{\xi})$. Then S^n can be formally considered as a local parametric domain for X^n with the coordinates $s^i = x^i$, $i = 1,\dots,n$. Let

$$\mathbf{r}(\mathbf{s}) : S^n \to R^n\,, \quad \mathbf{r} = (r^1,\dots,r^n)\,, \quad \mathbf{s} = (s^1,\dots,s^n)$$

be the mapping defined by

$$\mathbf{r}(\mathbf{s}) \equiv \boldsymbol{\xi}(\mathbf{s})\,, \quad \mathbf{s} \in S^n\,,$$

where $\boldsymbol{\xi}(\mathbf{s})$ is the inverse of $\mathbf{x}(\boldsymbol{\xi}) : \Xi^n \to S^n$. Now imposing in X^n a local metric in the coordinates s^i, $i = 1,\dots,n$, in the form (4.50), as

$$g^{\mathbf{s}}_{ij} = B^m_i B^m_j, \quad i,j,m = 1,\dots,n,$$

where

$$B^m_i = \frac{\partial \xi^m}{\partial s^i}, \quad i,m = 1,\dots,n,$$

we get

$$g^{\mathbf{s}}_{ij} = \frac{\partial \xi^l}{\partial s^i}\frac{\partial \xi^l}{\partial s^j} = \frac{\partial \boldsymbol{\xi}}{\partial s^i} \cdot \frac{\partial \boldsymbol{\xi}}{\partial s^j}\,, \quad i,j,l = 1,\dots,n\,. \tag{5.9}$$

For the contravariant metric elements we readily find the following formula

$$\begin{aligned} g^{ij}_{\mathbf{s}} &= \frac{\partial s^i}{\partial \xi^l}\frac{\partial s^j}{\partial \xi^l}\,, \quad i,j,l = 1,\dots,n\,, \\ g^{\mathbf{s}} &= \frac{1}{J^2}\,, \quad J = \det\Big(\frac{\partial s^i}{\partial \xi^j}\Big)\,, \quad i,j = 1,\dots,n\,. \end{aligned} \tag{5.10}$$

Applying the equations in (5.4) with metric (5.9) to the components of the function $\boldsymbol{\xi}(\mathbf{s})$, which is the inverse of (5.8) with $\mathbf{s} = \mathbf{x}$, and taking into account (5.10) and (2.48), we obtain

$$\begin{aligned} \frac{\partial}{\partial s^j}\Big(\sqrt{g^{\mathbf{s}}}g^{jk}_{\mathbf{s}}\frac{\partial \xi^i}{\partial s^k}\Big) &= \frac{\partial}{\partial s^j}\Big(\sqrt{g^{\mathbf{s}}}\frac{\partial s^j}{\partial \xi^l}\frac{\partial s^k}{\partial \xi^l}\frac{\partial \xi^i}{\partial s^k}\Big) \\ = \frac{\partial}{\partial s^j}\Big(\frac{1}{J}\frac{\partial s^j}{\partial \xi^i}\Big) &= 0\,, \quad i,j,k,l = 1,\dots,n\,. \end{aligned}$$

As a solution of the Dirichlet problem (5.4) with the boundary condition specified by the given mapping $\boldsymbol{\xi}(\mathbf{s})$ is unique, we obtain that this solution is the transformation that is the inverse of the original intermediate transformation $\mathbf{x}(\boldsymbol{\xi})$.

Grids on Surfaces

Let now S^{x2} be a surface lying in R^3 and locally represented by a parametrization

$$\mathbf{x}(\mathbf{s}) : S^2 \to R^3\,, \quad \mathbf{x} = (x^1, x^2, x^3)\,, \quad \mathbf{s} = (s^1, s^2)\,,$$

where S^2 is a two-dimensional parametric domain with the Cartesian coordinates s^1, s^2, while $\mathbf{x}(\mathbf{s})$ is a smooth nondegenerate vector-valued function. Note the transformation $\mathbf{x}(\mathbf{s})$ engenders in S^{x2} a natural metric tensor in the coordinates s^1, s^2:

$$g^{xs}_{ij} = \mathbf{x}_{s^i} \cdot \mathbf{x}_{s^j} = \frac{\partial x^k}{\partial s^i}\frac{\partial x^k}{\partial s^j}\,, \quad i,j = 1,2\,, \quad k = 1,2,3\,.$$

If a grid on the surface S^{x2} is constructed with the aid of a transformation

$$\mathbf{x}_1(\boldsymbol{\xi}) : \Xi^2 \to R^3\,,$$

then there is a diffeomorphism

$$\boldsymbol{\xi}(\mathbf{s}) : S^2 \to \Xi^2$$

such that

$$\mathbf{x}(\mathbf{s}) = \mathbf{x}_1[\boldsymbol{\xi}(\mathbf{s})] : S^2 \to R^3\,.$$

It is easily shown, likewise to the case of the domain discussed above, that the components of the transformation $\boldsymbol{\xi}(\mathbf{s})$ are the solutions of the grid equations in (5.4) with respect to the covariant metric tensor specified in the coordinates s^1, s^2 as

$$g_{ij}^{\mathbf{s}} = \frac{\partial \boldsymbol{\xi}}{\partial s^i} \cdot \frac{\partial \boldsymbol{\xi}}{\partial s^j}\,, \quad i,j = 1,2\,. \tag{5.11}$$

It is shown, analogously, that a grid in an arbitrary n–dimensional surface S^{xn} can be obtained by the solution of the boundary-value problem (5.4). So we see that an arbitrary nondegenerate grid in a domain or surface found by the mapping approach is realized, with a proper choice of the metric, by the grid generator based on the solution of the Beltrami equations.

Formulas (5.9), (5.10), and (5.11) appear to be useful for three purposes. First, one can use them to verify the reliability of any numerical algorithm applied to the solution of the system in (5.4). Second, they prompt one on how to find a coordinate system with necessary local coordinates by specifying the mapping $\boldsymbol{\xi}(\mathbf{s})$ in selected zones. For example, they can help one build an orthogonal system near boundary segments or in the vicinity of interfaces of blocks. Third, the expressions (5.9), (5.10), and (5.11) show which forms may be used for formulating monitor metrics.

Examples of Metrics Deriving Classical Grid Coordinates

Here we find, using (5.9), the expressions for the metric elements by which polar and spherical coordinate systems are realized through the solution of the problem (5.4). For the two-dimensional polar system of coordinates

$$\begin{aligned} x &= \rho\cos\varphi\,, \\ y &= \rho\sin\varphi\,, \end{aligned} \tag{5.12}$$

we find, assuming $x = s^1$, $y = s^2$, $\rho = \xi^1$, $\varphi = \xi^2$,

$$\frac{\partial \mathbf{s}}{\partial \xi^1} = \left(\frac{s^1}{\rho}, \frac{s^2}{\rho}\right), \quad \frac{\partial \mathbf{s}}{\partial \xi^2} = (-s^2, s^1)\,,$$

$$\frac{\partial \boldsymbol{\xi}}{\partial s^1} = \left(\frac{s^1}{\rho}, -\frac{s^2}{\rho^2}\right), \quad \frac{\partial \boldsymbol{\xi}}{\partial s^2} = \left(\frac{s^2}{\rho}, \frac{s^1}{\rho^2}\right),$$

where $\rho = \sqrt{(s^1)^2 + (s^2)^2}$. So the elements of the corresponding metric covariant and contravariant tensors (5.9) and (5.10) in the coordinates s^1, s^2, are as follows:

$$\begin{aligned} g_{ij}^{\mathbf{s}} &= \frac{\partial \boldsymbol{\xi}}{\partial s^i} \cdot \frac{\partial \boldsymbol{\xi}}{\partial s^j} = g^{\mathbf{s}}[\delta_j^i + s^i s^j (1 - g^{\mathbf{s}})]\,, \quad i,j = 1,2\,, \\ g_{\mathbf{s}}^{ij} &= \frac{\partial s^i}{\partial \xi^k}\frac{\partial s^j}{\partial \xi^k} = \frac{1}{g^{\mathbf{s}}}\delta_j^i - s^i s^j (1 - g^{\mathbf{s}})\,, \quad i,j = 1,2\,, \end{aligned} \tag{5.13}$$

where $g^{\mathrm{s}} = \det(g^{\mathrm{s}}_{ij}) = 1/((s^1)^2 + (s^2)^2) = 1/\rho^2$. We readily see that this metric is considerably different from the Euclidean metric. The equations in (5.4) with respect to the metric (5.13) realizing the polar coordinate system have the form

$$\frac{\partial}{\partial s^j}\left\{\sqrt{g^{\mathrm{s}}}\left[\frac{1}{g^{\mathrm{s}}}\delta^k_j - (1-g^{\mathrm{s}})s^k s^j\right]\frac{\partial \xi^i}{\partial s^k}\right\} = 0\,, \quad i,j,k = 1,2\,. \tag{5.14}$$

Analogous equations realize the three-dimensional spherical coordinate system

$$\begin{aligned} x &= \rho\cos\varphi\cos\phi\,, \\ y &= \rho\sin\varphi\cos\phi\,, \\ z &= \rho\sin\phi\,. \end{aligned} \tag{5.15}$$

5.1.5 Extension to Diffusion Equations

By substituting $w(\mathrm{s})$ for $\sqrt{g^{\mathrm{s}}}$ the system of equations in (5.4) becomes a more general system of the following equations

$$\frac{\partial}{\partial s^j}\left(w(\mathrm{s})g^{jk}_{\mathrm{s}}\frac{\partial \xi^i}{\partial s^k}\right) = 0\,, \quad i,j,k = 1,\dots,n\,, \tag{5.16}$$

where g^{jk}_{s} are the contravariant components in the coordinates $s^1,\dots,s^n$ of a monitor metric, $w(\mathrm{s}) > 0$ is a weight function aimed at increasing or decreasing the effect of the metric in the necessary zones of S^{xn}. The equations (5.16) will be referred to as diffusion equations.

The diffusion equations (5.16) are equivalent to the Beltrami equations if $w(\mathrm{s}) = \sqrt{g^{\mathrm{s}}}$, $g^{\mathrm{s}} = det(g^{\mathrm{s}}_{ij})$. Moreover for $n \neq 2$ they are always equivalent to the Beltrami equations, with respect to the metric

$$g_{ij} = (g^{\mathrm{s}})^{\frac{1}{2-n}}[w(\mathrm{s})]^{\frac{2}{n-2}}g^{\mathrm{s}}_{ij}, \quad i,j = 1,\dots,n\,,$$

regardless of the weight function $w(\mathrm{s}) > 0$. Indeed, for this metric,

$$g = \det(g_{ij}) = (g^{\mathrm{s}})^{\frac{n}{2-n}}[w(\mathrm{s})]^{\frac{2n}{n-2}}g^{\mathrm{s}} = (g^{\mathrm{s}})^{\frac{2}{2-n}}[w(\mathrm{s})]^{\frac{2n}{n-2}}\,,$$

$$g^{ij} = (g^{\mathrm{s}})^{\frac{1}{n-2}}[w(\mathrm{s})]^{\frac{2}{2-n}}g^{ij}_{\mathrm{s}}, \quad i,j = 1,\dots,n\,,$$

so

$$\sqrt{g}g^{ij} = w(\mathrm{s})g^{ij}_{\mathrm{s}}, \quad i,j = 1,\dots,n\,,$$

i.e. (5.16) are the Beltrami equations.

Though the Beltrami equations are comprehensive, i.e. an arbitrary nondegenerate intermediate transformation (5.2) can be computed as the inverse of the solution of these equations, the form (5.16) of the diffusion equations appears to be simpler for formulating, especially for $n = 2$, in order to realize the necessary requirements for the grid properties in different zones of S^{xn}. However one is to remember that for another coordinate system $v^1, \dots, v^n$ the equations (5.16) become equivalent to

$$\frac{\partial}{\partial v^j}\left(w[\mathbf{s}(\mathbf{v})]Jg^{jk}_{\mathbf{v}}\frac{\partial \xi^i}{\partial v^k}\right) = 0, \quad i,j,k = 1,\dots n,$$

where $J = det\Big(\partial s^i/\partial v^j\Big)$. This very system in the parametric coordinates $v^1, \dots, v^n$ should be solved in order to obtain the same grid computed by the solution of the system (5.16) in the parametric coordinates $s^1, \dots, s^n$.

Analogously to the form (5.5) of the Beltrami equations in (5.4), diffusion equations (5.16) have also the following equivalent form

$$\frac{\partial}{\partial \xi^j}\{Jw[\mathbf{s}(\boldsymbol{\xi})]g^{ij}_{\boldsymbol{\xi}}\} = 0, \quad i,j = 1,\dots,n, \qquad (5.17)$$

obtained by applying the identity (2.48) to (5.16).

5.1.6 Familiar Grid Equations

Laplace System

If M^n is a domain S^n in R^n with the Euclidean metric, i.e. in the Cartesian coordinates $s^1, \dots, s^n$,

$$g^{\mathbf{s}}_{ij} = \delta^i_j\,, \quad i,j = 1,\dots,n\,,$$

then

$$g^{\mathbf{s}} = 1\,, \quad g^{ij}_{\mathbf{s}} = \delta^i_j\,, \quad i,j = 1,\dots,n\,,$$

and consequently the grid equations in (5.4) in this metric are the Laplace equations

$$\nabla^2[\xi^i] \equiv \frac{\partial}{\partial s^j}\frac{\partial}{\partial s^j}\xi^i(\mathbf{s}) = 0\,, \quad i,j = 1,\dots,n\,. \qquad (5.18)$$

So the equations in (5.4) being a generalization of the Laplace equations are also called generalized Laplace equations.

The ordinary Laplace equations (5.18) were proposed for $n = 2$ by Crowley (1962) and Winslow (1967) for generating fixed grids in two-dimensional domains. Therefore the method for generating grids on surfaces S^{xn} by solving the boundary value problem (5.4) can also be considered as an extension of the Crowley–Winslow approach.

Diffusive System

A generalization of the Euclidean metric in S^n to the spherical metric specified in the original parametric Cartesian coordinates $s^1, \ldots, s^n$ by

$$g^{\mathbf{s}}_{ij} = v(\mathbf{s})\delta^i_j \,, \quad i,j = 1,\ldots,n \,, \quad v(\mathbf{s}) > 0 \,, \tag{5.19}$$

yields the grid equations for (5.4) in the following equivalent form

$$\frac{\partial}{\partial s^j}\left[(v(\mathbf{s}))^{(n-2)/2}\frac{\partial \xi^i}{\partial s^j}\right] = 0 \,, \quad i,j = 1,\ldots,n \,, \tag{5.20}$$

since

$$g^{\mathbf{s}} = [v(\mathbf{s})]^n \,, \quad g^{ij}_{\mathbf{s}} = \frac{1}{v(\mathbf{s})}\delta^i_j \,, \quad i,j = 1,\ldots,n \,.$$

Thus the popular diffusive form of the elliptic grid generator, following from (5.16) when $g^{ij}_{\mathbf{s}} = \delta^i_j$,

$$\frac{\partial}{\partial s^j}\left(w(\mathbf{s})\frac{\partial \xi^i}{\partial s^j}\right) = 0 \,, \quad i,j = 1,\ldots,n \,, \tag{5.21}$$

for constructing adaptive grids in a domain S^n is, in fact, equivalent to the grid system (5.20) for $n \neq 2$ in the metric (5.19) with

$$v(\mathbf{s}) = w(\mathbf{s})^{2/(n-2)}$$

imposed in the domain S^n.

Mind equations (5.20) for $n = 2$ are equivalent to Laplace equations (5.18) regardless of the form of the function $v(\mathbf{s})$ in (5.19). Therefore equations (5.21), with $w(\mathbf{s}) \neq$ *const*, are not equivalent to the grid equations in (5.4) for $n = 2$.

Note the diffusive equations (5.21) for $n = 2$ were introduced for generating adaptive grids in domains by Danaev, Liseikin, and Yanenko (1980) and Winslow (1981). An extension of the diffusive form (5.4) by introducing an individual weight function $w^i(\mathbf{s})$ for each direction $\xi^i(\mathbf{s})$ was made by Eiseman (1987) and Reed, Hsu, and Shiau (1988).

For a more general monitor metric with a diagonal tensor in the parametric coordinates $s^1, \ldots, s^n$,

$$g^{\mathbf{s}}_{ij} = v^k(\mathbf{s})\delta^k_i\delta^k_j \,, \quad i,j,k = 1,\ldots,n \,, \quad v^k(\mathbf{s}) > 0 \,, \tag{5.22}$$

i.e.

$$g^{\mathbf{s}}_{ij} = \begin{pmatrix} v^1(\mathbf{s}) & 0 & \ldots & 0 \\ 0 & v^2(\mathbf{s}) & 0 \cdots & 0 \\ \ldots & \ldots & \ldots & \ldots \\ 0 & \ldots & 0 & v^n(\mathbf{s}) \end{pmatrix}$$

we get

$$g^{\mathbf{s}} = \prod_{k=1}^{n} v^k(\mathbf{s}) \,, \quad g_{\mathbf{s}}^{ij} = \frac{1}{v^k(\mathbf{s})}\delta_i^k \delta_j^k \,, \quad i,j,k = 1,\dots,n \,.$$

This metric is of the form (4.50) when

$$z(\mathbf{s}) = 0 \,,\; B_i^k(\mathbf{s}) = \sqrt{v^k(\mathbf{s})\delta_i^k} \,,\; i,k = 1,\dots,n \,, \quad k \text{ fixed} \,.$$

The equations in (5.4) with respect to the metric (5.22) are equivalent to the following equations

$$\frac{\partial}{\partial s^j}\Big(\frac{\sqrt{g^{\mathbf{s}}}}{v^j(\mathbf{s})}\frac{\partial \xi^i}{\partial s^j}\Big) = 0 \,, \quad i,j = 1,\dots,n \,. \tag{5.23}$$

In particular, if $n = 2$, equations (5.23) are as follows:

$$\frac{\partial}{\partial s^1}\Big(F(\mathbf{s})\frac{\partial \xi^i}{\partial s^1}\Big) + \frac{\partial}{\partial s^2}\Big(\frac{1}{F(\mathbf{s})}\frac{\partial \xi^i}{\partial s^2}\Big) = 0 \,, \quad i = 1,2 \,, \tag{5.24}$$

where

$$F(\mathbf{s}) = \sqrt{v^2(\mathbf{s})/v^1(\mathbf{s})} \,.$$

Surface Grid System

Let $S^{x2} \subset R^3$ be a surface represented locally by a parametrization

$$\mathbf{x}(\mathbf{s}) : S^2 \to R^3 \,, \quad \mathbf{s} = (s^1, s^2) \,, \quad \mathbf{x} = (x^1, x^2, x^3) \,.$$

The covariant metric elements g_{ij}^{xs} of S^{x2} in the coordinates s^1, s^2 are defined by

$$g_{ij}^{xs} = \frac{\partial \mathbf{x}}{\partial s^i} \cdot \frac{\partial \mathbf{x}}{\partial s^j} \,, \quad i,j = 1,2 \,. \tag{5.25}$$

The equations

$$\frac{\partial}{\partial s^j}\Big(\sqrt{g^{xs}}\, g_{sx}^{kj} \frac{\partial \xi^i}{\partial s^k}\Big) = 0 \,, \quad i,j,k = 1,2 \,, \tag{5.26}$$

where $g^{xs} = \det(g_{ij}^{xs})$ and (g_{sx}^{kj}) is the contravariant metric tensor of S^{x2}, are equivalent to the grid equations in (5.4) for $n = 2$ with respect to the metric (5.25).

These equations popular for generating fixed grids on boundaries of domains were proposed by Warsi [1981]. The property of invariancy guarantees that the distribution of the grid nodes found through (5.26) does not depend on a parametrization of S^{x2}.

5.2 Variational Formulations

Here we discuss a relation of the grid equations in (5.4) and (5.16) to a variational grid generation approach.

5.2.1 Functional of Grid Smoothness

Formulation of the Functional

The system of the Beltrami equations in (5.4) is equivalent to the Euler–Lagrange equations for the following functional

$$I[\boldsymbol{\xi}] = \int\limits_{S^n} \sqrt{g^{\mathrm{s}}} g_{\mathrm{s}}^{jk} \frac{\partial \xi^i}{\partial s^k} \frac{\partial \xi^i}{\partial s^j} d\mathrm{s} \,, \quad i,j,k = 1,\dots,n \,. \tag{5.27}$$

Indeed the Euler–Lagrange equations for the functional in a general form

$$I[\boldsymbol{\xi}] = \int\limits_{S^n} G\Big[\mathrm{s}, \boldsymbol{\xi}(\mathrm{s}), \frac{\partial \boldsymbol{\xi}}{\partial s^1}(\mathrm{s}), \dots, \frac{\partial \boldsymbol{\xi}}{\partial s^n}(\mathrm{s})\Big] d\mathrm{s} \tag{5.28}$$

are as follows:

$$G_{\xi^i} - \frac{\partial}{\partial s^j}\Big\{G_{\partial\xi^i/\partial s^j}\Big[\mathrm{s}, \boldsymbol{\xi}(\mathrm{s}), \frac{\partial \boldsymbol{\xi}}{\partial s^1}(\mathrm{s}), \dots, \frac{\partial \boldsymbol{\xi}}{\partial s^n}(\mathrm{s})\Big]\Big\} = 0 \,, \qquad i,j = 1,\dots,n \,. \tag{5.29}$$

The quantities g^{s} and g_{s}^{jk} in (5.29) are specified in the parametric coordinates $s^1,\dots,s^n$ therefore they remain unchanged when the functions $\xi^i(\boldsymbol{s})$ are varied. So the system of the Euler–Lagrange equations (5.29) derived from the functional (5.27) is readily obtained and has the following form

$$2\frac{\partial}{\partial s^j}\Big(\sqrt{g^{\mathrm{s}}} g_{\mathrm{s}}^{jk} \frac{\partial \xi^i}{\partial s^k}\Big) = 0 \,, \quad i,j,k = 1,\dots,n \,,$$

which is equivalent to the system of the grid equations in (5.4). Thus the technique based on the minimization of the functional (5.27) produces the very grids obtained by the differential grid generator using Beltrami equations.

Availing us of the relations (4.19) the functional (5.27) is also expressed as

$$I[\boldsymbol{\xi}] = \int\limits_{S^n} \sqrt{g^{\mathrm{s}}} \mathrm{tr}(g_{\boldsymbol{\xi}}^{ij}) d\boldsymbol{s} \,, \tag{5.30}$$

where

$$\mathrm{tr}(g_{\boldsymbol{\xi}}^{ij}) = \sum_{i=1}^{n} g_{\boldsymbol{\xi}}^{ii} \,.$$

Taking into account that

$$\mathrm{d}M^n = \sqrt{g^{\mathrm{s}}} \mathrm{d}\mathrm{s} = \sqrt{g^{\boldsymbol{\xi}}} \mathrm{d}\boldsymbol{\xi}$$

we also obtain the following form of the functional (5.27)

$$I[\boldsymbol{\xi}] = \int\limits_{M^n} \mathrm{tr}(g_{\boldsymbol{\xi}}^{ij}) dM^n$$

and

$$I[\boldsymbol{\xi}] = \int\limits_{\Xi^n} \sqrt{g^{\boldsymbol{\xi}}} \mathrm{tr}(g_{\boldsymbol{\xi}}^{ij}) d\boldsymbol{\xi} \,.$$

Thus for $n = 1, 2$, and 3 we have, from (5.30),

$$I[\boldsymbol{\xi}] = \begin{cases} \int\limits_{S^1} \sqrt{g^{\mathrm{s}}} g_{\boldsymbol{\xi}}^{11} \mathrm{d}s \,, & n = 1 \,, \\ \int\limits_{S^2} \sqrt{g^{\mathrm{s}}} (g_{\boldsymbol{\xi}}^{11} + g_{\boldsymbol{\xi}}^{22}) \mathrm{d}s^1 \mathrm{d}s^2 \,, & n = 2 \,, \\ \int\limits_{S^3} \sqrt{g^{\mathrm{s}}} (g_{\boldsymbol{\xi}}^{11} + g_{\boldsymbol{\xi}}^{22} + g_{\boldsymbol{\xi}}^{33}) \mathrm{d}s^1 \mathrm{d}s^2 \mathrm{d}s^3 \,, & n = 3 \,, \end{cases} \tag{5.31}$$

with the corresponding contravariant metric elements $g_{\boldsymbol{\xi}}^{ij}$ and determinants $g^{\boldsymbol{s}}$ for each $n = 1, 2, 3$. Analogously, using in (5.31) suitable formulas for the elements of inverse matrices, we obtain the formulation of the functional with respect to intermediate transformations $\boldsymbol{s}(\boldsymbol{\xi})$ in terms of the covariant metric elements $g_{ij}^{\boldsymbol{\xi}}$:

$$I[\boldsymbol{s}] = \begin{cases} \int\limits_{\Xi^1} \frac{1}{J\sqrt{g^{s}}} \mathrm{d}\xi \,, & n = 1 \,, \\ \int\limits_{\Xi^2} \frac{1}{J\sqrt{g^{\boldsymbol{s}}}} (g_{11}^{\boldsymbol{\xi}} + g_{22}^{\boldsymbol{\xi}}) \mathrm{d}\xi^1 \mathrm{d}\xi^2 \,, & n = 2 \,, \\ \int\limits_{\Xi^3} \frac{1}{J\sqrt{g^{\boldsymbol{s}}}} [g_{11}^{\boldsymbol{\xi}} g_{22}^{\boldsymbol{\xi}} + g_{11}^{\boldsymbol{\xi}} g_{33}^{\boldsymbol{\xi}} + g_{22}^{\boldsymbol{\xi}} g_{33}^{\boldsymbol{\xi}} & \\ \quad -(g_{12}^{\boldsymbol{\xi}})^2 - (g_{13}^{\boldsymbol{\xi}})^2 - (g_{23}^{\boldsymbol{\xi}})^2] \mathrm{d}\xi^1 \mathrm{d}\xi^2 \mathrm{d}\xi^3 \,, & n = 3 \,, \end{cases} \tag{5.32}$$

where $J = det(\partial s^i / \partial \xi^j)$. Note that the functional $I[\mathrm{s}]$, in this formulation, is defined on the set of invertible transformations $\boldsymbol{s}(\boldsymbol{\xi}) \in C^2(\Xi^n)$.

When a monitor manifold M^n is an n-dimensional domain X^n, the functional (5.27) is the functional referred to as a functional of grid smoothness on X^n,

$$I[\boldsymbol{\xi}] = \int\limits_{X^n} \Big(\sum_{i=1}^{n} g_{\xi x}^{ii} \Big) \mathrm{d}\boldsymbol{x} \,, \tag{5.33}$$

where

$$g_{\xi x}^{ij} = \frac{\partial \xi^i}{\partial x^m} \frac{\partial \xi^j}{\partial x^m} \,, \quad i, j, m = 1, \dots, n \,,$$

introduced by Brackbill and Saltzman (1988). Therefore it is reasonable to call the functional (5.27), being the generalization of (5.33), as the functional of grid smoothness on the manifold M^n. It was shown by Liseikin (1999) that such a generalization of the functional (5.27) to n-dimensional regular surfaces preserves all salient features of grids obtained by applying the smoothness functional (5.33) to domains.

Relation to Harmonic Functions

This paragraph discusses an interpretation of the smoothness functional (5.27), which is related to the harmonic-functions approach for generating adaptive grids. For this purpose we express the smoothness functional (5.27) in the following form

$$I[\boldsymbol{\xi}] = \int\limits_{M^n} g_{\mathbf{s}}^{ml} \frac{\partial \xi^i}{\partial s^m} \frac{\partial \xi^i}{\partial s^l} \, \mathrm{d}M^n \,, \qquad i,l,m = 1,\cdots,n \,, \tag{5.34}$$

since

$$\sqrt{g^{\mathbf{s}}} d\mathbf{s} = dM^n \,.$$

Now we describe the definition of a harmonic map between two general n-dimensional Riemannian manifolds M^n and Z^n with covariant metric tensors g_{ij} and D_{ij} in some local coordinates x^i, $i = 1,\cdots,n$, of M^n and z^i, $i = 1,\cdots,n$, of Z^n, respectively.

Every $C^1(M^n)$ map $\boldsymbol{z}(\boldsymbol{x}) : M^n \to Z^n$ defines an energy density by the following formula:

$$e(\boldsymbol{z}) = \frac{1}{2} g^{ij}(\boldsymbol{x}) D_{kl}(\boldsymbol{z}) \frac{\partial z^k}{\partial x^i} \frac{\partial z^l}{\partial x^j} \,, \qquad i,j,k,l = 1,\cdots,n \,, \tag{5.35}$$

where (g^{ij}) is the contravariant metric tensor of M^n in the coordinates $x^1,\ldots,x^n$, i.e. $g_{ij}g^{jk} = \delta_i^k$. The total energy associated with the mapping $\boldsymbol{z}(\boldsymbol{x})$ is then defined as the integral of (5.35) over the manifold M^n:

$$E(\boldsymbol{z}) = \int_{M^n} e(\boldsymbol{z}) \mathrm{d}M^n \,. \tag{5.36}$$

A transformation $\boldsymbol{z}(\boldsymbol{x})$ of class $C^2(M^n)$ is referred to as a harmonic mapping if it is a critical point of the functional of the total energy (5.36). The Euler–Lagrange equations whose solution minimizes the energy functional (5.36) are given by

$$\frac{\partial}{\partial x^k}\Big(\sqrt{g} g^{kj} \frac{\partial z^l}{\partial x^j}\Big) + \sqrt{g} g^{kj} \Gamma^l_{mp} \frac{\partial z^m}{\partial x^k} \frac{\partial z^p}{\partial x^j} = 0 \,, \tag{5.37}$$

where $g = \det(g_{ij})$ and Γ^l_{mp} are Christoffel symbols of the second kind on the manifold Z^n :

$$\Upsilon^l_{mp} = \frac{1}{2} D^{lj} \left(\frac{\partial D_{jm}}{\partial z^p} + \frac{\partial D_{jp}}{\partial z^m} - \frac{\partial D_{mp}}{\partial z^j} \right) . \tag{5.38}$$

The following theorem guarantees the uniqueness of the harmonic mapping.

Theorem 1. *Let M^n, with metric g_{ij}, and Z^n, with metric D_{ij}, be two Riemannian manifolds with boundaries ∂M^n and ∂Z^n and let $\boldsymbol{\phi} : M^n \to Z^n$ be a diffeomorphism. If the curvature of Z^n is nonpositive and ∂Z^n is convex (with respect to the metric D_{ij}) then there exists a unique harmonic map $\boldsymbol{z}(\boldsymbol{x}) : M^n \to Z^n$ such that $\boldsymbol{z}(\boldsymbol{x})$ is a homotopy equivalent to $\boldsymbol{\phi}$. In other words, one can deform $\boldsymbol{z}$ to $\boldsymbol{\phi}$ by constructing a continuous family of maps $\boldsymbol{z}_t : M^n \to Z^n$, $t \in [0,1]$, such that $\boldsymbol{z}_0(\boldsymbol{x}) = \boldsymbol{\phi}(\boldsymbol{x})$, $\boldsymbol{z}_1(\boldsymbol{x}) = \boldsymbol{z}(\boldsymbol{x})$, and $\boldsymbol{z}_t(\boldsymbol{x}) = \boldsymbol{z}(\boldsymbol{x})$ for all $\boldsymbol{x} \in \partial M^n$.*

In application of the harmonic theory to grid generation the manifold Z^n is assumed to correspond to the logical domain Ξ^n with the Euclidean metric $D_{ij} = \delta^i_j$ in the coordinates $\xi^1, \dots, \xi^n$. Since the Euclidean space Ξ^n is flat, i.e. it has zero curvature, and the domain Ξ^n is constructed by the user, both requirements of the above theorem can be satisfied. For the manifold M^n, one uses a set of points of a physical geometry S^{xn} with an introduced Riemannian metric $g^{\mathbf{x}}_{ij}$. The functional of the total energy (5.36) has then the form

$$E(\boldsymbol{\xi}) = \frac{1}{2} \int\limits_{M^n} g^{kl}_{\mathbf{x}} \frac{\partial \xi^i}{\partial x^k} \frac{\partial \xi^i}{\partial x^l} \mathrm{d}M^n , \qquad i,k,l = 1, \cdots, n . \tag{5.39}$$

Thus, assuming in (5.35) $\mathbf{s} = \mathbf{x}$, we obtain in the case of the Euclidean metric in Ξ^n that the integral (5.27) is twice the total energy associated with the mapping $\boldsymbol{\xi}(\boldsymbol{s}) : S^n \to \Xi^n$ representing a transformation between the manifold M^n, with its metric tensor $g^{\boldsymbol{s}}_{ij}$ in the coordinates s^i, and the computational domain Ξ^n, with the Cartesian coordinates ξ^i:

$$I[\boldsymbol{\xi}] = 2E[\boldsymbol{\xi}] .$$

As for the Euler–Lagrange equations (5.37) for the functional (5.39) we have

$$\frac{\partial}{\partial x^k} \left(\sqrt{g^{\mathbf{x}}} g^{kj}_{\mathbf{x}} \frac{\partial \xi^i}{\partial x^j} \right) = 0 , \qquad i,j,k = 1, \cdots, n , \tag{5.40}$$

since, from (5.38), $\Upsilon^l_{mp} = 0$. So the equations (5.40) with $\mathbf{x} = \mathbf{s}$ are equivalent to the grid equations in (5.4).

Equations (5.40), in contrast to (5.37), are linear and have a conservative form. Therefore being elliptic they satisfy the maximum principle, and the Dirichlet boundary value problem is a well-posed problem for this system of equations, i.e. the above theorem is obvious for the functional (5.39). Thus the functions $\xi^i(\mathrm{s})$ obtained from (5.4) compose a harmonic transformation

$$\boldsymbol{\xi}(\mathrm{s}) : S^n \to \Xi^n , \quad \boldsymbol{\xi}(\mathrm{s}) = [\xi^1(\mathrm{s}), \dots, \xi^n(\mathrm{s})] .$$

Note the value of any grid generation method is commonly judged by its ability to rule out the construction of folded grids in domains or on surfaces with arbitrary geometry. In the case $n = 2$ the mathematical foundation of this requirement for the technique based on the generalized Laplace system in (5.4) is solid when the computation domain Ξ^2 is convex. It is founded on the following result, derived from a theorem of Rado.

Theorem 2. *Let M^2 be a simply connected bounded Riemannian manifold. In this case, the Jacobian of the transformation $\boldsymbol{\xi}(\mathbf{s})$ generated by the boundary value problem (5.4) does not vanish in the interior of S^2, if Ξ^2 is a convex domain and $\boldsymbol{\xi}(\mathbf{s}) : \partial S^2 \to \partial \Xi^2$ is a homeomorphism.*

Geometric Interpretation

This paragraph describes a geometric meaning of the smoothness functional which justifies to some extent its expression for the generation of adaptive grids in domains and on surfaces. For this purpose we consider for a monitor manifold over a physical geometry S^{xn} an n-dimensional regular surface represented in the form (5.1) as

$$\mathbf{r}(\mathbf{s}) : S^n \to R^{n+k+l} , \quad \mathbf{r}(\mathbf{s}) = [\mathbf{x}(\mathbf{s}), f^1(\mathbf{s}), \dots , f^l(\mathbf{s})] , \tag{5.41}$$

where $\mathbf{x}(\mathbf{s})$ is a parametric transformation (5.1), $\mathbf{f}(\mathbf{s}) = (f^1(\mathbf{s}), \dots , f^l(\mathbf{s}))$ is a vector-valued function. This surface designated by S^{rn} is referred to as a monitor surface over S^{xn}. Its covariant metric elements $g^{\mathbf{s}}_{ij}$ in the coordinates $s^1, \dots , s^n$ are computed by the formula

$$g^{\mathbf{s}}_{ij} = \boldsymbol{r}_{s^i} \cdot \boldsymbol{r}_{s^j} = g^{xs}_{ij} + f^k_{s^i} \cdot f^k_{s^j}, \quad i, j = 1, \dots , n, \quad k = 1, \dots , l. \tag{5.42}$$

The monitor surface is naturally parametrized in the grid coordinates $\xi^1, \dots , \xi^n$ as

$$\boldsymbol{r}[\mathbf{s}(\boldsymbol{\xi})] : \Xi^n \to R^{n+k+l} ,$$

while its covariant metric elements in these coordinates are specified by

$$g^{\boldsymbol{\xi}}_{ij} = \boldsymbol{r}_{\xi^i} \cdot \boldsymbol{r}_{\xi^j} = g^{\mathbf{s}}_{km} \frac{\partial s^k}{\partial \xi^i} \frac{\partial s^m}{\partial \xi^j} , \quad i, j, k, m = 1, \dots , n ,$$

where

$$\boldsymbol{r}_{\xi^i} = \frac{\partial}{\partial \xi^i} \boldsymbol{r}[\mathbf{s}(\boldsymbol{\xi})] , \quad i = 1, \dots , n .$$

The functional of grid smoothness (5.27) with respect to the metric of the monitor surface S^{rn} is as follows:

$$I[\boldsymbol{\xi}] = \int\limits_{S^{rn}} \Big(\sum_{i=1}^{n} g^{ii}_{\boldsymbol{\xi}}\Big) dS^{rn} = \int\limits_{S^{rn}} \mathrm{tr}(g^{ij}_{\boldsymbol{\xi}}) dS^{rn} . \tag{5.43}$$

First, note that the trace of any contravariant n-dimensional tensor $(g^{ij}_{\boldsymbol{\xi}})$ can be expressed through the invariants I_{n-1} and I_n of the orthogonal transforms of the covariant tensor $(g^{\boldsymbol{\xi}}_{ij})$, namely

$$\operatorname{tr}(g^{ij}_{\boldsymbol{\xi}}) = \frac{I_{n-1}}{I_n} , \tag{5.44}$$

where I_i, $i = 1, \dots , n$, is the sum of the principal minors of order i of the matrix $(g^{\boldsymbol{\xi}}_{ij})$. Therefore the functional (5.43) can also be expressed through these invariants:

$$I[\boldsymbol{\xi}] = \int\limits_{S^{rn}} \Big(\frac{I_{n-1}}{I_n}\Big) \mathrm{d}S^{rn} . \tag{5.45}$$

The expression (5.45) of the smoothness functional with respect to the metric (5.42) through the invariants of orthogonal transformations was given by Liseikin (1991). A generalization of this expression by substituting the quantity $(I_{n-1}/I_n)^q$, $0 \le q \le 1$ for I_{n-1}/I_n in (5.45) was suggested by Liseikin (1991, 1999) and Huang (2001).

Now, for the purpose of simplicity, we restrict our consideration to three dimensions. The functional (5.45) then has the form

$$I[\boldsymbol{\xi}] = \int\limits_{S^{r3}} \Big(\frac{I_2}{I_3}\Big) \mathrm{d}S^{r3} . \tag{5.46}$$

In three dimensions the invariant I_3 of the covariant metric tensor (5.42) in the coordinates ξ^1, ξ^2, ξ^3 is the Jacobian $g^{\boldsymbol{\xi}}$ of the matrix $(g^{\boldsymbol{\xi}}_{ij})$ and it represents the volume V^3 of the three-dimensional basic parallelepiped $P^3 \subset R^{n+k+l}$ formed by the basic tangent vectors $\boldsymbol{r}_{\xi^i}$, $i = 1, 2, 3$ (see Fig. 2.5). The invariant I_2 of the matrix $(g^{\boldsymbol{\xi}}_{ij})$ is the sum of its principal minors of order 2. Every principal minor of order 2 equals the Jacobian of the two-dimensional matrix A^2 obtained from $(g^{\boldsymbol{\xi}}_{ij})$ by crossing out a row and a column which intersect on the diagonal. Since the edges of the basic parallelepiped are the tangential vectors $\mathbf{r}_{\xi^i}$ we find that each element of the matrix A^2 is a dot product of two such vectors and, consequently, the Jacobian of A^2 equals the square of the area of the parallelogram formed by these two vectors. So the invariants I_2 and I_3 can be expressed as

$$I_2 = \sum_{m=1}^{3} \left(V^2_m\right)^2 , \qquad I_3 = \left(V^3\right)^2 , \tag{5.47}$$

where V^2_m is the area of the boundary face of the basic parallelepiped P^3 formed by the basic tangent vectors $\boldsymbol{r}_{\xi^i}$, $i = 1, 2, 3$, except for $\boldsymbol{r}_{\xi^m}$, and V^3 is the volume of P^3. Therefore

$$\frac{I_2}{I_3} = \sum_{m=1}^{3} \left(V^2_m\right)^2 / \left(V^3\right)^2 . \tag{5.48}$$

It is obvious that

$$V^3 = d_m V_m^2 , \qquad m = 1, 2, 3 ,$$

where d_m is the distance between the vertex of the vector $\boldsymbol{r}_{\xi^m}$ and the plane spanned by the vectors $\boldsymbol{r}_{\xi^i}$, $i \neq m$. Hence, from (5.48)

$$\frac{I_2}{I_3} = \sum_{m=1}^{3} (1/d^m)^2 . \tag{5.49}$$

Now let us consider two grid hypersurfaces $\xi^m = c$ and $\xi^m = c + h$ in S^{r3} obtained by mapping a uniform rectangular grid with a step size h in the computational domain Ξ^3 onto the monitor surface S^{r3}. The distance l_m between a node on the surface $\xi^m = c$ and the nearest node on the surface $\xi^m = c + h$ equals $d_m h + O(h)^2$. Therefore (5.49) is equivalent to

$$\frac{I_2}{I_3} = \sum_{m=1}^{3} (h/l_m)^2 + O(h) .$$

The quantity $(h/l_m)^2$ increases as the nodes of a coordinate grid cluster in the direction normal to the hypersurface $\xi^m = c$, and therefore it can be considered as some measure of the grid nonuniformity in this direction; consequently, the functional (5.46) defines an integral measure of the grid nonuniformity in all directions. So the problem of minimizing the functional of smoothness (5.43) for $n = 3$ can be interpreted as a problem of finding a grid with a minimum of nonuniform clustering, namely a quasiuniform grid on the monitor surface S^{r3}. Consequently, the minimization of the functional (5.43) makes the grid in S^{xn} be adaptive to the gradient of the function $\mathbf{f}(\mathbf{s}) = (f^1(\mathbf{s}), \dots, f^l(\mathbf{s}))$. □

Analogous interpretations of the functional of grid smoothness with respect to the metric (5.42) are valid for arbitrary dimensions, i.e. the integrand

$$\sigma(\mathbf{s}) = g_{\boldsymbol{s}}^{km} \frac{\partial \xi^i}{\partial s^k} \frac{\partial \xi^i}{\partial s^m}, \qquad i, k, m = 1, \dots, n \tag{5.50}$$

in the functional (5.43) can be considered as a relative measure of the grid nonuniformity on the monitor surface S^{rn} represented by (5.41). The measure (5.50) with respect to the metric (5.42) can also be considered as a measure of departure of the grid in the physical geometry S^{xn} from an adaptive grid with node clustering in the zones of large variation of the function $\mathbf{f}(\mathbf{s})$.

The interpretation of the smoothness functional considered above justifies, to some extent, its potential to generate grids adapting to the gradient of a function specified at the points of a domain or surface by projecting onto the domain or surface quasiuniform grids built on monitor surfaces (see Fig. 5.5) by the minimization of the functional.

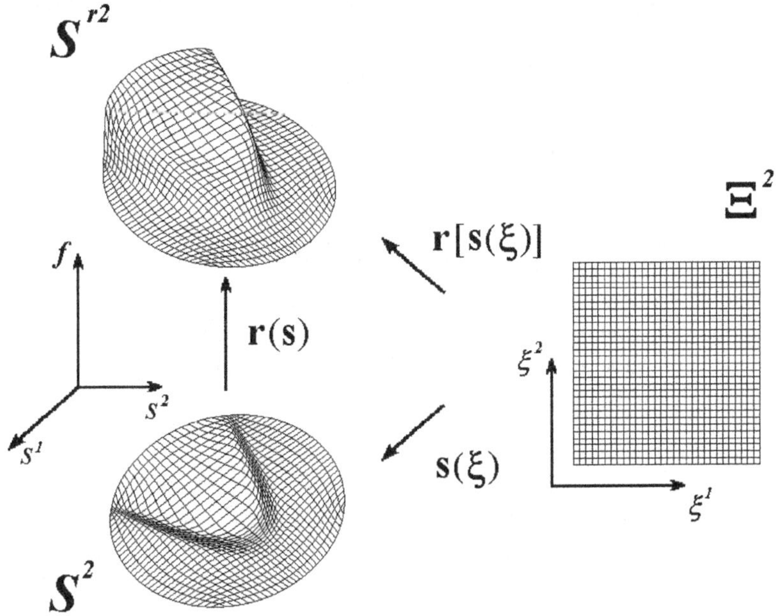

Fig. 5.5. Illustration for grid adaptation

5.2.2 Diffusion Functional

Similarly to the Beltrami equations in (5.4) the diffusive equations (5.16) are equivalent to the Euler-Lagrange equations for the diffusive functional

$$I[\boldsymbol{\xi}] = \int\limits_{S^n} w(\mathbf{s}) g_{\mathbf{s}}^{jk} \frac{\partial \xi^i}{\partial s^k} \frac{\partial \xi^i}{\partial s^j} d\mathbf{s} \equiv \int\limits_{S^n} w(\mathbf{s}) g_{\boldsymbol{\xi}}^{ii} d\mathbf{s} \equiv \int\limits_{\Xi^n} w[\mathbf{s}(\boldsymbol{\xi})] J g_{\boldsymbol{\xi}}^{ii} d\boldsymbol{\xi}, \tag{5.51}$$
$$i, j, k = 1, \ldots, n.$$

Analogously to the functional of grid smoothness (5.27) the diffusion functional (5.51) yields the following functional $I[\mathbf{s}]$ with respect to the intermediate transformations $\mathbf{s}(\boldsymbol{\xi})$:

$$
I[s] = \begin{cases} \displaystyle\int_{\Xi^1} \frac{w[s(\xi)]}{Jg^s} \mathrm{d}\xi \,, & n = 1\,, \\ \displaystyle\int_{\Xi^2} \frac{w[\mathbf{s}(\boldsymbol{\xi})]}{Jg^{\mathbf{s}}} (g^{\boldsymbol{\xi}}_{11} + g^{\boldsymbol{\xi}}_{22}) \mathrm{d}\xi^1 \mathrm{d}\xi^2 \,, & n = 2\,, \\ \displaystyle\int_{\Xi^3} \frac{w[\mathbf{s}(\boldsymbol{\xi})]}{Jg^{\mathbf{s}}} [g^{\boldsymbol{\xi}}_{11} g^{\boldsymbol{\xi}}_{22} + g^{\boldsymbol{\xi}}_{11} g^{\boldsymbol{\xi}}_{33} + g^{\boldsymbol{\xi}}_{22} g^{\boldsymbol{\xi}}_{33} & \\ \qquad -(g^{\boldsymbol{\xi}}_{12})^2 - (g^{\boldsymbol{\xi}}_{13})^2 - (g^{\boldsymbol{\xi}}_{23})^2] \mathrm{d}\xi^1 \mathrm{d}\xi^2 \mathrm{d}\xi^3 \,, & n = 3\,, \end{cases} \tag{5.52}
$$

The expression of functional (5.51) promts one what the monitor metric should be to provide the generation of the numerical grid with a required property. For this purpose the metric is to have such a form so that the integrand of the functional (5.51)

$$
\sigma(\mathbf{s}) = w(\mathbf{s}) g^{jk}_{\mathbf{s}} \frac{\partial \xi^i}{\partial s^j} \frac{\partial \xi^i}{\partial s^k}, \quad i, j, k = 1, \dots, n \tag{5.53}
$$

describes a measure of departure of the grid from the necessary grid at the point $\mathbf{s} \in S^n$. If such a metric is found then it can be expected that the minimization of the functional (5.51) will produce the grid with the required property.

Similarly, the metric elements should be specified for the functional of grid smoothness (5.27). In particular, formula (5.50) with respect to the metric (5.42) of the monitor surface S^{rn} represented by (5.41) can be considered as a measure of departure of a grid from the reference grid with node clustering in the zones of large variations of the function $\mathbf{f}(\mathbf{s})$.

5.3 Formulation of Monitor Metrics

The Dirichlet boundary value problem for both the inverted Beltrami and diffusion equations is well posed for an arbitrary monitor metric. Nevertheless for the purpose of better handling grid control it is reasonable to restrict the whole set of the monitor metrics to a basic subset which, however, can be adequate for realizing the necessary grid properties. Also these basic monitor metrics are to be described by simple formulas which allow one to establish readily the relations between them and the grid characteristics. Further, one of the natural ways to satisfy balanced grid properties, each of which is realized by an individual monitor metric, is to combine linearly these metrics. Therefore the basic metric tensors formulated should be subject to the operation of summation in sense that the sum of two monitor metrics from the subset is also the metric (nonsingular tensor). Note, in general, the sum of two metrics may not be a metric since the sum of two nonsingular matrices may be a singular matrix. This section describes an approach for formulating such basic monitor metrics.

5.3.1 General Formulas for Covariant Elements

The expressions (5.9) and (5.13) for the metrics realizing both arbitrary and particular polar coordinate systems, respectively, through the solution of the grid equations in (5.4) prompt some ways on how to design general formulas for specifying necessary metrics in general geometries in order to obtain suitable intermediate transformations for generating grids. For example, let g^{xs}_{ij}, $i,j=1,\dots,n$, and g^{ij}_{sx}, $i,j=1,\dots,n$, be the covariant and contravariant metric tensor, respectively, of a physical geometry (in general of a Riemannian manifold) S^{xn} in the coordinates $s^1,\dots,s^n$. Then the metric g^{xs}_{ij} and three smooth functions $z(\mathbf{s})>0$, $v(\mathbf{s})$, and $f(\mathbf{s})$ define the following new covariant and contravariant monitor metric tensors $g^{\mathbf{s}}_{ij}$ and $g^{ij}_{\mathbf{s}}$, respectively, in the coordinates $s^1,\dots,s^n$:

$$\begin{aligned} g^{\mathbf{s}}_{ij} &= z(\mathbf{s})g^{xs}_{ij} + v(\mathbf{s})\frac{\partial f}{\partial s^i}\frac{\partial f}{\partial s^j}\,, \quad i,j=1,\dots,n\,, \\ g^{ij}_{\mathbf{s}} &= \frac{1}{z(\mathbf{s})}g^{ij} - d(\mathbf{s})g^{im}_{sx}\frac{\partial f}{\partial s^m}g^{jp}_{sx}\frac{\partial f}{\partial s^p}\,, \quad i,j,m,p=1,\dots,n\,, \end{aligned} \tag{5.54}$$

where

$$\begin{aligned} d(\mathbf{s}) &= \frac{v(\mathbf{s})}{z(\mathbf{s})[z(\mathbf{s})+v(\mathbf{s})\nabla(f)]}\,, \\ \nabla(f) &= g^{ij}_{sx}\frac{\partial f}{\partial s^i}\frac{\partial f}{\partial s^j}\,, \quad i,j=1,\dots,n\,. \end{aligned}$$

The points of the geometry S^{xn} with this metric comprise a Rimennian manifold M^n called a monitor manifold over S^{xn}.

The formula in (5.54) for the contravariant metric tensor of M^n will be proved below. Note the functions $z(\mathbf{s})$, $v(\mathbf{s})$, and $f(\mathbf{s})$ in (5.54) must be such that $\det(g^{\mathbf{s}}_{ij})>0$.

The covariant $g^{\mathbf{s}}_{ij}$ and contravariant $g^{ij}_{\mathbf{s}}$ metric tensors in the coordinates s^1,s^2 considered in (5.13) for generating the classical polar coordinate system (5.12) are realized by the metric tensors (5.54) with

$$\begin{aligned} &g_{ij}=g^{ij}=\delta^i_j\,, \quad i,j=1,2\,, \\ &z(\mathbf{s})=\frac{1}{\rho^2}\,, \quad v(\mathbf{s})=\frac{1}{\rho^2}\Big(1-\frac{1}{\rho^2}\Big)\,, \quad f(\mathbf{s})=\frac{1}{2}\rho^2\,, \end{aligned} \tag{5.55}$$

where $\rho^2=(s^1)^2+(s^2)^2=|\mathbf{s}|^2$.

A coordinate transformation $\mathbf{s}(\boldsymbol{\xi})$ representing a more general polar coordinate system

$$s^1=b(\xi^1)\cos\xi^2\,, \quad s^2=b(\xi^1)\sin\xi^2,$$

when $b'(\xi^1)>0$ is realized as the inverse of the solution of (5.4) in the metric (5.54) if

$$g^{xs}_{ij} = g^{ij}_{sx} = \delta^i_j , \quad i,j = 1,2 ,$$

$$z(\mathbf{s}) = \frac{1}{\rho^2} , \qquad f(\mathbf{s}) = \frac{1}{2}\rho^2 ,$$

$$v(\mathbf{s}) = z(\mathbf{s})\Big[\frac{1}{[b'(b^{-1}(\rho))]^2} - z(\mathbf{s})\Big] ,$$

where $\rho^2 = (s^1)^2 + (s^2)^2$.

The functions $z(\mathbf{s})$ and $v(\mathbf{s})$ in (5.54) can be interpreted as weight functions which control the influence of the original metric g^{xs}_{ij} and the monitor function $f(\mathbf{s})$ on the grid behavior. A generalization of the metric (5.54) is naturally carried out by the following formula

$$g^{\mathbf{s}}_{ij} = z(\mathbf{s})g^{xs}_{ij} + v^k(\mathbf{s})f^k_{s^i}(\mathbf{s})f^k_{s^j}(\mathbf{s}) , \; i,j = 1,\dots,n , \; k = 1,\dots,l , \qquad (5.56)$$

with weight functions $z(\mathbf{s}) > 0$ and $v^k(\mathbf{s}) \geq 0$, $k = 1,\dots,l$, and monitor functions $f^k(\mathbf{s})$, $k = 1,\dots,l$. Remind it is assumed that a summation is carried out in (5.56) over the index k.

Note when

$$g^{xs}_{ij} = \mathbf{x}_{s^i} \cdot \mathbf{x}_{s^j} , \quad i,j = 1,\dots,n ,$$

where $\mathbf{x}(\mathbf{s})$ is the parametrization (5.1) of S^{xn} then the metric (5.56) for $z(\mathbf{s}) = v^k(\mathbf{s}) = 1$, $k = 1,\dots,l$, $\mathbf{f}(\mathbf{s}) = (f^1(\mathbf{s}),\dots,f^l(\mathbf{s}))$ is the metric

$$g^{\mathbf{s}}_{ij} = g^{xs}_{ij} + \mathbf{f}_{s^i} \cdot \mathbf{f}_{s^j}, \quad i,j = 1,\dots,n, \qquad (5.57)$$

of the monitor surface S^{rn} represented by (5.41).

The general monitor metrics in the forms (5.57) and (5.56) were introduced by Liseikin (1991) and Liseikin (2002a, 2003), respectively.

The approach for formulating monitor metrics in the form (5.56) is somewhat similar to the classical variational approaches in which the grid functionals are formulated as a combination of several functionals with weights (see (1.24)) each of which is responsible for providing some individual grid property. Note the boundary value problem (5.4) with respect to an arbitrary metric is well-posed. Moreover it is well-posed in the monitor metric (5.56) with arbitrary weight and monitor functions contrary to the problem of the minimization of the combined functional that is well-posed only for special weight functions. Besides this the grid obtained from the minimization of the combined functional may be dependent on a parametrization of the physical geometry.

The monitor metric (5.56) is extended to an even more general but to a more convenient metric through a set of covariant tensors of the first rank

$$\mathbf{F}^k(\mathbf{s}) = [F^k_1(\mathbf{s}),\dots,F^k_n(\mathbf{s})] , \quad k = 1,\dots,l ,$$

by the following formula (see Liseikin (2004)) in the parametric coordinates $s^1,\dots,s^n$:

$$g^{\mathbf{s}}_{ij} = z(\mathbf{s})g^{xs}_{ij} + F^k_i(\mathbf{s})F^k_j(\mathbf{s}) \ , \ i,j = 1,\dots,n \ , \ k = 1,\dots,l \ , \tag{5.58}$$

where $z(\mathbf{s}) \geq 0$ is a weight function specifying the contribution of the metric of S^{xn} in the monitor metric. Of course it is assumed that the function $z(\mathbf{s})$ and the vectors $\mathbf{F}^k(\mathbf{s})$, $k = 1,\dots,l$, are subject to the relation

$$g^{\mathbf{s}} = det(g^{\mathbf{s}}_{ij}) > 0,$$

at each point $\mathbf{s} \in S^n$.

Note, if we introduce in R^{n+k+l} vectors $\mathbf{w}_i(\mathbf{s})$, $i = 1,\dots,n$ by the formula

$$\mathbf{w}_i(\mathbf{s}) = [\sqrt{z(\mathbf{s})}\frac{\partial \mathbf{x}}{\partial s^i}, F^1_i, \dots, F^l_i], \quad i = 1,\dots,n,$$

where $\mathbf{x}(\mathbf{s})$ is the parametrization (5.1) of the physical geometry S^{xn}, then it is obvious that

$$g^{\mathbf{s}}_{ij} = \mathbf{w}_i \cdot \mathbf{w}_j, \quad i,j = 1,\dots,n.$$

So for nonsingularity of the monitor metric tensor (5.58) the vectors $\mathbf{w}_i(\mathbf{s})$, $i = 1,\dots,n$, must be independent. In particular, as vectors $\partial \mathbf{x}/\partial s^i$, $\dots,n$, are independent, the vectors $\mathbf{w}_i(\mathbf{s})$, $i = 1,\dots,n$, will be independent if $z(\mathbf{s}) > 0$ at all points $\mathbf{s} \in S^n$.

It is evident that the linear combination of two metric tensors of the form (5.58) with corresponding nonnegative coefficients $\varepsilon_1(\mathbf{s})$ and $\varepsilon_2(\mathbf{s})$ is the matrix of the same form (5.58) and it is nonsingular (metric tensor) if $\varepsilon_1(\mathbf{s}) + \varepsilon_2(\mathbf{s}) > 0$ at all points of S^n.

The monitor metric (5.56) is realized by the metric (5.58) if

$$F^k_i(\mathbf{s}) = \sqrt{v^k(\mathbf{s})}\frac{\partial f^k}{\partial s^i} \ , \ i = 1,\dots,n \ , \ k = 1,\dots,l \ , \ k \text{ fixed} \ .$$

The form (5.58) of the monitor metric allows one to formulate more easily metrics for generating grids adapting to arbitrary vector fields.

This form may also be useful for generating a family of grid coordinates orthogonal to a boundary segment. For example, let a boundary curve of a two-dimensional domain S^2 be defined from the equation $\varphi(s^1, s^2) = 0$. If we assume the variable $\varphi(\mathbf{s})$ as a logical grid coordinate then the second grid coordinate $\psi(\mathbf{s})$ which is orthogonal to the curve $\varphi(\mathbf{s}) = 0$ is subject to the relation

$$\text{grad}\varphi \cdot \text{grad}\psi = 0 \ ,$$

i.e.

$$(\psi_{s^1}, \psi_{s^2}) = w(\mathbf{s})(\varphi_{s^2}, -\varphi_{s^1}) \ , \tag{5.59}$$

at the points of the curve. As the coordinate system φ, ψ is realized by the solution of the Beltrami's equations in (5.4) with respect to the metric defined by (5.9) for $n = 2$, $\xi^1 = \varphi$, $\xi^2 = \psi$, we find, availing us of (5.59) that this metric should be as follows:

$$g_{ij}^{\mathbf{s}} = \varphi_{s^i}\varphi_{s^j} + \psi_{s^i}\psi_{s^j} = \varphi_{s^i}\varphi_{s^j} + (-1)^{i+j}\varphi_{s^{3-i}}\varphi_{s^{3-j}} \,, \qquad i,j = 1,2 \,, \quad i,j \text{ fixed} \,. \tag{5.60}$$

Since

$$(-1)^{i+j}\varphi_{s^{3-i}}\varphi_{s^{3-j}} = [(\varphi_{s^1})^2 + (\varphi_{s^2})^2]\delta_j^i - \varphi_{s^i}\varphi_{s^j} \,, \quad i,j = 1,2 \,, \quad i,j \text{ fixed} \,,$$

so the expression (5.60) has the following equivalent form

$$g_{ij}^{\mathbf{s}} = [(\varphi_{s^1})^2 + (\varphi_{s^2})^2]w^2(\mathbf{s})\delta_j^i + [1 - w^2(\mathbf{s})]\varphi_{s^i}\varphi_{s^j} \,, \quad i,j = 1,2 \,, \tag{5.61}$$

which is, in fact, of the form (5.54).

Computation of Contravariant Metric Components

We are to know in the equations (5.4) and (5.16) as well as in the functionals (5.27) and (5.51) the contravariant metric components of any specified monitor metric. These quantities are readily computed for the most general metric (5.58) in the case $z(\mathbf{s}) > 0$.

General Formula.

Let (d_{ab}) and (d^{ab}), $a, b = 1, \dots, l$ be two mutually inverse matrices, where

$$d^{ab} = \delta_b^a + \frac{1}{z(\mathbf{s})}\nabla(\mathbf{F}^a, \mathbf{F}^b) \,, \quad a, b = 1, \dots, l \,, \tag{5.62}$$

here $\nabla(,)$ is the mixed parameter of Beltrami in the metric of the physical surface S^{xn} and $\mathbf{F}^a$, $a = 1, ..., l$, is a covariant vector. The mixed parameter with respect to the covariant tensors $\mathbf{F}^a$ and $\mathbf{F}^b$ of the first rank is defined by the formula

$$\nabla(\mathbf{F}^a, \mathbf{F}^b) = g_{sx}^{ij} F_i^a F_j^b \,, \quad i,j = 1, \dots, n \,, \quad a, b = 1, \dots, l \,, \tag{5.63}$$

where F_i^a and F_j^b, $i, j = 1, \dots, n$, are the components of the corresponding tensors $\mathbf{F}^a$ and $\mathbf{F}^b$ in the coordinates $s^1, \dots, s^n$, while g_{sx}^{ij} are the contravariant metric components of S^{xn} in the same coordinates.

Theorem 3. *The contravariant metric components $g_{\mathbf{s}}^{ij}$ of the monitor manifold M^n over S^{xn} with the metric (5.58) for $z(\mathbf{s}) > 0$ are computed by the following formulas*

$$g_{\mathbf{s}}^{ij} = \frac{1}{z(\mathbf{s})} g_{sx}^{ij} - \frac{1}{[z(\mathbf{s})]^2} d_{ab} \nabla(\mathbf{F}^a, \mathbf{e}^i)\nabla(\mathbf{F}^b, \mathbf{e}^j) \,, \qquad i,j,k,m = 1, \dots, n \,, \quad a, b = 1, \dots, l \,, \tag{5.64}$$

where $\mathbf{e}^k$ is a covariant vector whose i-th component in the coordinates $s^1, \dots, s^n$ equals to δ_i^k.

Proof. For proving the theorem it is sufficient to show that the matrix $(g^{ij}_{\mathbf{s}})$ compiled by the quantities from (5.64) is inverse to the matrix $(g^{\mathbf{s}}_{ij})$, i.e. their elements are subject to the relations

$$g^{\mathbf{s}}_{kj} g^{ji}_{\mathbf{s}} = \delta^i_k \ , \quad i,j,k = 1,\dots,n \ , \tag{5.65}$$

where $g^{\mathbf{s}}_{kj}$ are covariant metric components (5.58). As

$$g^{\mathbf{s}}_{kj} = z(\mathbf{s}) g^{xs}_{kj} + F^a_k F^a_j \ , \quad j,k = 1,\dots,n \ , \quad a = 1,\dots,l \ ,$$

we have, using (5.63) in (5.64)

$$\begin{aligned} g^{\mathbf{s}}_{kj} g^{ji}_{\mathbf{s}} &= [(z(\mathbf{s}) g^{xs}_{kj} + F^a_k F^a_j)] \Big(\frac{1}{z(\mathbf{s})} g^{ji}_{sx} - \frac{1}{[z(\mathbf{s})]^2} g^{jm}_{sx} g^{ip}_{sx} d_{bc} F^b_m F^c_p \Big) = \\ &= \delta^k_i + \frac{1}{z(\mathbf{s})} F^a_j g^{ji}_{sx} F^a_k - \frac{1}{z(\mathbf{s})} F^c_p g^{ip}_{sx} d_{bc} F^b_k - \frac{1}{[z(\mathbf{s})]^2} F^c_p g^{ip}_{sx} d_{bc} \nabla(\mathbf{F}^a, \mathbf{F}^b) F^a_k \ , \\ & i,j,k,m,p = 1,\dots,n \ , \quad a,b,c = 1,\dots,l \ . \end{aligned} \tag{5.66}$$

Since (5.62) yields

$$\frac{1}{z(\mathbf{s})} \nabla(\mathbf{F}^a, \mathbf{F}^b) = d^{ab} - \delta^a_b \ , \quad a,b = 1,\dots,l \ ,$$

therefore

$$\begin{aligned} & \frac{1}{z(\mathbf{s})} F^a_j g^{ji}_{sx} F^a_k - \frac{1}{z(\mathbf{s})} F^c_p g^{ip}_{sx} d_{bc} F^b_k - \frac{1}{[z(\mathbf{s})]^2} F^c_p g^{ip}_{sx} d_{bc} \nabla(\mathbf{F}^a, \mathbf{F}^b) F^a_k = \\ &= \frac{1}{z(\mathbf{s})} F^a_j g^{ji}_{sx} F^a_k - \frac{1}{z(\mathbf{s})} F^c_p g^{ip}_{sx} d_{bc} F^b_k - \frac{1}{z(\mathbf{s})} F^c_p g^{ip}_{sx} d_{bc} (d^{ab} - \delta^a_b) F^a_k = 0 \ , \\ & i,j,k,p = 1,\dots,n \ , \quad a,b,c = 1,\dots,l \ . \end{aligned}$$

Thus the relations (5.65) follow from (5.66). □

In patrticular, for the metric (5.57) of the monitor surface S^{rn} over S^{xn} we get

$$g^{ij}_{\mathbf{s}} = g^{ij}_{sx} - g^{ik}_{sx} g^{jm}_{sx} d_{ab} \frac{\partial f^a}{\partial s^k} \frac{\partial f^b}{\partial s^m}, \quad i,j,k,m = 1,\dots,n, \quad a,b,c = 1,\dots,l, \tag{5.67}$$

taking into account that this metric is of the form (5.58) with

$$z(\mathbf{s}) = 1, \ \mathbf{F}^a = grad \ f^a, \ a = 1,\dots,l.$$

Formulas for Domains.

Let S^{xn} be a domain identified with the parametric area S^n. Then in the coordinates $s^1, \dots, s^n$ we have

$$g_{ij}^{xs} = g_{sx}^{ij} = \delta_j^i, \quad i,j = 1, \dots, n,$$

while the formulas (5.58) and (5.62) come to

$$g_{ij}^{\mathbf{s}} = z(\mathbf{s})\delta_j^i + F_i^a F_j^a, \quad i,j = 1, \dots, n, \quad a = 1, \dots, l,$$

$$d^{ab} = \delta_b^a + \frac{1}{z(\mathbf{s})} F_i^a F_i^b, \quad i = 1, \dots, n, \quad a,b = 1, \dots, l.$$

So the formula (5.64), in this case, is as follows:

$$g_{\mathbf{s}}^{ij} = \frac{1}{z(\mathbf{s})}\Big[\delta_j^i - \frac{1}{z(\mathbf{s})} d_{ab} F_i^a F_j^b\Big], \quad i,j = 1, \dots, n, \quad a,b = 1, \dots, l.$$

Analogously to (5.67) we obtain for the metric

$$g_{ij}^{\mathbf{s}} = \delta_j^i + \frac{\partial f^a}{\partial s^i}\frac{\partial f^a}{\partial s^j}, \quad i,j = 1, \dots, n, \quad a = 1, \dots, l,$$

of the monitor surface S^{rn} over the domain S^n

$$g_{\mathbf{s}}^{ij} = \delta_j^i - d_{ab}\frac{\partial f^a}{\partial s^i}\frac{\partial f^b}{\partial s^j}, \quad i,j = 1, \dots, n, \quad a,b,c = 1, \dots, l. \tag{5.68}$$

Computation of the Metric Jacobian

For computing the equations in (5.4) one also needs to know the determinant of the monitor tensor $(g_{ij}^{\mathbf{s}})$.

Theorem 4. *Let $g_{ij}^{\mathbf{s}}$ be the monitor metric (5.58) with $z(\mathbf{s}) > 0$. Then there is valid the following formula*

$$g^{\mathbf{s}} = [z(\mathbf{s})]^n g^{xs} \det(d^{ab}), \tag{5.69}$$

where $g^{xs} = \det(g_{ij}^{xs})$, while (d^{ab}) is the matrix whose elements are specified by (5.62).

Proof. First we prove the theorem for $l = 1$. Let

$$g_{ij}^1 = z(\mathbf{s}) g_{ij}^{xs}, \quad g_1^{ij} = \frac{1}{z(\mathbf{s})} g_{sx}^{ij}, \quad i,j = 1, \dots, n,$$

then formula (5.58) is as follows:

$$g^{\mathbf{s}}_{ij} = g^1_{ij} + F^1_i F^1_j \,, \quad i,j = 1,\dots,n \,. \tag{5.70}$$

Notice the rank of the matrix $(F^1_i F^1_j)$ is not more than 1, therefore we get from (5.70)

$$g^{\mathbf{s}} = g^1 + \sum_{i=1}^{n} \det(\bar{g}^i_{km}) \,, \tag{5.71}$$

where $g^1 = \det(g^1_{ij})$, while $(\bar{g}^i_{km})$ is the matrix derived from the matrix (g^1_{km}) by replacing its i-th row with

$$(F^1_i F^1_1, \dots, F^1_i F^1_n) = F^1_i \mathbf{F}^1 \,, \quad i = 1,\dots,n \,,$$

i. e.

$$(\bar{g}^i_{km}) = \begin{pmatrix} g^1_{11} & \cdots & g^1_{1n} \\ \cdots & \cdots & \cdots \\ g^1_{i-11} & \cdots & g^1_{i-1n} \\ F^1_i F^1_1 & \dots & F^1_i F^1_n \\ g^1_{i+11} & \cdots & g^1_{i+1n} \\ \cdots & \cdots & \cdots \\ g^1_{n1} & \cdots & g^1_{nn} \end{pmatrix} .$$

It is evident that

$$\det(\bar{g}^i_{km}) = F^1_i F^1_k G^{ik} \,, \quad i,k = 1,\dots,n \,, \quad i \text{ fixed} \,,$$

where

$$G^{ik} = g^1 g^{ik}_1 \,, \quad i,k = 1,\dots,n \,,$$

here g^{ik}_1 is the (ik)-th element of the matrix (g^{ij}_1) inverse to (g^1_{ij}). So equation (5.71) gives

$$g^{\mathbf{s}} = g^1(1 + g^{ik}_1 F^1_i F^1_k) = g^1\Big(1 + \frac{1}{z(\mathbf{s})} g^{jk}_{sx} F^1_i F^1_k\Big) = [z(\mathbf{s})]^n g^{xs} d^{11} \,,$$
$$i,k = 1,\dots,n \,,$$

i.e. formula (5.69) is valid for $l = 1$.

Let now $l > 1$. Assuming that the theorem is valid for $l-1$, we represent the monitor metric (5.58) as

$$g^{\mathbf{s}}_{ij} = g^2_{ij} + F^l_i F^l_j \,, \quad i,j = 1,\dots,n \,,$$

where

$$g^2_{ij} = z(\mathbf{s}) g^{xs}_{ij} + F^k_i F^k_j \,, \quad i,j = 1,\dots,n \,, \quad k = 1,\dots,l-1 \,.$$

Identifying g^2_{ij} with g^{xs}_{ij} we obtain that this monitor metric $g^{\mathbf{s}}_{ij}$ has the form (5.58) for $l = 1$ and $z(\mathbf{s}) \equiv 1$. By induction we get from (5.69) for $z(\mathbf{s}) \equiv 1$ and $l = 1$

$$g^{\mathbf{s}} = g^2(1 + \nabla(\mathbf{F}^l)) , \tag{5.72}$$

where

$$g^2 = det(g^2_{ij}), \quad \nabla(\mathbf{F}^l) = g_2^{ij} F^l_i F^l_j , \quad i,j = 1, \dots , n ,$$

g_2^{ij} is the (ij)-th element of the matrix inverse to the matrix (g^2_{mp}). Let (d_2^{ab}), $a, b = 1, \dots , l-1$ be a $(l-1) \times (l-1)$ matrix whose elements are determined as

$$d_2^{ab} = \delta^a_b + \frac{1}{z(\mathbf{s})} \nabla(\mathbf{F}^a, \mathbf{F}^b) , \quad a, b = 1, \dots , l-1 ,$$

here $\nabla(\mathbf{F}^a, \mathbf{F}^b)$ is defined by (5.63). It is obvious that (d_2^{ab}) is the first principal minor of order $l-1$ of the matrix (d^{ab}) whose elements are specified by (5.62). Availing us of (5.64) and (5.69), valid in accordance with the induction assumption, gives

$$\begin{aligned} g_2^{ij} &= \frac{1}{z(\mathbf{s})} g^{ij}_{sx} - \frac{1}{[z(\mathbf{s})]^2} d^2_{ab} \nabla(\mathbf{F}^a, \mathbf{e}^i) \nabla(\mathbf{F}^b, \mathbf{e}^j) , \\ i,j &= 1, \dots , n , \quad a, b = 1, \dots , l-1 , \\ g^2 &= [z(\mathbf{s})]^n g^{xs} \det(d^2_{ab}) , \end{aligned}$$

where g_2^{ij} is the (ij)-th element of matrix inverse to (g^2_{ij}), $g^2 = \det(g^2_{ij})$, while (d^2_{ab}) is the matrix inverse to (d_2^{ab}). Substituting these expressions in (5.72) yields

$$\begin{aligned} g^{\mathbf{s}} &= g^2\Big(1 + \frac{1}{z(\mathbf{s})} g^{ij}_{sx} F^l_i F^l_j - \frac{1}{[z(\mathbf{s})]^2} d^2_{ab} g^{ik}_{sx} F^a_k g^{jm}_{sx} F^b_m F^l_i F^l_j\Big) = \\ &= g^2\Big(1 + \frac{1}{z(\mathbf{s})} \nabla(\mathbf{F}^l) - \frac{1}{[z(\mathbf{s})]^2} d^2_{ab} \nabla(\mathbf{F}^a, \mathbf{F}^l) \nabla(\mathbf{F}^b, \mathbf{F}^l)\Big) = \\ &= [z(\mathbf{s})]^n g^{xs} \det(d^{cd}) , \\ i,j,k,m &= 1, \dots , n , \quad a, b = 1, \dots , l-1 , \quad c, d = 1, \dots , l . \end{aligned}$$

This proves the theorem. □

If S^{xn} is a domain S^n then $g^{xs} = 1$ and equation (5.69), in this case, comes to

$$g^{\mathbf{s}} = [z(\mathbf{s})]^n det(d^{ab}).$$

5.3.2 Formulations of Contravariant Elements

General Formulas

Notice both the functionals of grid smoothness (5.27) and diffusion (5.51) as well as the Beltrami equations in (5.4) and diffusion equations (5.16) are formulated through the contravariant metric components $g^{ij}_{\mathbf{s}}$ in the coordinates

$s^1, \dots, s^n$. Therefore instead of the covariant metric components $g^{\mathbf{s}}_{ij}$ one may originally, in particular in order to define a measure of grid departure from a required grid, formulate the contravariant components $g^{ij}_{\mathbf{s}}$ of the monitor metric, for example, like the most general form (5.58),

$$g^{ij}_{\mathbf{s}} = \epsilon(\mathbf{s}) g^{ij}_{sx} + B^i_k B^j_k, \quad i,j = 1, \dots, n\,, \ k = 1, \dots, l, \tag{5.73}$$

where $B^i_k, i = 1, \dots, n$, are the components of a contravariant vector $\mathbf{B}_k = (B^1_k, \dots, B^n_k)$, $k = 1, \dots, l$.

Then the quantity $g^{\mathbf{s}}$ in (5.4) and (5.27) is defined from the equation

$$g^{\mathbf{s}} = 1/g_{\mathbf{s}}, \quad g_{\mathbf{s}} = det(g^{ij}_{\mathbf{s}}).$$

Consequently if $\epsilon(\mathbf{s}) > 0$ then the covariant metric components can be computed using an analog of formula (5.64). For this purpose we, similarly to (5.62), introduce two mutually inverse matrices (c_{ab}) and (c^{ab}), $a,b = 1, \dots, l$, assuming

$$c_{ab} = \delta^a_b + \frac{1}{\epsilon(\mathbf{s})} g^{xs}_{ij} B^i_a B^j_b, \quad i,j = 1, \dots, n, \quad a,b = 1, \dots, l. \tag{5.74}$$

Then, in the same way as for (5.64) and (5.69), the following formulas are proved

$$\begin{aligned} g^{\mathbf{s}}_{ij} &= \frac{1}{\epsilon(\mathbf{s})} g^{xs}_{ij} - \frac{1}{[\epsilon(\mathbf{s})]^2} g^{xs}_{mi} g^{xs}_{pj} c^{ab} B^m_a B^p_b, \\ & i,j,m,p = 1, \dots, n, \quad a,b = 1, \dots, l, \end{aligned} \tag{5.75}$$

$$g_{\mathbf{s}} = [\epsilon(\mathbf{s})]^n g_{sx} det(c_{ab}), \tag{5.76}$$

where $g_{sx} = det(g^{ij}_{sx})$.

Formulas for Domains.

Let S^{xn} be a domain S^n and $s^1, \dots, s^n$ be the Cartesian coordinates. Then, in the coordinates $s^1, \dots, s^n$,

$$g^{ij}_{sx} = g^{xs}_{ij} = \delta^i_j, \quad i,j = 1, \dots, n,$$

so the contravariant components (5.73) of the monitor metric become as

$$g^{ij}_{\mathbf{s}} = \epsilon(\mathbf{s}) \delta^i_j + B^i_k B^j_k, \quad i,j = 1, \dots, n\,, \ k = 1, \dots, l. \tag{5.77}$$

Thus for the elements of the matrix (5.74) we have

$$c_{ab} = \delta^a_b + \frac{1}{\epsilon(\mathbf{s})} B^m_a B^m_b, \quad m = 1, \dots, n, \quad a,b = 1, \dots, l.$$

Therefore equation (5.76) yields

$$g_{\mathbf{s}} = [\epsilon(\mathbf{s})]^n det(c_{ab}), \quad a, b = 1, \dots, l, \tag{5.78}$$

in particular,

$$\begin{aligned} g_{\mathbf{s}} &= [\epsilon(\mathbf{s})]^{n-1}[\epsilon(\mathbf{s}) + |\mathbf{B}_1|^2], \quad l = 1, \\ g_{\mathbf{s}} &= [\epsilon(\mathbf{s})]^{n-2}\Big\{[\epsilon(\mathbf{s}) + |\mathbf{B}_1|^2][\epsilon(\mathbf{s}) + |\mathbf{B}_2|^2] - (\mathbf{B}_1 \cdot \mathbf{B}_2)^2\Big\}, \quad l = 2. \end{aligned} \tag{5.79}$$

Here and further

$$\mathbf{B}_a \cdot \mathbf{B}_b = B_a^i B_b^i, \quad i = 1, \dots, n,$$

$$|\mathbf{B}_a|^2 = \mathbf{B}_a \cdot \mathbf{B}_a, \quad a \text{ fixed}.$$

Consequently formula (5.75) gives

$$g_{ij}^{\mathbf{s}} = \frac{1}{\epsilon(\mathbf{s})}\delta_j^i - \frac{1}{[\epsilon(\mathbf{s})]^2}c^{ab}B_a^i B_b^j, \quad i, j = 1, \dots, n, \quad a, b = 1, \dots, l, \tag{5.80}$$

thus, for $l = 1$,

$$g_{ij}^{\mathbf{s}} = \frac{1}{\epsilon(\mathbf{s})}\Big[\delta_j^i - \frac{1}{\epsilon(\mathbf{s}) + |\mathbf{B}_1|^2}B_1^i B_1^j\Big], \quad i, j = 1, \dots, n. \tag{5.81}$$

Analogously, for $l = 2$,

$$g_{ij}^{\mathbf{s}} = \frac{1}{\epsilon(\mathbf{s})}\delta_j^i - \frac{1}{d(\mathbf{s})}(c_{22}B_1^i B_1^j - 2c_{12}B_1^i B_2^j + c_{11}B_2^i B_2^j), \quad i, j = 1, \dots, n, \tag{5.82}$$

where

$$d(\mathbf{s}) = [\epsilon(\mathbf{s}) + |\mathbf{B}_1|^2][\epsilon(\mathbf{s}) + |\mathbf{B}_2|^2] - (\mathbf{B}_1 \cdot \mathbf{B}_2)^2.$$

5.3.3 Specification of Individual Monitor Metrics

Generation of Vector Field-Aligned Grids

Contravariant metric tensor in the form (5.73) can be used to define a measure of departure of the grid from a grid aligned to a vector-field $\mathbf{B}$ in a domain S^n, in particular, to control the angle between a normal to the grid coordinate hypersurface and the vector-field $\mathbf{B}$ at the points of the domain S^n. As a tensor of the first rank in the formulas (5.73) one may take either the same or a transformed vector field. The generation of grids through such metric is helpful for solving numerically problems with strong anisotropy, in particular, problems of magnetically-confined plasmas. For example, the condition of orthogonality between a vector field $\mathbf{B} = (B^1, \dots, B^n)$ specified at the points of the domain S^n and the vector *grad* ξ^1 normal to the coordinate hypersurface $\xi^1 = const$ can be described as an equation for a quadratic form

$$(\mathbf{B} \cdot grad\ \xi^1)^2 \equiv B^i B^j \frac{\partial \xi^1}{\partial s^i} \frac{\partial \xi^1}{\partial s^j} = 0, \quad i,j = 1,\dots,n.$$

This quadratic form as a measure of the grid departure from field-alignment at a point of the domain S^n was proposed by Glasser and Tang (2004). The integral measure of the grid departure can be expressed, for example, in the form

$$L = \int\limits_{S^n} w(\mathbf{s}) B^i B^j \frac{\partial \xi^1}{\partial s^i} \frac{\partial \xi^1}{\partial s^j} d\mathbf{s}\,, \quad i,j = 1,\dots,n. \tag{5.83}$$

However the problem of minimizing this functional is ill-posed for $n \geq 2$ since the matrix $(B^i B^j)$ is singular.

The integrands in the functionals of grid smoothness (5.27) and diffusion (5.51) are formulated as the sum over the index i of quadratic forms

$$g_{\mathbf{s}}^{jk} \frac{\partial \xi^i}{\partial s^k} \frac{\partial \xi^i}{\partial s^j}, \quad i,j,k = 1,\dots,n, \quad i \text{ fixed},$$

multiplied by $\sqrt{g^{\mathbf{s}}}$ and $w(\mathbf{s})$, respectively, and contrary to the integrand in (5.83) with a nondegenerate matrix $(g_{\mathbf{s}}^{jk})$. The condition of non-degeneracy is indispensable for well-posedness of the corresponding variational and differential problems and for obtaining unfolded grids through the minimization of the functional or through the solution of the Euler-Lagrange equations. Therefore in order to get the grids that are both nearly field-aligned (*grad* ξ^i for some i is nearly orthogonal to the vector field) and unfolded we have to change slightly the matrix $(B^i B^j)$ in the functional (5.83) to make it nondegenerate and to replace the expressions $(\partial \xi^1/\partial s^i)$ $(\partial \xi^1/\partial s^j)$ by $(\partial \xi^k/\partial s^i)$ $(\partial \xi^k/\partial s^j)$. The matrix $(g_{\mathbf{s}}^{ij})$ whose elements are specified in the form (5.73) is nondegenerate for an arbitrary $\epsilon(\mathbf{s}) > 0$ (see formula (5.76)), in addition this matrix is close to the matrix $(B^i B^j)$ when both $\epsilon(\mathbf{s})$ and $\mathbf{B}_k$, $k = 2,\dots,l$ are small and $\mathbf{B}_1 = \mathbf{B}$. Assuming this matrix as a contravariant metric tensor of a monitor manifold M^n over S^n yields, in accordance with (5.53), the following measure of grid nonalignment with the vector field $\boldsymbol{B}$

$$\sigma(\boldsymbol{s}) = w(\boldsymbol{s})[\epsilon(\boldsymbol{s})\delta_k^i + B_a^j B_a^k] \frac{\partial \xi^i}{\partial s^j} \frac{\partial \xi^i}{\partial s^k}, \qquad i,j,k = 1,\dots,n, \quad a = 1,\dots,l. \tag{5.84}$$

Consequently the functional of diffusion (5.51) in such monitor metric over S^n is as follows:

$$
\begin{aligned}
I[\boldsymbol{\xi}] &= \int\limits_{S^n} w(\mathbf{s}) \Big[\epsilon(\mathbf{s}) \frac{\partial \xi^i}{\partial s^m} \frac{\partial \xi^i}{\partial s^m} + B^j_k B^p_k \frac{\partial \xi^i}{\partial s^j} \frac{\partial \xi^i}{\partial s^p} \Big] ds \,, \\
& i,j,m,p = 1,\dots,n, \quad k = 1,\dots,l.
\end{aligned}
\tag{5.85}
$$

Substituting here $\sqrt{g^{\mathbf{s}}}$ for $w(\mathbf{s})$, where $g^{\mathbf{s}} = 1/\sqrt{g_{\mathbf{s}}}$, $\quad g_{\mathbf{s}} = det(g^{ij}_{\mathbf{s}})$, produces the functional of grid smoothness (5.27). Availing us of functional (5.85) yields that the diffusion equations (5.16) aimed at the generation of grids providing that the angle between **B** and a normal to a coordinate hypersurface is close to $\pi/2$ have the following form

$$
\frac{\partial}{\partial s^j} \Big\{ w(\mathbf{s}) [\epsilon(\mathbf{s}) \delta^j_k + B^j_a B^k_a] \frac{\partial \xi^i}{\partial s^k} \Big\} = 0, \quad i,j,k = 1,\dots,n, \quad a = 1,\dots,l.
\tag{5.86}
$$

Substituting $\sqrt{g^{\mathbf{s}}}$ for $w(\mathbf{s})$ in (5.86) yields the Beltrami equations in (5.4) in the metric (5.77).

Figure 5.6 demonstrates integral lines of a vector field (left-hand) and a corresponding field-aligned quadrilateral domain grid obtained through the solution of Beltrami equations in the metric (5.77).

Similarly, using general formula (5.73) one can formulate a measure of grid nonalignment for an arbitrary physical geometry S^{xn} as

$$
\begin{aligned}
\sigma(s) &= w(s) [\epsilon(s) g^{jk}_{sx} + B^j_a B^k_a] \frac{\partial \xi^i}{\partial s^j} \frac{\partial \xi^i}{\partial s^k}, \\
& i,j,k = 1,\dots,n, \quad a = 1,\dots,l.
\end{aligned}
\tag{5.87}
$$

and corresponding functionals and equations for generating field-aligned grids in S^{xn}.

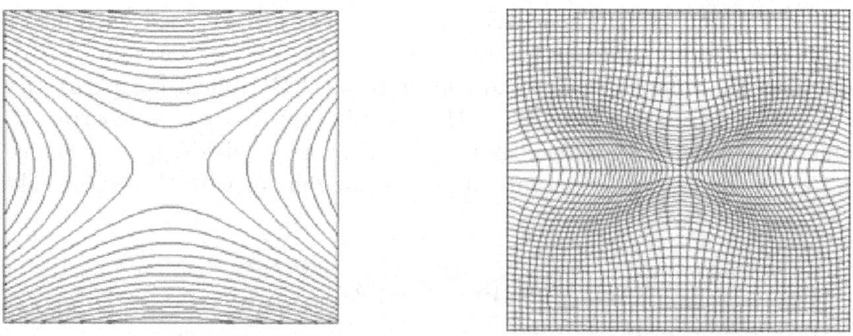

Fig. 5.6. Flux lines of a vector field (left-hand) and a field-aligned grid (right-hand).

Monitor Metric for Generating Grids Adapting to the Gradient of a Function

In accordance with a concept of Eiseman (1987) grid clustering in a physical geometry S^{xn} in the zones of large variation of a function $\mathbf{f}(\mathbf{s}) = (f^1(\mathbf{s}), \dots, f^l(\mathbf{s}))$ can be provided by projecting a quasiuniform grid from a regular monitor surface defined as the graph of the values of $\mathbf{f}(\mathbf{s})$. This function $\mathbf{f}(\mathbf{s})$ referred to as a monitor function can be a solution to the problem of interest, a combination of its components or derivatives, or any other variable scalar– or vector–valued quantity that suitably monitors the features of the physical solution and/or of the geometry of the physical domain or surface which significantly affect the accuracy of the calculations. The monitor functions provide an efficient opportunity to control the grid quality, in particular, the concentration of grid nodes and the size of angles between grid lines.

One of the techniques to generating quasiuniform grids on the monitor surface is based on the use of the smoothness functional (5.27) proposed by Liseikin (1991) which generalizes the functional introduced by Brackbill and Saltzman (1982) for generating fixed grids in domains.

Generation of Adaptive Grids in Domains.

In the case, important for the generation of adaptive grids in a physical domain $X^n \subset R^n$, the monitor manifold can be defined as an n-dimensional monitor surface S^{rn} formed by the values of some monitor vector-valued function

$$\mathbf{f}(\mathbf{x}) : X^n \to R^l\,, \quad \mathbf{x} = (x^1, \dots, x^n)\,, \quad \mathbf{f} = [f^1(\mathbf{x}), \dots, f^l(\mathbf{x})]\,, \tag{5.88}$$

over X^n. Thus the monitor surface S^{rn}, whose points are

$$(x^1, \dots, x^n, f^1(\mathbf{x}), \dots, f^l(\mathbf{x})), \ \mathbf{x} = (x^1, \dots, x^n) \in X^n,$$

is the subset of the $(n+l)$-dimensional space R^{n+l}. It is apparent that for the parametric domain S^n there may be taken the domain X^n, so the parametric mapping $\mathbf{r}(\mathbf{s}) : S^n \to R^{n+l}$ is defined as

$$\mathbf{r}(\mathbf{s}) = [\mathbf{s}, \mathbf{f}(\mathbf{s})] = [s^1, \dots, s^n, f^1(\mathbf{s}), \dots, f^l(\mathbf{s})]\,, \qquad \mathbf{s} = \mathbf{x}\,. \tag{5.89}$$

Consequently the covariant metric elements of this surface are specified in the coordinates $s^1, \dots, s^n$ by

$$g^{\mathbf{s}}_{ij} = \delta^i_j + \frac{\partial \mathbf{f}}{\partial s^i} \cdot \frac{\partial \mathbf{f}}{\partial s^j}, \quad i,j = 1, \dots, n, \tag{5.90}$$

i.e. in the form (5.58) with $z(\mathbf{s}) = 1$,

$$g^{xs}_{ij} = \delta^i_j, \quad F^k_i = \frac{\partial f^k}{\partial s^i}, \quad i,j = 1, \dots, n, \quad k = 1, \dots, l.$$

According to Sect. 5.2.1. the numerical solution of the problem (5.4) with respect to the metric (5.90) makes the grid become uniform in S^{rn}. Therefore the intermediate transformation $\mathbf{s}(\boldsymbol{\xi})$ that is the inverse of the map $\boldsymbol{\xi}(\mathbf{s})$ found by the solution of the problem (5.4) in the metric (5.90) produces, in fact, the very adaptive grid in S^n obtained by projecting the quasiuniform grid from S^{rn}. This adaptive grid provides node concentration in the zones of large variation of $\mathbf{f}(\mathbf{s})$ (Fig. 5.5).

Formulas (5.64) and (5.69) give the following expressions for the contravariant elements of the monitor metric (5.90) and its Jacobian in the parametric coordinates $s^1, \dots, s^n$

$$\begin{aligned} g_{\mathbf{s}}^{ij} &= \delta_j^i - d_{ab}\frac{\partial f^a}{\partial s^i}\frac{\partial f^b}{\partial s^j}, \quad i,j = 1,\dots,n, \quad a,b = 1,\dots,l, \\ g^{\mathbf{s}} &= det(d^{ab}), \quad a,b = 1,\dots,l, \end{aligned} \tag{5.91}$$

where (d_{ab}) is an $l \times l$ matrix inverse to the matrix (d^{ab}) with

$$d^{ab} = \delta_b^a + \frac{\partial f^a}{\partial s^i}\frac{\partial f^b}{\partial s^i}, \quad i = 1,\dots,n, \quad a,b = 1,\dots,l,$$

Thus one measure of grid nonuniformity, given by (5.50), i.e. departure of the grid from a quasiuniform grid in S^{rn} and consequently from an adaptive grid in S^n has the following form

$$\sigma(\mathbf{s}) = \Big(\delta_k^j - d_{ab}\frac{\partial f^a}{\partial s^j}\frac{\partial f^b}{\partial s^k}\Big)\frac{\partial \xi^i}{\partial s^j}\frac{\partial \xi^i}{\partial s^k}$$

$$i,j,k = 1,\dots,n, \quad a,b = 1,\dots,l.$$

In the case of a scalar monitor function $f(\mathbf{s})$ we have

$$\begin{aligned} d^{aa} &= 1 + \nabla(f), \quad a = 1, \\ g_{\mathbf{s}}^{ij} &= \delta_j^i - \frac{1}{1+\nabla(f)}\frac{\partial f}{\partial s^i}\frac{\partial f}{\partial s^j}, \quad i,j = 1,\dots,n, \\ g^{\mathbf{s}} &= 1 + \nabla(f), \quad \nabla(f) = \frac{\partial f}{\partial s^i}\frac{\partial f}{\partial s^i} = |grad\ f|^2, \quad i = 1,\dots,n. \end{aligned} \tag{5.92}$$

These expressions substituted in the Beltrami equations in (5.4) give the following equations for determining the components $\xi^i(\mathbf{s})$ of the transformation $\boldsymbol{\xi}(\mathbf{s}) : S^n \to \Xi^n$:

$$\frac{\partial}{\partial s^j}\Big[\sqrt{1+\nabla(f)}\Big(\delta_k^j - \frac{1}{1+\nabla(f)}\frac{\partial f}{\partial s^j}\frac{\partial f}{\partial s^k}\Big)\frac{\partial \xi^i}{\partial s^k}\Big] = 0\,,$$

$$i,j,k = 1,\dots,n\,.$$

The inverse mapping $\mathbf{s}(\boldsymbol{\xi})$ forms an adaptive grid in S^n. An example of such domain grid adapting to the gradient of a scalar-valued function $f(\mathbf{s})$ is demonstrated by Fig. 5.5.

Analogously, the grid equations are written out with a vector-valued monitor function $\mathbf{f}(\mathbf{s})$ for generating adaptive grids in domains.

Note the popular equations for generating adaptive grids in domains are based on the numerical solution of the Poisson system

$$\frac{\partial}{\partial x^j}\left(\frac{\partial \xi^i}{\partial x^j}\right) = P^i , \qquad i,j, = 1,\cdots,n , \tag{5.93}$$

where P^i are the control functions. These equations are not equivalent to the generalized Laplace equations in (5.4) if $P^i \neq 0$, $i = 1,\cdots,n$. In the case where S^{rn} is a monitor surface, the system in (5.4) can also be interpreted as a system of elliptic equations with a control function. The control function is the monitor mapping $\boldsymbol{f}(\boldsymbol{s})$ whose values over the physical domain or surface form the monitor hypersurface S^{rn}. The influence of the control function $\boldsymbol{f}(\boldsymbol{s})$ is realized through the magnitudes $\boldsymbol{f}_{s^i} \cdot \boldsymbol{f}_{s^j}$ in the terms $g^{\mathbf{s}}$ and $g^{ml}_{\mathbf{s}}$ in (5.4) and (5.27). These terms are determined by the tensor elements $g^{\mathbf{s}}_{ij}$ which define the covariant metric tensor (5.90) of the surface S^{rn} in the coordinates s^i represented by the parametrization (5.89). The system in (5.4), in contrast to that of (5.93), has a divergent form and its solution is a harmonic function, as was mentioned above. Besides this, its solution is independent of parametrizations of X^n; as a rule, this is not valid for solutions of the system (5.93).

Generation of Adaptive Grids on Surfaces.

When the monitor surface is formed by the values of a function $\mathbf{f}(\mathbf{x})$ over a general n-dimensional surface S^{xn} lying in the space R^{n+k} and represented by the parametrization

$$\mathbf{x}(\mathbf{s}) : S^n \to R^{n+k} , \quad \mathbf{x}(\mathbf{s}) = [x^1(\mathbf{s}),\ldots,x^{n+k}(\mathbf{s})]$$

from an n-dimensional parametric domain $S^n \in R^n$ then the monitor surface S^{rn} over S^{xn} can be described by a parametrization from S^n in the form

$$\mathbf{r}(\mathbf{s}) : S^n \to R^{n+l+k} , \qquad \mathbf{r}(\mathbf{s}) = \{\mathbf{x}(\mathbf{s}), \mathbf{f}[\mathbf{x}(\mathbf{s})]\} . \tag{5.94}$$

In particular, a one-dimensional monitor surface S^{r1} (in fact curve) over a curve S^{x1} lying in R^n and represented by $\mathbf{x}(s) : [a,b] \to R^n$, has the following parametrization

$$\mathbf{r}(s) : [a,b] \to R^{n+l} , \qquad \mathbf{r}(s) = \{\mathbf{x}(s), \mathbf{f}[\mathbf{x}(s)]\} .$$

It is evident that the adaptive grid on the surface S^{xn} obtained by projecting the quasiuniform grid from S^{rn} is formed, in fact, by mapping a reference grid in Ξ^n with a composition of $\mathbf{x}(\mathbf{s})$ and the intermediate grid transformation $\mathbf{s}(\boldsymbol{\xi})$, i.e. with $\mathbf{x}(\mathbf{s}(\boldsymbol{\xi})) : \Xi^n \to S^{xn}$.

If the monitor surface over S^{xn} is formed by a scalar-valued monitor function $f(\mathbf{s})$, then for the covariant metric tensor of S^{rn}, designated by $g^{\mathbf{s}}_{ij}$ in the coordinates $s^1, \dots, s^n$, we have

$$g^{\mathbf{s}}_{ij} = g^{xs}_{ij} + f_{s^i} f_{s^j} \ , \quad i,j = 1, \dots, n \ ,$$

where

$$g^{xs}_{ij} = \mathbf{x}_{s^i} \cdot \mathbf{x}_{s^j} \ , \quad i,j = 1, \dots, n \ ,$$

is the metric of S^{xn}. This monitor metric is a particular case of the metric (5.57) and (5.58). Computing the elements $g^{ij}_{\mathbf{s}}$ of the contravariant metric tensor of S^{rn} and the Jacobian of the monitor metric by formulas (5.64) and (5.69) gives

$$\begin{aligned} & g^{ij}_{\mathbf{s}} = g^{ij}_{sx} - g^{ik}_{sx} \frac{\partial f}{\partial s^k} g^{jm}_{sx} \frac{\partial f}{\partial s^m}, \quad i,j,k,m = 1, \dots, n, \\ & g^{\mathbf{s}} = g^{xs}[1 + \nabla(f)], \end{aligned} \tag{5.95}$$

where

$$\nabla(f) = g^{ij}_{sx} \frac{\partial f}{\partial s^i} \frac{\partial f}{\partial s^j}, \quad i,j = 1, \dots, n.$$

Substituting these expressions in (5.4) yields the following system for generating adaptive grids on surfaces

$$\begin{aligned} & \frac{\partial}{\partial s^j} \Big[\sqrt{g^{xs}[1 + \nabla(f)]} \\ & \times \Big(g^{jk}_{sx} - \frac{1}{1 + \nabla(f)} g^{jl}_{sx} g^{km}_{sx} \frac{\partial f}{\partial s^l} \frac{\partial f}{\partial s^m} \Big) \frac{\partial \xi^i}{\partial s^k} \Big] = 0 \ , \\ & i,j,k,l,m = 1, \dots, n \ . \end{aligned}$$

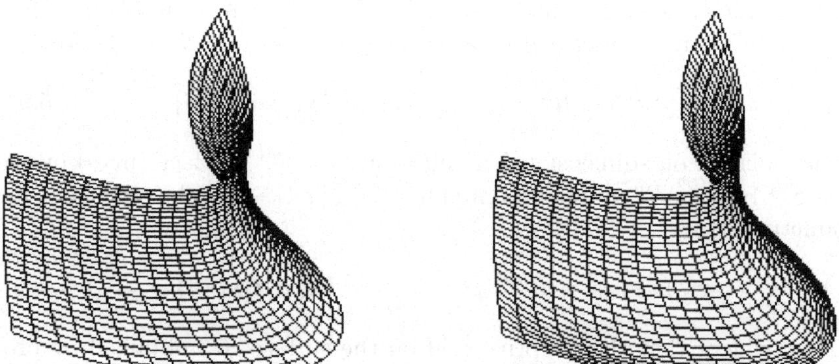

Fig. 5.7. Fixed (left-hand) and adaptive (right-hand) surface grids.

Figure 5.7 demonstrates fixed (left-hand) and adaptive (right-hand) surface grids. The adaptive grid is subject to the Beltrami equations in the monitor metric (5.95).

Similarly to the Poisson equations (5.93), for generating adaptive grids on two-dimensional surfaces $S^{x2} \subset R^3$ there can also be used a combination of the Beltramian operator and forcing terms:

$$\Delta_B^x[\xi^i] = P^i\ , \quad i = 1, 2\ . \tag{5.96}$$

Here the Beltramian operator is defined through the metric of S^{x2}, i.e.

$$\Delta_B^x[\xi^i] = \frac{1}{\sqrt{g^{xs}}}\frac{\partial}{\partial s^j}\left(\sqrt{g^{xs}}g_{sx}^{kj}\frac{\partial \xi^i}{\partial s^k}\right), \quad i, j = 1, 2\ . \tag{5.97}$$

If the forcing terms P^i, $i = 1, 2$, are specified as functions of the coordinates s^1, s^2 then the system (5.96) is independent of parametrizations of S^{x2}. However, the equations (5.96) are not the generalized Laplace equations and their solution $\boldsymbol{\xi}(\mathbf{s})$ is not a harmonic function. So the theorem of Rado is not held and consequently $\boldsymbol{\xi}(\mathbf{s})$ may not be a one-to-one mapping. Thus the grid cells obtained from (5.96) may be folded.

Monitor Metric for Generating Grids Adapting to the Values of a Function.

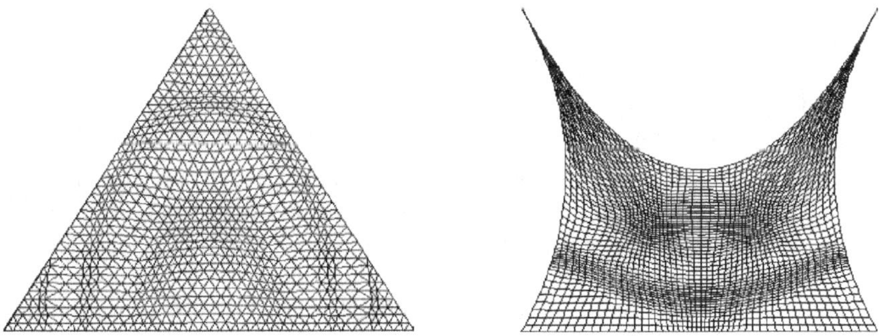

Fig. 5.8. Examples of adaptive domain grids.

For generating a numerical grid in a physical geometry S^{xn} with node clustering in the zones of the large values of a function $v(\mathbf{s})$ the measure of departure from the necessary grid can be expressed in the form

$$\sigma(\mathbf{s}) = Z[v](\mathbf{s})g_{sx}^{km}\frac{\partial \xi^i}{\partial s^k}\frac{\partial \xi^i}{\partial s^m}, \quad i, j, k, m = 1, \dots, n, \tag{5.98}$$

where $Z[v] > 0$ is a positive operator such that $Z[v](\mathbf{s})$ is large (small) where $v(\mathbf{s})$ is small (large). This measure for generating adaptive grids in domains was introduced by Danaev, Liseikin, and Yanenko (1980) and Winslow (1981). Consequently the contravariant elements of the monitor metric are as follows:

$$g^{ij}(\mathbf{s}) = Z[v](\mathbf{s})g^{ij}_{sx}, \quad i,j = 1,\dots,n. \tag{5.99}$$

This contravariant metric tensor can also be used for providing node clustering in the zones of large variations of a function $\mathbf{f}(\mathbf{s})$, introducing for this purpose an operator $Z[grad\ \mathbf{f}]$ such that $Z[grad\ \mathbf{f}](\mathbf{s})$ is large where $|grad\ \mathbf{f}|(\mathbf{s})$ is small and vice versa.

Figure 5.8 illustrates adaptive domain grids generated through diffusion equations (5.16) in the monitor metric (5.99) with a function $v(\mathbf{s})$ that has large values in the zones of grid clustering.

5.3.4 Monitor Metrics for Generating Balanced Grids.

For computing balanced numerical grids, that are field-aligned and adaptive to the values of one function and/or to the variations of another function, a natural way for defining a monitor metric consists in combining the covariant or contravariant elements of the corresponding individual monitor metrics, i.e. as

$$g^{\mathbf{s}}_{ij}(\mathbf{s}) = \varepsilon_1(\mathbf{s})g^{al}_{ij} + \varepsilon_2(\mathbf{s})g^{adg}_{ij} + \varepsilon_3(\mathbf{s})g^{adv}_{ij}, \quad i,j = 1,\dots,n. \tag{5.100}$$

or

$$g^{ij}_{\mathbf{s}}(\mathbf{s}) = w_1(\mathbf{s})g^{ij}_{al} + w_2(\mathbf{s})g^{ij}_{adg} + w_3(\mathbf{s})g^{ij}_{adv}, \quad i,j = 1,\dots,n. \tag{5.101}$$

where $\epsilon_i(\mathbf{s}) \geq 0$ and $w_i(\mathbf{s}) \geq 0, i = 1,2,3$ are weight functions specifying the contribution of the covariant elements g^{al}_{ij}, g^{adg}_{ij}, and g^{adv}_{ij} or contravariant elements g^{ij}_{al}, g^{ij}_{adg}, and g^{ij}_{adv}, respectively. The marks *al*, *adg* and *adv* in these formulas mean that the corresponding metric elements are chosen to grid alignment, adaptation to gradients, and adaptation to values, respectively. These metric elements were formulated above.

Figure 5.9 exhibits balanced grids adapting to the gradients of a function and to a vector field (left-hand), and to the values of a function and to a vector field (right-hand).

There may be other effective ways for combining the corresponding metric components, in particular, for generating grids that are field-aligned and adaptive to the values of a function $v(\mathbf{s})$ good results are demonstrated by the following contravariant elements of the monitor metric in the form (5.99):

$$g^{ij}_{\mathbf{s}} = Z[v](\mathbf{s})g^{ij}_{al}, \quad i,j = 1,\dots,n. \tag{5.102}$$

Figure 5.10 illustrates both the integral lines of a two-dimensional magnetic field (left-hand) and such a balanced grid aligned to the magnetic field

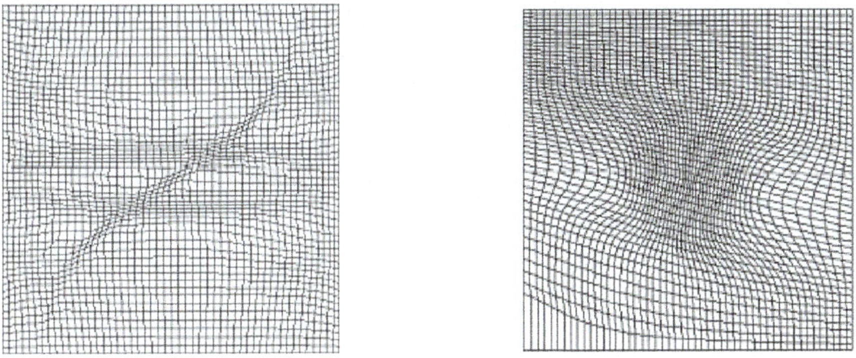

Fig. 5.9. Examples of balanced grids.

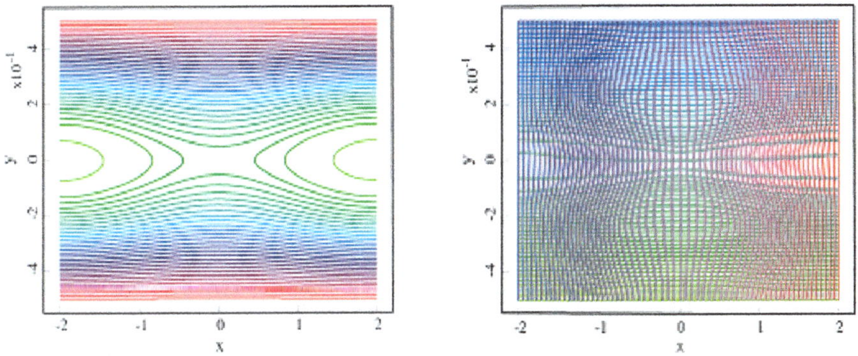

Fig. 5.10. Magnetic vector field flux (left-hand) and a balanced grid (right-hand).

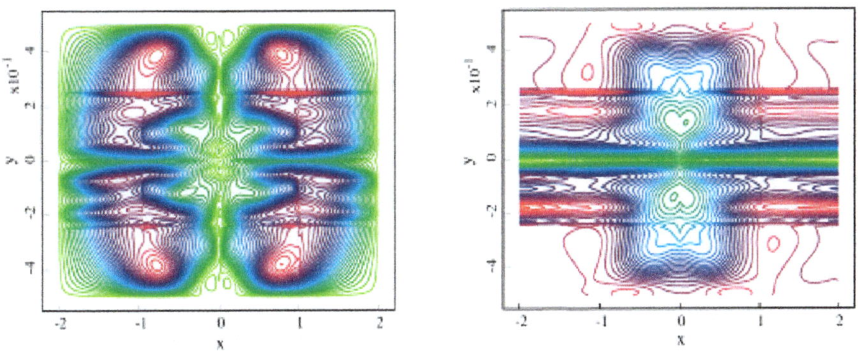

Fig. 5.11. Alignment and adaptation (left-hand) and scaled grid density (right-hand).

and adapted to the numerical error (right-hand) via the metric (5.102). Figure 5.11 exhibits a contour plot of alignment error for both alignment and adaptation (left-hand) and scaled grid density (right-hand). The pictures of Fig. 5.10 and Fig. 5.11 were performed by A. Glasser who used a spectral element method for computing plasmas and diffusion grid equations (5.17) in the metric (5.102) (see Glasser, Kitaeva, and Liseikin (2005), and Glasser et al. (2005)).

6 Inverted Equations

In this chapter we establish some equivalent forms of the grid equations with respect to the components $s^i(\boldsymbol{\xi})$ of intermediate transformations (5.2) using for this purpose the relations outlined in Chaps. 4–5. The equations are obtained by inverting the comprehensive grid equations specified in (5.4) and (5.16). The chapter also reviews the role of the mean curvature in the grid equations and in measure of grid clustering.

6.1 General Forms of Equations

6.1.1 Relations to Beltrami Equations

Let s^i, $i = 1, \dots, n$, be a local coordinate system in a Riemannian manifold M^n and g^{s}_{ij} be its covariant metric tensor. Expanding the differentiation in $\Delta_B[s^i]$ (formula (5.3)), we have

$$\begin{aligned}\Delta_B[s^i] &= \frac{1}{\sqrt{g^{\mathrm{s}}}}\frac{\partial}{\partial s^j}(\sqrt{g^{\mathrm{s}}}g^{ji}_{\mathrm{s}}) \\ &= \frac{g^{ji}_{\mathrm{s}}}{\sqrt{g^{\mathrm{s}}}}\frac{\partial}{\partial s^j}\sqrt{g^{\mathrm{s}}} + \frac{\partial}{\partial s^j}g^{ji}_{\mathrm{s}} , \quad i, j = 1, \dots, n ,\end{aligned} \tag{6.1}$$

where $g^{\mathrm{s}} = det(g^{\mathrm{s}}_{ij})$, g^{ji}_{s} is the (ji)-th contravariant metric element of M^n in the coordinates $s^1, \dots, s^n$.

The application of (4.26) to the first item in the second line of this system of equations yields

$$\frac{g^{ji}_{\mathrm{s}}}{\sqrt{g^{\mathrm{s}}}}\frac{\partial}{\partial s^j}\sqrt{g^{\mathrm{s}}} = g^{ji}_{\mathrm{s}}\Upsilon^k_{kj} , \quad i, j, k = 1, \dots, n ,$$

while for the second item we find from (4.25)

$$\frac{\partial}{\partial s^j}g^{ji}_{\mathrm{s}} = -g^{ik}_{\mathrm{s}}\Upsilon^j_{kj} - g^{lj}_{\mathrm{s}}\Upsilon^i_{lj} , \quad i, j, k, l = 1, \dots, n .$$

Substituting these relations in (6.1) gives the following expression of $\Delta_B[s^i]$

$$\Delta_B[s^i] \equiv -g_{\mathbf{s}}^{kj}\Upsilon_{kj}^i\,, \quad i,j,k=1,\dots,n\,. \tag{6.2}$$

Thus, assuming in (6.2) that $\boldsymbol{\xi}=\mathbf{s}$, the Beltrami equations in (5.4) formulated for generating grids are as follows:

$$\Delta_B[\xi^i] \equiv -g_{\boldsymbol{\xi}}^{kj}\Upsilon_{kj}^i = 0\,, \quad i,j,k=1,\dots,n\,. \tag{6.3}$$

Note, the quantities in this formula are in the grid coordinates ξ^i, $i=1,\dots,n$, in particular, the Christoffel symbols of the second kind are, in accordance with (4.24), as

$$\Upsilon_{kj}^i = {}^{\boldsymbol{\xi}}\Upsilon_{kj}^i = g_{\boldsymbol{\xi}}^{im}[kj,m]^{\boldsymbol{\xi}}\,, \quad i,j,k,m=1,\dots,n\,.$$

Thus the grid equations (6.3) also have the following equivalent form

$$g_{\boldsymbol{\xi}}^{kj}g_{\boldsymbol{\xi}}^{im}[kj,m]^{\boldsymbol{\xi}} = 0\,, \quad i,j,k,m=1,\dots,n\,.$$

Multiplying these equations by $g_{ip}^{\boldsymbol{\xi}}$ gives the following equivalent grid system

$$g_{\boldsymbol{\xi}}^{kj}[kj,p]^{\boldsymbol{\xi}} = 0\,, \quad j,k,p=1,\dots,n\,. \tag{6.4}$$

Let M^n be a regular surface $S^{rn}\subset R^{n+l}$ represented in the parametric coordinates $s^1,\dots,s^n$ by

$$\mathbf{r}(\mathbf{s}): S^n \to R^{n+l}\,, \quad \mathbf{r}(\mathbf{s}) = [r^1(\mathbf{s}),\dots,r^{n+l}(\mathbf{s})]\,, \quad \mathbf{s}=(s^1,\dots,s^n)\,,$$

then, using in (6.4) the relations (4.27) with $\mathbf{s}=\boldsymbol{\xi}$, we find that the Beltrami comprehensive grid equations in (5.4) with respect to the metric of the surface S^{rn} are equivalent to

$$g_{\boldsymbol{\xi}}^{kj}(\mathbf{r}_{\xi^k\xi^j}\cdot\mathbf{r}_{\xi^i}) = 0\,, \quad i,j,k=1,\dots,n\,, \tag{6.5}$$

where

$$\mathbf{r}_{\xi^k\xi^j} = \frac{\partial^2\mathbf{r}[\mathbf{s}(\boldsymbol{\xi})]}{\partial\xi^k\partial\xi^j}\,, \quad j,k=1,\dots,n\,, \quad \mathbf{r}_{\xi^i} = \frac{\partial\mathbf{r}[\mathbf{s}(\boldsymbol{\xi})]}{\partial\xi^i}\,, \quad i=1,\dots,n\,.$$

Note equations (6.5) mean that in the grid coordinates $\xi^1,\dots,\xi^n$ the vector

$$g_{\boldsymbol{\xi}}^{ij}\mathbf{r}_{\xi^i\xi^j}\,, \quad i,j=1,\dots,n\,,$$

is orthogonal to the n-dimensional plane formed by the basic tangent vectors $\mathbf{r}_{\xi^i}$, $i=1,\dots,n$ consequently, it is orthogonal to the surface $S^{rn}\subset R^{n+l}$.

As

$$\begin{aligned} &g_{\boldsymbol{\xi}}^{ij}\mathbf{r}_{\xi^i\xi^j} + \frac{1}{\sqrt{g^{\boldsymbol{\xi}}}}\frac{\partial}{\partial\xi^j}(\sqrt{g^{\boldsymbol{\xi}}}g_{\boldsymbol{\xi}}^{ij})\mathbf{r}_{\xi^i} \\ &= \frac{1}{\sqrt{g^{\boldsymbol{\xi}}}}\frac{\partial}{\partial\xi^j}(\sqrt{g^{\boldsymbol{\xi}}}g_{\boldsymbol{\xi}}^{ij}\mathbf{r}_{\xi^i}) = \Delta_B[\mathbf{r}]\,, \quad i,j=1,\dots,n\,, \end{aligned}$$

we obtain, taking advantage of (5.5) that in the grid coordinates $\xi^1, \dots, \xi^n$

$$g_{\boldsymbol{\xi}}^{ij} \mathbf{r}_{\xi^i \xi^j} = \Delta_B[\mathbf{r}] \,, \quad i,j = 1, \dots, n \,, \tag{6.6}$$

i.e. the vector $\Delta_B[\mathbf{r}]$ being the same in arbitrary coordinates is also orthogonal to the surface $S^{rn} \subset R^{n+l}$. If $l = 1$, i.e. $S^{rn} \subset R^{n+1}$ then there is defined the mean curvature of S^{rn} by the formula

$$K_m = \frac{1}{n} g_{\boldsymbol{\xi}}^{ij} \mathbf{r}_{\xi^i \xi^j} \cdot \mathbf{n} \,, \quad i,j = 1, \dots, n \,,$$

where $\mathbf{n}$ is a unit normal to S^{rn} (see (4.66)). As $\Delta_B[\mathbf{r}]$ is a normal to S^{rn} we can assume

$$\mathbf{n} = \frac{1}{|\Delta_B[\mathbf{r}]|} \Delta_B[\mathbf{r}] \,, \quad |\Delta_B[\mathbf{r}]| = \sqrt{\Delta_B[\mathbf{r}] \cdot \Delta_B[\mathbf{r}]} \,,$$

hence with respect to this unit normal

$$K_m = \frac{1}{n|\Delta_B[\mathbf{r}]|} \Delta_B[\mathbf{r}] \cdot \Delta_B[\mathbf{r}] = \frac{1}{n} |\Delta_B[\mathbf{r}]| \,. \tag{6.7}$$

Notice $\Delta_B[\mathbf{r}]$ is an invariant of parametrizations of an arbitrary regular surface $S^{rn} \subset R^{n+l}, l \geq 1$, so formula (6.7) can be considered as the mean curvature of this surface as well. Thus using this formula (6.7) for the definition of the mean curvature of the regular surface $S^{rn} \subset R^{n+l}$ represented by (4.1) we obtain from (6.6) that in the grid coordinates $\xi^1, \dots, \xi^n$, satisfying (5.4)

$$g_{\boldsymbol{\xi}}^{ij} \mathbf{r}_{\xi^i \xi^j} = n K_m \mathbf{n} \,, \quad i,j = 1, \dots, n \,, \tag{6.8}$$

where $\mathbf{n} = \Delta_B[\mathbf{r}] / |\Delta_B[\mathbf{r}]|$.

6.1.2 Resolved Grid Equations

In general the systems of equations (6.4) and (6.5) are not resolved with respect to the components $s^i(\boldsymbol{\xi})$ of the intermediate transformation $\mathbf{s}(\boldsymbol{\xi})$. In this section we establish some resolved forms of the grid equations convenient for the implementation into numerical codes. Crucial for this purpose is a basic elliptic operator.

Basic Elliptic Operator

The basic elliptic grid operator designated in local coordinates $\mathbf{v} = (v^1, \dots, v^n)$ of the Riemannian manifold M^n by $L^{\mathbf{v}}$ is specified at an arbitrary twice-differential function $y(\mathbf{v})$ as

$$L^{\mathbf{v}}[y] = g_{\mathbf{v}}^{ij} \frac{\partial^2 y}{\partial v^i \partial v^j} \,, \quad i,j = 1, \dots, n \,, \tag{6.9}$$

where $g^{ij}_{\mathbf{v}}$ is the (ij)th element of the contravariant metric tensor of M^n in the coordinates $v^1, \dots, v^n$. Note the value of the operator at $y(\boldsymbol{v})$ is not an invariant of parametrizations of M^n. The first line of equation (4.58) yields the following relation between the basic elliptic and Beltrami operators:

$$L^{\mathbf{v}}[y] = \Delta_B[y] + y_{v^k} g^{ij}_{\mathbf{v}} \Upsilon^k_{ij} \,, \quad i,j,k = 1, \dots, n \,,$$

while the application of (6.2) to this formula gives

$$L^{\mathbf{v}}[y] = \Delta_B[y] - y_{v^k} \Delta_B[v^k] \,, \quad k = 1, \dots, n \,.$$

Assuming here that the coordinates $v^1, \dots, v^n$ are the grid coordinates $\xi^1, \dots, \xi^n$ satisfying (5.4) we obtain

$$L^{\boldsymbol{\xi}}[y] = \Delta_B[y] \,, \tag{6.10}$$

i.e. in the grid coordinates the value of the basic elliptic operator at an arbitrary function $y(\boldsymbol{\xi})$ coincides with Beltrami's second differential parameter of this function.

Inverted Beltrami Equations

A general form of the inverted Beltrami grid equations in (5.4) resolved with respect to the intermediate transformation $\mathbf{s}(\boldsymbol{\xi})$ is obtained after substituting in (6.10) $s^i(\boldsymbol{\xi})$ for $y(\boldsymbol{\xi})$. Thus we have the following system of the inverted Beltrami grid equations with respect to the components $s^i(\boldsymbol{\xi})$, $i = 1, \dots, n$,

$$L^{\boldsymbol{\xi}}[s^i] = \Delta_B[s^i] \,, \quad i = 1, \dots, n \,,$$

i.e., using (6.9) and (5.3)

$$g^{km}_{\boldsymbol{\xi}} \frac{\partial^2 s^i}{\partial \xi^k \partial \xi^m} = \frac{1}{\sqrt{g^{\mathbf{s}}}} \frac{\partial}{\partial s^j} (\sqrt{g^{\mathbf{s}}} g^{ji}_{\mathbf{s}}), \quad i,j,k,m = 1, \dots, n. \tag{6.11}$$

As the value of the Beltramian operator is independent of the choice of parametrizations we also find from (6.11)

$$L^{\boldsymbol{\xi}}[s^i] = \frac{1}{\sqrt{g^{\boldsymbol{\xi}}}} \frac{\partial}{\partial \xi^k} \Big(\sqrt{g^{\boldsymbol{\xi}}} g^{kp}_{\boldsymbol{\xi}} \frac{\partial s^i}{\partial \xi^p} \Big) \,, \quad i,k,p = 1, \dots, n \,.$$

Another inference of equations (6.11) is carried out by multiplying the system in (5.4) with $\partial s^l / \partial \xi^i$. Indeed this multiplication yields

$$\begin{aligned} &\frac{\partial}{\partial s^j} \Big(\sqrt{g^{\mathbf{s}}} g^{jk}_{\mathbf{s}} \frac{\partial \xi^i}{\partial s^k} \Big) \frac{\partial s^l}{\partial \xi^i} \\ &= \frac{\partial}{\partial s^j} (\sqrt{g^{\mathbf{s}}} g^{jl}_{\mathbf{s}}) - \sqrt{g^{\mathbf{s}}} g^{jk}_{\mathbf{s}} \frac{\partial^2 s^l}{\partial \xi^i \partial \xi^m} \frac{\partial \xi^i}{\partial s^k} \frac{\partial \xi^m}{\partial s^j} \\ &= \sqrt{g^{\mathbf{s}}} \Big(\Delta_B[s^l] - g^{im}_{\boldsymbol{\xi}} \frac{\partial^2 s^l}{\partial \xi^i \partial \xi^m} \Big) = 0 \,, \quad i,j,k,l,m = 1, \dots, n \,, \end{aligned}$$

since (5.4), (4.15) and (4.18). We see that the equations in the last line are identical to (6.11).

Inverted Diffusion Equations

Similarly multiplying the system (5.16) by $\partial s^k/\partial \xi^i$ yields the following inverted diffusion grid equations resolved with respect to $s^k(\boldsymbol{\xi})$:

$$w[\mathbf{s}(\boldsymbol{\xi})]L^{\boldsymbol{\xi}}[s^k] = \frac{\partial}{\partial s^j}[w(\mathbf{s})g_{\mathbf{s}}^{jk}], \quad j,k = 1,\dots,n, \tag{6.12}$$

i.e. this system can be obtained from (6.11) after substituting $w(\mathbf{s})$ for $\sqrt{g^{\mathbf{s}}}$, and vice-versa, assuming in (6.12) $w(\mathbf{s}) = \sqrt{g^{\mathbf{s}}}$ yields equations (6.11).

6.1.3 Fluxes-Sources Equations

Finite-element and spectral-element methods are typically applied to the solution of equations having fluxes-sources forms. This section establishes some of these forms to the inverted grid equations.

As a system of grid equations with respect to the components $s^i(\boldsymbol{\xi})$ of the intermediate transformation (5.2) there can also be the system of the Euler-Lagrange equations of the functional of grid smoothness (5.27) or diffusion (5.51) rewritten with respect to the function $\mathbf{s}(\boldsymbol{\xi})$, i.e. in the form

$$I[\mathbf{s}] = \int_{\Xi^n} F[\mathbf{s}(\boldsymbol{\xi}), \frac{\partial \mathbf{s}}{\partial \xi^1},\dots,\frac{\partial \mathbf{s}}{\partial \xi^n}]d\boldsymbol{\xi}, \tag{6.13}$$

where, for the functional (5.51),

$$F[\mathbf{s}(\boldsymbol{\xi}), \frac{\partial \mathbf{s}}{\partial \xi^1},\dots,\frac{\partial \mathbf{s}}{\partial \xi^n}] = Jw[\mathbf{s}(\boldsymbol{\xi})]g_{\mathbf{s}}^{jm}[\mathbf{s}(\boldsymbol{\xi})]\frac{\partial \xi^k_{\mathbf{s}}}{\partial s^j}\frac{\partial \xi^k_{\mathbf{s}}}{\partial s^m}, \quad j,k,m = 1,\dots,n, \tag{6.14}$$

Of course we assume that $w[\mathbf{s}(\boldsymbol{\xi})] = \sqrt{g^{\mathbf{s}}[\mathbf{s}(\boldsymbol{\xi})]}$ for the functional (5.27). In (6.14) the Jacobian $J = det(\partial s^i/\partial \xi^j)$ and the elements $\partial \xi^l/\partial s^p$, $l,p = 1,\dots,n$, of the matrix inverse to the Jacobi matrix $(\partial s^i/\partial \xi^j)$ are the well-known functions of the variables $\partial s^k/\partial \xi^m, k,m = 1,\dots,n$.

In accordance with (5.29) the components $s^i(\boldsymbol{\xi})$ of the transformation $\mathbf{s}(\boldsymbol{\xi})$ optimal to the functional (6.13) are subject to the Euler-Lagrange equations

$$F_{s^i} - \frac{\partial}{\partial \xi^j}\Big(\frac{\partial}{\partial(\partial s^i/\partial \xi^j)}F\Big) = 0, \quad i,j = 1,\dots,n. \tag{6.15}$$

Availing us of the formulas of the tensor analysis yields

$$\frac{\partial}{\partial(\partial s^k/\partial \xi^j)}J = J\frac{\partial \xi^j}{\partial s^k}, \quad j,k = 1,\dots,n, \tag{6.16}$$

and

$$\frac{\partial}{\partial(\partial s^k/\partial \xi^j)}\frac{\partial \xi^l}{\partial s^m} = -\frac{\partial \xi^j}{\partial s^m}\frac{\partial \xi^l}{\partial s^k}, \quad j,k,l,m = 1,\dots,n. \tag{6.17}$$

In particular, the formula (6.16) follows from the relations

$$\frac{\partial s^k}{\partial \xi^m}\Big(J\frac{\partial \xi^m}{\partial s^k}\Big) = J, \quad k,m = 1,\dots,n, \quad k \text{ fixed},$$

and

$$\frac{\partial}{\partial(\partial s^k/\partial \xi^j)}\Big(J\frac{\partial \xi^m}{\partial s^k}\Big) = 0, \quad k,m = 1,\dots,n, \quad k \text{ fixed},$$

For obtaining formula (6.17) we see that

$$\frac{\partial}{\partial(\partial s^m/\partial \xi^i)}\Big(\frac{\partial s^l}{\partial \xi^p}\frac{\partial \xi^p}{\partial s^j}\Big) = \delta^m_l \frac{\partial \xi^i}{\partial s^j} + \frac{\partial s^l}{\partial \xi^p}\frac{\partial}{\partial(\partial s^m/\partial \xi^i)}\frac{\partial \xi^p}{\partial s^j} = 0,$$

$$i,j,l,m,p = 1,\dots,n.$$

Multiplying these equations by $\partial \xi^k/\partial s^l$ yields

$$\frac{\partial \xi^k}{\partial s^m}\frac{\partial \xi^i}{\partial s^j} = -\frac{\partial}{\partial(\partial s^m/\partial \xi^i)}\frac{\partial \xi^k}{\partial s^j}, \quad i,j,k,m = 1,\dots,n,$$

i.e. the equations (6.17).

Using the relations (6.16) and (6.17) we obtain for the function (6.14)

$$\begin{aligned}\frac{\partial}{\partial(\partial s^i/\partial \xi^j)}F &= F\frac{\partial \xi^j}{\partial s^i} - w(\mathbf{s})Jg^{km}_{\mathbf{s}}\frac{\partial \xi^l}{\partial s^i}\Big(\frac{\partial \xi^l}{\partial s^m}\frac{\partial \xi^j}{\partial s^k} + \frac{\partial \xi^l}{\partial s^k}\frac{\partial \xi^j}{\partial s^m}\Big) = \\ &= F\frac{\partial \xi^j}{\partial s^i} - 2w(\mathbf{s})Jg^{jl}_{\boldsymbol{\xi}}\frac{\partial \xi^l}{\partial s^i}, \quad i,j,k,l,m = 1,\dots,n.\end{aligned} \tag{6.18}$$

Thus, taking advantage of (6.18) in (6.15) we find the following fluxes-sources diffusion grid equations for the functional (6.13) with the integrand (6.14)

$$\begin{aligned}J\frac{\partial \xi^p}{\partial s^k}\frac{\partial \xi^p}{\partial s^m}\frac{\partial}{\partial s^i}[w(\mathbf{s})g^{km}_{\mathbf{s}}] &= \frac{\partial}{\partial \xi^j}\Big[F\frac{\partial \xi^j}{\partial s^i} - 2w(\mathbf{s})Jg^{jm}_{\boldsymbol{\xi}}\frac{\partial \xi^m}{\partial s^i}\Big], \\ &i,j,k,m,p = 1,\dots,n. \ \square\end{aligned} \tag{6.19}$$

It is easily verified that the relations (6.18) for $n = 1$ and $n = 2$ become as follows:

$$\frac{\partial}{\partial(\mathrm{d}s/\mathrm{d}\xi)}F = -\frac{w(s)}{g^s(\mathrm{d}s/\mathrm{d}\xi)^2}, \quad n = 1. \tag{6.20}$$

$$\frac{\partial}{\partial(\partial s^i/\partial \xi^j)}F = 2\frac{w(\mathbf{s})}{Jg^{\mathbf{s}}}g^{\mathbf{s}}_{ki}\frac{\partial s^k}{\partial \xi^j} - F\frac{\partial \xi^j}{\partial s^i}, \quad i,j = 1,2, \quad n = 2. \tag{6.21}$$

Therefore equation (6.19) for $n = 1$ is as

$$\begin{aligned}\frac{\mathrm{d}}{\mathrm{d}s}\Big[\frac{w(s)}{g^s}\Big]\frac{\mathrm{d}\xi}{\mathrm{d}s} + \frac{\mathrm{d}}{\mathrm{d}\xi}\Big[\frac{w[s(\xi)]}{g^s}\Big(\frac{\mathrm{d}\xi}{\mathrm{d}s}\Big)^2\Big] &= 2\frac{\mathrm{d}\xi}{\mathrm{d}s}\frac{\mathrm{d}}{\mathrm{d}\xi}\Big[\frac{w[s(\xi)]}{g^s}\frac{\mathrm{d}\xi}{\mathrm{d}s}\Big] = \\ = -2\Big(\frac{g^s}{w[s(\xi)]}\Big)^2\frac{\partial \xi}{\partial s}\frac{\partial}{\partial \xi}\Big[\frac{g^s}{w[s(\xi)]}\frac{\partial s}{\partial \xi}\Big] &= 0.\end{aligned} \tag{6.22}$$

Analogously substituting (6.21) in (6.15) yields the following fluxes-sources grid equations for $n=2$

$$J\frac{\partial\xi^p}{\partial s^k}\frac{\partial\xi^p}{\partial s^m}\frac{\partial}{\partial s^i}[w(\mathbf{s})g_{\mathbf{s}}^{km}] = \frac{\partial}{\partial\xi^j}\Big(2w[\mathbf{s}(\boldsymbol{\xi})]\frac{g^{\mathbf{s}}_{ki}}{Jg^{\mathbf{s}}}\frac{\partial s^k}{\partial\xi^j} - F\frac{\partial\xi^j}{\partial s^i}\Big), \qquad i,j,k,l,m,p=1,2. \tag{6.23}$$

The corresponding fluxes-sources Beltrami grid equations for the functional of the grid smoothness (5.27) written out with respect to the intermediate transformation $\mathbf{s}(\boldsymbol{\xi})$, i.e.

$$I[\mathbf{s}] = \int\limits_{\Xi^n}\Big(\sqrt{g^{\mathbf{s}}}g_{\mathbf{s}}^{jk}\frac{\partial\xi^i}{\partial s^j}\frac{\partial\xi^i}{\partial s^k}\Big)J\mathrm{d}\boldsymbol{\xi},$$

are obtained by substituting $\sqrt{g^{\mathbf{s}}}$ for $w(\mathbf{s})$ in the equations (6.19), (6.22), and (6.23).

Note the inverted equations (6.12) and the fluxes-sources equations (6.19) are equivalent if the transformation $\mathbf{s}(\boldsymbol{\xi})$ is not singular. This fact follows from a relation connecting the Euler-Lagrange equations for a general functional

$$I[\boldsymbol{\xi}] = \int\limits_{S^n} F\Big[\mathbf{s},\frac{\partial\boldsymbol{\xi}}{\partial s^1},\dots,\frac{\partial\boldsymbol{\xi}}{\partial s^n}\Big]\mathrm{d}\mathbf{s}, \tag{6.24}$$

and for the same functional with respect to the inverted transformations $\mathbf{s}(\boldsymbol{\xi})$

$$I[\mathbf{s}] = \int\limits_{\Xi^n} F\Big[\mathbf{s}(\boldsymbol{\xi}),\frac{\partial\boldsymbol{\xi}}{\partial s^1},\dots,\frac{\partial\boldsymbol{\xi}}{\partial s^n}\Big]J\mathrm{d}\boldsymbol{\xi}. \tag{6.25}$$

Indeed, using the relations (6.16) and (6.17) gives the following Euler-Lagrange equations for the functional (6.25)

$$JF_{s^i} - \frac{\partial}{\partial\xi^j}\Big(FJ\frac{\partial\xi^j}{\partial s^i}\Big) + \frac{\partial}{\partial\xi^j}\Big(J\frac{\partial\xi^j}{\partial s^m}\frac{\partial\xi^k}{\partial s^i}\frac{\partial}{\partial(\partial\xi^k/\partial s^m)}F\Big) = 0, \qquad i,j,k,m=1,\dots,n. \tag{6.26}$$

Since

$$\frac{\partial}{\partial\xi^j}\Big(J\frac{\partial\xi^j}{\partial s^m}\Big) = 0, \quad j,m=1,\dots,n,$$

we find from (6.26)

$$JF_{s^i} - J\frac{\partial F}{\partial s^i} + J\frac{\partial}{\partial s^m}\Big(\frac{\partial\xi^k}{\partial s^i}\frac{\partial}{\partial(\partial\xi^k/\partial s^m)}F\Big) = J\frac{\partial\xi^k}{\partial s^i}\frac{\partial}{\partial s^m}\Big(\frac{\partial}{\partial(\partial\xi^k/\partial s^m)}F\Big) = 0.$$

Multiplying this system by $(\partial s^i/\partial\xi^l)/J$ gives the Euler-Lagrange equations for the functional (6.24). □

One more form of the fluxes-sources diffusion grid equations is obtained if we use the following relation of the tensor analysis

$$\frac{\partial A^i}{\partial s^i} = \frac{1}{J}\frac{\partial}{\partial \xi^j}(J\tilde{A}^j), \quad i,j = 1,\dots,n, \tag{6.27}$$

where

$$\tilde{A}^j = A^k \frac{\partial \xi^j}{\partial s^k}, \quad j,k = 1,\dots,n.$$

Assuming

$$A^m = w(\mathbf{s}) g_{\mathbf{s}}^{mk} \frac{\partial \xi^i}{\partial s^k}, \quad i,k,m = 1,\dots,n,$$

we find

$$\tilde{A}^j = w[\mathbf{s}(\boldsymbol{\xi})] g_{\mathbf{s}}^{mk} \frac{\partial \xi^i}{\partial s^k}\frac{\partial \xi^j}{\partial s^m} = w[\mathbf{s}(\boldsymbol{\xi})] g_{\boldsymbol{\xi}}^{ij}, \quad i,j,k,m = 1,\dots,n.$$

So, in accordance with (6.27), the diffusion grid equations (5.16), as well as (6.19), are equivalent to the following fluxes-sources diffusion grid equations

$$\frac{\partial}{\partial \xi^j}\{Jw[\mathbf{s}(\boldsymbol{\xi})] g_{\boldsymbol{\xi}}^{ij}\} = 0, \quad i,j = 1,\dots,n, \tag{6.28}$$

(see also (5.17)).

Substituting here $\sqrt{g^{\mathbf{s}}}$ for $w(\mathbf{s})$ yields the fluxes-sources Beltrami grid equations noted above by the formula (5.5).

In the two-dimensional case the equations (6.28) are as follows:

$$\frac{\partial}{\partial \xi^j} F_i^j = 0, \quad i,j = 1,2, \tag{6.29}$$

where

$$F_1^1 = \frac{w[\mathbf{s}(\boldsymbol{\xi})]}{J}\Big[g_{\mathbf{s}}^{11}\Big(\frac{\partial s^2}{\partial \xi^2}\Big)^2 + g_{\mathbf{s}}^{22}\Big(\frac{\partial s^1}{\partial \xi^2}\Big)^2 - 2g_{\mathbf{s}}^{12}\frac{\partial s^1}{\partial \xi^2}\frac{\partial s^2}{\partial \xi^2}\Big],$$

$$F_2^2 = \frac{w[\mathbf{s}(\boldsymbol{\xi})]}{J}\Big[g_{\mathbf{s}}^{11}\Big(\frac{\partial s^2}{\partial \xi^1}\Big)^2 + g_{\mathbf{s}}^{22}\Big(\frac{\partial s^1}{\partial \xi^1}\Big)^2 - 2g_{\mathbf{s}}^{12}\frac{\partial s^1}{\partial \xi^1}\frac{\partial s^2}{\partial \xi^1}\Big],$$

$$F_1^2 = F_2^1 = \frac{w[\mathbf{s}(\boldsymbol{\xi})]}{J}\Big[-g_{\mathbf{s}}^{11}\frac{\partial s^2}{\partial \xi^1}\frac{\partial s^2}{\partial \xi^2} - g_{\mathbf{s}}^{22}\frac{\partial s^1}{\partial \xi^1}\frac{\partial s^1}{\partial \xi^2} + g_{\mathbf{s}}^{12}\Big(\frac{\partial s^1}{\partial \xi^1}\frac{\partial s^2}{\partial \xi^2} + \frac{\partial s^1}{\partial \xi^2}\frac{\partial s^2}{\partial \xi^1}\Big)\Big].$$

6.2 Equations for Classical Monitor Metrics

In this section the inverted grid equations are written for some special monitor metrics realizing the popular grid equations discussed in Sect. 5.1.6.

6.2.1 Domain Grid Equations for a Diagonal Monitor Metric

Inverted Beltrami Equations

Euler Metric.

Let both S^{xn} and the monitor manifold be a domain S^n, i.e.

$$g^{\mathbf{s}}_{ij} = g^{ij}_{\mathbf{s}} = \delta^i_j, \quad i,j = 1,\dots,n. \tag{6.30}$$

In this case the inverted Beltrami grid equations (6.11) become the well-known Winslow grid equations

$$g^{km}_{\xi s} \frac{\partial^2 s^i}{\partial \xi^k \partial \xi^m} = 0, \quad i,k,m = 1,\dots,n, \tag{6.31}$$

where

$$g^{km}_{\xi s} = \frac{\partial \xi^k}{\partial s^l} \frac{\partial \xi^m}{\partial s^l}, \quad k,l,m = 1,\dots,n,$$

are the elements of the contravariant Euclidean metric tensor of S^n in the grid coordinates $\xi^i, \dots, \xi^n$.

Spherical Metric.

If the monitor metric $g^{\mathbf{s}}_{ij}$ of a monitor manifold M^n over a domain S^n is a spherical one, i.e. in the Cartesian coordinates $s^1, \dots, s^n$ of S^n it is of the form

$$g^{\mathbf{s}}_{ij} = v(\mathbf{s})\delta^i_j\,, \quad i,j = 1,\dots,n\,, \quad v(\mathbf{s}) > 0\,, \tag{6.32}$$

then

$$g^{\mathbf{s}} = [v(\mathbf{s})]^n, \quad g^{ij}_{\mathbf{s}} = \frac{1}{v(\mathbf{s})}\delta^i_j\,, \quad i,j = 1,\dots,n\,,$$

so, in accordance with (5.3),

$$\Delta_B[s^k] = \frac{1}{[v(\mathbf{s})]^{n/2}} \frac{\partial}{\partial s^j}\left\{\delta^j_k [v(\mathbf{s})]^{n/2-1}\right\} = \frac{n-2}{2(v(\mathbf{s}))^2} \frac{\partial v}{\partial s^k}\,,$$
$$i,j,k = 1,\dots,n\,.$$

Therefore the inverted Beltrami grid equations (6.11), in the spherical metric (6.32), have the following form

$$L^{\boldsymbol{\xi}}[s^i] = \frac{n-2}{2[v(\mathbf{s})]^2} \frac{\partial v}{\partial s^k}\,, \quad k = 1,\dots,n\,. \tag{6.33}$$

Since the rule (4.15)

$$g^{im}_{\boldsymbol{\xi}} = g^{kj}_{\mathbf{s}} \frac{\partial \xi^i}{\partial s^k} \frac{\partial \xi^m}{\partial s^j} = \frac{1}{v(\mathbf{s})} g^{im}_{\xi s}\,, \quad i,j,k,m = 1,\dots,n\,, \tag{6.34}$$

therefore we readily see that the basic operator specified by (6.9) in the metric (6.32) is as follows:

$$L^{\boldsymbol{\xi}}[y] = \frac{1}{v(\mathbf{s})} g^{im}_{\xi s} \frac{\partial^2 y}{\partial \xi^i \partial \xi^m}\,, \qquad i,m = 1,\dots,n\,.$$

So (6.33) is also expressed as follows:

$$g^{im}_{\xi s} \frac{\partial^2 s^k}{\partial \xi^i \partial \xi^m} = \frac{n-2}{2v(\mathbf{s})} \frac{\partial v}{\partial s^k}\,, \qquad i,k,m = 1,\dots,n\,. \tag{6.35}$$

As

$$\frac{n-2}{2v(\mathbf{s})} \frac{\partial v}{\partial s^i} = \frac{1}{f(\mathbf{s})} \frac{\partial f}{\partial s^i}\,, \qquad i = 1,\dots,n\,,$$

where

$$f(\mathbf{s}) = [v(\mathbf{s})]^{(n-2)/2}\,,$$

we find, using (6.35),

$$g^{ij}_{\xi s} \frac{\partial^2 s^k}{\partial \xi^i \partial \xi^j} - \frac{1}{f} \frac{\partial f}{\partial s^k} = 0\,, \quad i,j,k = 1,\cdots,n\,. \tag{6.36}$$

Availing us of the relation (2.24), we have

$$\frac{\partial f}{\partial s^k} = \frac{\partial f}{\partial \xi^i} \frac{\partial \xi^i}{\partial s^k} = g^{ij}_{\xi s} \frac{\partial f}{\partial \xi^i} \frac{\partial s^k}{\partial \xi^j}\,.$$

Therefore we obtain that the system (6.36) has also the following equivalent form

$$g^{ij}_{\xi s}\left(\frac{\partial^2 s^k}{\partial \xi^i \partial \xi^j} - \frac{1}{f} \frac{\partial f}{\partial \xi^i} \frac{\partial s^k}{\partial \xi^j}\right) = f g^{ij}_{\xi s} \frac{\partial}{\partial \xi^i}\left(\frac{1}{f} \frac{\partial s^k}{\partial \xi^j}\right) = 0\,.$$

Thus we have a more compact form in comparison with (6.36) of the inverted Beltrami grid equations in the spherical metric (6.32):

$$g^{ij}_{\xi s} \frac{\partial}{\partial \xi^i}\left(\frac{1}{f} \frac{\partial s^k}{\partial \xi^j}\right) = 0\,, \quad i,j,k = 1,\dots,n\,. \tag{6.37}$$

Note $f(\mathbf{s}) \equiv 1$ for n=2 hence these equation for n=2 are equivalent to the Winslow equations (6.31) regardless of the form of $v(\mathbf{s})$.

General Diagonal Metric.

For the monitor metric (5.22) we get

$$\Delta_B[s^i] = \frac{1}{\sqrt{g^{\mathbf{s}}}} \frac{\partial}{\partial s^i}\left(\frac{\sqrt{g^{\mathbf{s}}}}{v^i(\mathbf{s})}\right), \quad i = 1,\dots,n\,, \quad i \text{ fixed.}$$

Therefore equations (6.11) become the following inverted Beltrami grid equations with respect to the monitor metric (5.22)

$$g_{\boldsymbol{\xi}}^{pm}\frac{\partial^2 s^i}{\partial\xi^p\partial\xi^m}=\frac{1}{\sqrt{g^{\mathrm{s}}}}\frac{\partial}{\partial s^i}\left(\frac{\sqrt{g^{\mathrm{s}}}}{v^i(\mathrm{s})}\right),\quad i,m,p=1,\dots,n\,,\quad i\ \text{fixed}\,,\tag{6.38}$$

where

$$g_{\boldsymbol{\xi}}^{pm}=\frac{1}{v^k(\mathrm{s})}\delta_i^k\delta_j^k\frac{\partial\xi^p}{\partial s^i}\frac{\partial\xi^m}{\partial s^j}=\frac{1}{v^k(\mathrm{s})}\frac{\partial\xi^p}{\partial s^k}\frac{\partial\xi^m}{\partial s^k}\,,\quad i,j,k,m,p=1,\dots,n\,.\tag{6.39}$$

In the two-dimensional case we have from (6.38)

$$g_{\boldsymbol{\xi}}^{pm}\frac{\partial^2 s^i}{\partial\xi^p\partial\xi^m}=\frac{1}{\sqrt{v^1(\mathrm{s})v^2(\mathrm{s})}}\frac{\partial}{\partial s^i}G_i(\mathrm{s})\,,\quad i,m,p=1,2\,,\quad i\ \text{fixed}\,,\tag{6.40}$$

where

$$G_1(\mathrm{s})=\sqrt{v^2(\mathrm{s})/v^1(\mathrm{s})}\,,\quad G_2(\mathrm{s})=1/G_1(\mathrm{s})\,.$$

As

$$g^{\boldsymbol{\xi}}=det(g_{ij}^{\boldsymbol{\xi}})=g^{\mathrm{s}}J^2,\ J=\det(\partial s^i/\partial\xi^k),$$

and in the two-dimensional case

$$g_{\boldsymbol{\xi}}^{pm}=(-1)^{p+m}\frac{1}{g^{\boldsymbol{\xi}}}g_{3-p3-m}^{\boldsymbol{\xi}}\,,\quad m,p=1,2\,,\quad m,p\ \text{fixed}\,,$$

so the system (6.40) is equivalent to

$$g_{22}^{\boldsymbol{\xi}}\frac{\partial^2 s^i}{\partial\xi^1\partial\xi^1}-2g_{12}^{\boldsymbol{\xi}}\frac{\partial^2 s^i}{\partial\xi^1\partial\xi^2}+g_{11}^{\boldsymbol{\xi}}\frac{\partial^2 s^i}{\partial\xi^2\partial\xi^2}=J^2\sqrt{g^{\mathrm{s}}}\frac{\partial}{\partial s^i}G_i(\mathrm{s})\,,\quad i=1,2\,,\quad i\ \text{fixed}\,.\tag{6.41}$$

where

$$g_{ij}^{\boldsymbol{\xi}}=g_{lm}^{\mathrm{s}}\frac{\partial s^l}{\partial\xi^i}\frac{\partial s^m}{\partial\xi^j}=v^k(\mathrm{s})\delta_l^k\delta_m^k\frac{\partial s^l}{\partial\xi^i}\frac{\partial s^m}{\partial\xi^j}=v^k(\mathrm{s})\frac{\partial s^k}{\partial\xi^i}\frac{\partial s^k}{\partial\xi^j},\quad i,j,k,l,m=1,2.$$

The system (6.11) in the metric (5.22) can also be transformed to a more compact form, similarly to the form of equations (6.37) in the metric (6.32). For finding this form we note that

$$\frac{\partial\xi^i}{\partial s^m}=g_{\mathrm{s}}^{pt}\frac{\partial\xi^i}{\partial s^p}\frac{\partial\xi^j}{\partial s^t}\frac{\partial s^b}{\partial\xi^j}g_{bm}^{\mathrm{s}}=g_{\boldsymbol{\xi}}^{ij}g_{bm}^{\mathrm{s}}\frac{\partial s^b}{\partial\xi^j}\,,\quad b,i,j,m,p,t=1,\dots,n\,,$$

therefore

$$\frac{1}{\sqrt{g^{\mathrm{s}}}}\frac{\partial}{\partial\xi^i}(\sqrt{g^{\mathrm{s}}}g_{\mathrm{s}}^{mk})\frac{\partial\xi^i}{\partial s^m}=g_{\boldsymbol{\xi}}^{ij}\frac{1}{\sqrt{g^{\mathrm{s}}}}g_{bm}^{\mathrm{s}}\frac{\partial s^b}{\partial\xi^j}\frac{\partial}{\partial\xi^i}(\sqrt{g^{\mathrm{s}}}g_{\mathrm{s}}^{mk})\quad b,i,j,k,m=1,\dots,n\,.$$

Availing us of these relations in (6.11) yields the following equivalent inverted Beltrami equations

$$g^{ij}_{\boldsymbol{\xi}}\Big[\frac{\partial^2 s^k}{\partial\xi^i\partial\xi^j}-\frac{1}{\sqrt{g^{\mathbf{s}}}}g^{\mathbf{s}}_{bm}\frac{\partial}{\partial\xi^i}(\sqrt{g^{\mathbf{s}}}g^{mk}_{\mathbf{s}})\frac{\partial s^b}{\partial\xi^j}\Big]=0\,,\quad b,i,j,k,m=1,\dots,n\,, \tag{6.42}$$

Now if the monitor metric $g^{\mathbf{s}}_{bm}$ is diagonal, i.e. $g^{\mathbf{s}}_{bm}=0$, $b\neq m$, then

$$g^{bm}_{\mathbf{s}}=0\quad b\neq m\quad g^{mm}_{\mathbf{s}}=1/g^{\mathbf{s}}_{mm},\ m\text{ fixed},$$

so equations (6.42) become

$$g^{ij}_{\boldsymbol{\xi}}\Big[\frac{\partial^2 s^k}{\partial\xi^i\partial\xi^j}-\frac{1}{\sqrt{g^{\mathbf{s}}}g^{kk}_{\mathbf{s}}}\frac{\partial}{\partial\xi^i}(\sqrt{g^{\mathbf{s}}}g^{kk}_{\mathbf{s}})\frac{\partial s^k}{\partial\xi^j}\Big]=$$

$$\sqrt{g^{\mathbf{s}}}g^{kk}_{\mathbf{s}}g^{ij}_{\boldsymbol{\xi}}\frac{\partial}{\partial\xi^i}\Big(\frac{g^{\mathbf{s}}_{kk}}{\sqrt{g^{\mathbf{s}}}}\frac{\partial s^k}{\partial\xi^j}\Big)=0\,,\quad i,j,k=1,\dots,n\,,\ k\text{ fixed}\,.$$

Therefore the inverted Beltrami grid equations (6.11) with respect to the metric (5.22) are also equivalent to the following equations

$$g^{ij}_{\boldsymbol{\xi}}\frac{\partial}{\partial\xi^i}\Big(\frac{v_k(\mathbf{s})}{\sqrt{g^{\mathbf{s}}}}\frac{\partial s^k}{\partial\xi^j}\Big)=0\,,\quad i,j,k=1,\dots,n\,,\ k\text{ fixed}\,, \tag{6.43}$$

while in the two-dimensional case they, analogously to (6.41), have the following form

$$\begin{aligned}&g^{\boldsymbol{\xi}}_{22}\frac{\partial}{\partial\xi^1}\Big(\frac{1}{G_i(\mathbf{s})}\frac{\partial s^i}{\partial\xi^1}\Big)-g^{\boldsymbol{\xi}}_{12}\Big[\frac{\partial}{\partial\xi^1}\Big(\frac{1}{G_i(\mathbf{s})}\frac{\partial s^i}{\partial\xi^2}\Big)+\frac{\partial}{\partial\xi^2}\Big(\frac{1}{G_i(\mathbf{s})}\frac{\partial s^i}{\partial\xi^1}\Big)\Big]+\\&+g^{\boldsymbol{\xi}}_{11}\frac{\partial}{\partial\xi^2}\Big(\frac{1}{G_i(\mathbf{s})}\frac{\partial s^i}{\partial\xi^2}\Big)=0\,,\quad i=1,2\,,\ i\text{ fixed}\,.\end{aligned} \tag{6.44}$$

Inverted Diffusion Equations

Availing us of (6.34) we obtain that the inverted diffusion grid equations (6.12) with respect to the spherical monitor metric (6.32) (in particular (6.30)) have the form (6.37) with $f(\mathbf{s})=v(\mathbf{s})/w(\mathbf{s})$, i.e.

$$g^{ij}_{\xi s}\frac{\partial}{\partial\xi^i}\Big(\frac{w(\mathbf{s})}{v(\mathbf{s})}\frac{\partial s^k}{\partial\xi^j}\Big)=0,\quad i,j,k=1,\dots,n, \tag{6.45}$$

while for the general diagonal monitor metric (5.22) they, similarly to (6.43), are as follows:

$$g^{ij}_{\boldsymbol{\xi}}\frac{\partial}{\partial\xi^i}\Big(\frac{w(\mathbf{s})}{v_k(\mathbf{s})}\frac{\partial s^k}{\partial\xi^j}\Big)=0,\quad i,j,k=1,\dots,n,\quad k\text{ fixed}. \tag{6.46}$$

Fluxes-Sources Equations

The fluxes-sources diffusion grid equations (6.19) with respect to the spherical metric (6.32) have the following form

$$Jg^{kk}_{\xi s}\frac{\partial}{\partial s^i}\left[\frac{w(\mathrm{s})}{v(\mathrm{s})}\right]=\frac{\partial}{\partial\xi^j}\left\{\frac{w[\mathrm{s}(\boldsymbol{\xi})]}{v[\mathrm{s}(\boldsymbol{\xi})]}J\left(g^{ll}_{\xi s}\frac{\partial\xi^j}{\partial s^i}-2g^{jm}_{\xi s}\frac{\partial\xi^m}{\partial s^i}\right)\right\}, \quad (6.47)$$
$$i,j,k,l,m=1,\dots,n.$$

In the two-dimensional case equations (6.23) with respect to the metric (6.32) are transformed to

$$Jg^{kk}_{\xi s}\frac{\partial}{\partial s^i}\left[\frac{w(\mathrm{s})}{v(\mathrm{s})}\right]=\frac{\partial}{\partial\xi^j}\left\{\frac{w[\mathrm{s}(\boldsymbol{\xi})]}{v[\mathrm{s}(\boldsymbol{\xi})]}\left(\frac{2}{J}\frac{\partial s^i}{\partial\xi^j}-Jg^{ll}_{\xi s}\frac{\partial\xi^j}{\partial s^i}\right)\right\}, \quad (6.48)$$
$$i,j,k,l=1,2.$$

Substituting in (6.47) and (6.48) $\sqrt{g^{\mathrm{s}}}=[v(\mathrm{s})]^{n/2}$ for $w(\mathrm{s})$ yields the corresponding fluxes-sources Beltrami grid equations in the metric (6.32). Also assuming in (6.47) and (6.48) $v(\mathrm{s})=1$ gives the corresponding fluxes-sources diffusion grid equations with respect to the metric (6.32). When additionally $w(\mathrm{s})=1$ these equations become the fluxes-sources Winslow equations, equivalent to (6.31).

The fluxes-sources equations (6.28) in the metric (6.32) are as follows:

$$\frac{\partial}{\partial\xi^j}\left(J\frac{w(\mathrm{s})}{v(\mathrm{s})}g^{ij}_{\xi s}\right)=0,\quad i,j=1,\dots,n. \quad (6.49)$$

In the general monitor metric (5.22) the fluxes-sources grid equations (6.19) and (6.28) are

$$J\frac{\partial\xi^p}{\partial s^k}\frac{\partial\xi^p}{\partial s^k}\frac{\partial}{\partial s^i}\left[\frac{w(\mathrm{s})}{v_k(\mathrm{s})}\right]=\frac{\partial}{\partial\xi^j}\left[J\frac{w(\mathrm{s})}{v_m(\mathrm{s})}\frac{\partial\xi^k}{\partial s^m}\left(\frac{\partial\xi^k}{\partial s^m}\frac{\partial\xi^j}{\partial s^i}-2\frac{\partial\xi^j}{\partial s^m}\frac{\partial\xi^k}{\partial s^i}\right)\right],$$
$$i,j,k,m,p=1,\dots,n, \quad (6.50)$$

and

$$\frac{\partial}{\partial\xi^j}\left[J\frac{w(\mathrm{s})}{v_k(\mathrm{s})}\frac{\partial\xi^i}{\partial s^k}\frac{\partial\xi^i}{\partial s^k}\right]=0,\quad i,j,k=1,\dots,n, \quad (6.51)$$

respectively.

6.2.2 Domain Grid Equations with Respect to the Metric of a Monitor Surface

Inverted Beltrami Equations

Let the physical domain X^n be identified with the parametric domain S^n while the intermediate transformation

$$\mathbf{s}(\boldsymbol{\xi}) : \Xi^n \to S^n$$

for generating a grid in X^n be determined as the inverse of the map

$$\boldsymbol{\xi}(\mathbf{s}) : S^n \to \Xi^n$$

satisfying (5.4) with respect to the metric of a monitor surface S^{rn} over the domain S^n. If the monitor surface S^{rn} over the domain X^n is formed by a vector-valued monitor function $\mathbf{f}(\mathbf{x}) = [f^1(\mathbf{x}), \dots, f^l(\mathbf{x})]$ then, assuming $\mathbf{s} = \mathbf{x}$, the parametrization of S^{rn} is determined by

$$\mathbf{r}(\mathbf{s}) : S^n \to R^{n+l}, \quad \mathbf{r}(\mathbf{s}) = \{\mathbf{s}, \mathbf{f}(\mathbf{s})\} .$$

Hence in the grid coordinates $\xi^1, \dots, \xi^n$ we obtain

$$\begin{aligned} g^{\boldsymbol{\xi}}_{ij} &= \mathbf{s}_{\xi^i} \cdot \mathbf{s}_{\xi^j} + \mathbf{f}_{\xi^i} \cdot \mathbf{f}_{\xi^j} , \quad i,j = 1, \dots, n , \\ \mathbf{r}_{\xi^m \xi^j} &= (\mathbf{s}_{\xi^m \xi^j}, \mathbf{f}_{\xi^m \xi^j}) , \quad j,m = 1, \dots, n , \\ \mathbf{r}_{\xi^k} &= (\mathbf{s}_{\xi^k}, \mathbf{f}_{\xi^k}) , \quad k = 1, \dots, n , \end{aligned}$$

where for a function $\mathbf{v}(\mathbf{s})$

$$\mathbf{v}_{\xi^i} = \frac{\partial \mathbf{v}[\mathbf{s}(\boldsymbol{\xi})]}{\partial \xi^i} , \quad \mathbf{v}_{\xi^m \xi^j} = \frac{\partial^2 \mathbf{v}[\mathbf{s}(\boldsymbol{\xi})]}{\partial \xi^m \partial \xi^j} , \quad j,m = 1, \dots, n .$$

Therefore, in this case, the grid system (6.5) is as follows:

$$g^{mj}_{\boldsymbol{\xi}} (\mathbf{s}_{\xi^m \xi^j} \cdot \mathbf{s}_{\xi^k} + \mathbf{f}_{\xi^m \xi^j} \cdot \mathbf{f}_{\xi^k}) = 0 , \quad j,k,m = 1, \dots, n , \tag{6.52}$$

and the multiplication of this system by $\partial \xi^k / \partial s^i$ yields the following inverted grid equations, with respect to $s^i(\boldsymbol{\xi})$, $i = 1, \dots, n$,

$$g^{mj}_{\boldsymbol{\xi}} \left(\frac{\partial^2 s^i}{\partial \xi^m \partial \xi^j} + \mathbf{f}_{\xi^m \xi^j} \cdot \mathbf{f}_{s^i} \right) = 0 , \quad i,j,m = 1, \dots, n , \tag{6.53}$$

where

$$\mathbf{f}_{s^i} = \frac{\partial \mathbf{f}[\mathbf{s}]}{\partial s^i} , \quad i = 1, \dots, n .$$

The coefficients $g^{mj}_{\boldsymbol{\xi}}$ in (6.53) are computed by the formula

$$g^{mj}_{\boldsymbol{\xi}} = g^{ik}_{\mathbf{s}} \frac{\partial \xi^m}{\partial s^i} \frac{\partial \xi^j}{\partial s^k}, \quad i,j,k,m = 1, \dots, n,$$

where the contravariant metric elements $g^{ij}_{\mathbf{s}}$ of the monitor surface S^{rn} are described by formula (5.68).

Using the definition of the basic operator $L^{\boldsymbol{\xi}}$, specified by (6.9), the system (6.53) is rewritten in the form

$$L^{\boldsymbol{\xi}}[s^i] + f^k_{s^i} L^{\boldsymbol{\xi}}[f^k] = 0\,, \quad i = 1, \ldots, n\,, \quad k = 1, \ldots, l\,. \tag{6.54}$$

Note, if $\xi^1, \ldots, \xi^n$ are the coordinates satisfying (5.4) and consequently (5.5) then for a function $v(\boldsymbol{\xi})$

$$L^{\boldsymbol{\xi}}[v] \equiv g^{mj}_{\boldsymbol{\xi}} \frac{\partial^2 v}{\partial \xi^m \partial \xi^j} = \frac{1}{\sqrt{g^{\boldsymbol{\xi}}}} \frac{\partial}{\partial \xi^j} \left(\sqrt{g^{\boldsymbol{\xi}}} g^{jm}_{\boldsymbol{\xi}} \frac{\partial v}{\partial \xi^m} \right) = \Delta_B[v]\,,$$
$$j, m = 1, \ldots, n\,,$$

so the system (6.54) also has the following equivalent forms

$$L^{\boldsymbol{\xi}}[s^i] + f^k_{s^i} \Delta_B[f^k] = 0\,, \quad i = 1, \ldots, n\,, \quad k = 1, \ldots, l\,. \tag{6.55}$$

Note $\Delta_B[f^p]$ is independent of a parametrization of S^{rn} over S^n therefore it can be computed in an arbitrary coordinate system, in particular, in the parametric coordinates $s^1, \ldots, s^n$.

If we consider in S^n new curvelinear coordinates $\boldsymbol{v} = (v^1, \ldots, v^n)$, $\boldsymbol{v} \in V^n$ connected with the Cartesian coordinates $\boldsymbol{s} = (s^1, \ldots, s^n)$ by the relations

$$\boldsymbol{v}(\boldsymbol{s}) : S^n \to V^n\,, \quad \boldsymbol{s}(\boldsymbol{v}) : V^n \to S^n\,,$$

then the parametrization of the monitor surface S^{rn} in the coordinates $v^1, \ldots, v^n$ is as follows:

$$\mathbf{r}_1(\boldsymbol{v}) : V^n \to R^{n+l}\,, \quad \mathbf{r}_1(\boldsymbol{v}) = \{\mathbf{s}(\boldsymbol{v}), \mathbf{f}[\mathbf{s}(\boldsymbol{v})]\}\,.$$

Therefore the grid equations equivalent to (6.53) and resolved with respect to the functions $v^i(\boldsymbol{\xi})$ can be obtained from (6.11). Using (6.2), (4.24), and (4.27) in (6.11) gives these equations the form

$$g^{ij}_{\boldsymbol{\xi}} \frac{\partial^2 v^p}{\partial \xi^i \partial \xi^j} = -g^{ij}_{\boldsymbol{v}} g^{pm}_{\boldsymbol{v}} (\boldsymbol{s}_{v^i v^j} \cdot \boldsymbol{s}_{v^m} + \boldsymbol{f}_{v^i v^j} \cdot \boldsymbol{f}_{v^m})\,,$$
$$i, j, m, p = 1, \ldots, n\,, \tag{6.56}$$

where

$$\boldsymbol{f}_{v^i v^j} = \frac{\partial^2 \boldsymbol{f}[\boldsymbol{s}(\boldsymbol{v})]}{\partial v^i \partial v^j}\,, \quad \boldsymbol{f}_{v^m} = \frac{\partial \boldsymbol{f}[\boldsymbol{s}(\boldsymbol{v})]}{\partial v^m}\,.$$

Availing us of the notion of the basic elliptic operator (6.9) this system is transformed to

$$L^{\boldsymbol{\xi}}[v^p] = -g^{mp}_{\boldsymbol{v}} (s^i_{v^m} L^{\boldsymbol{v}}[s^i] + f^k_{v^m} L^{\boldsymbol{v}}[f^k])\,,$$
$$i, m, p = 1, \ldots, n\,, \quad k = 1, \ldots, l\,.$$

Note, in general, equations (6.56) are more complicated than (6.53) and (6.54).

The system (6.53) is a special equivalent form of the general inverted Beltrami grid equations (6.11), valid for the metric of the monitor surface S^{rn} over the domain S^n. The general equations (6.11) for this metric are as follows:

$$g_{\boldsymbol{\xi}}^{mj}\frac{\partial^2 s^i}{\partial\xi^m\partial\xi^j} = \frac{1}{\sqrt{g^{\mathbf{s}}}}\frac{\partial}{\partial s^p}\left(\sqrt{g^{\mathbf{s}}}g_{\mathbf{s}}^{pi}\right), \quad i,j,m,p = 1,\dots,n, \tag{6.57}$$

where, in accordance with (5.68) and (5.69)

$$g_{\mathbf{s}}^{ij} = \delta_j^i - d_{ab}f_{s^i}^a f_{s^j}^b, \quad i,j = 1,\dots,n, \quad a,b = 1,\dots,l,$$

$$g^{\mathbf{s}} = \det(d^{ab}), \quad d^{ab} = \delta_b^a + f_{s^i}^a f_{s^i}^b, \quad i = 1,\dots,n, \quad a,b = 1,\dots,l,$$

(d_{ab}) is the matrix inverse to the matrix (d^{ab}).

Inverted Diffusion Equations

The inverted diffusion grid equations with respect to the metric of the monitor surface S^{rn} over a domain S^n are described by the formula (6.12) where the contravariant metric elements are computed by (5.68).

Fluxes-Sources Equations

Analogously the fluxes-sources grid equations for the monitor metric (6.87) are of the form (6.19) and (6.28) with the contravariant metric elements (5.68).

6.2.3 Surface Grid Equations for Some Special Monitor Metrics

Inverted Beltrami Equations

For generating a fixed grid on the surface S^{xn} represented by (5.1) there are usually applied the Beltrami equations in (5.4) with respect to the metric of S^{xn}, i.e. $g_{ij}^{\mathbf{s}} = g_{ij}^{xs}$, where

$$g_{ij}^{xs} = \mathbf{x}_{s^i}\cdot\mathbf{x}_{s^j}, \quad i,j = 1,\dots,n. \tag{6.58}$$

Consequently the inverted Beltrami equations (6.11) with respect to the metric (6.58) are as follows:

$$g_{\xi x}^{ij}\frac{\partial^2 s^k}{\partial\xi^i\partial\xi^j} = \frac{1}{\sqrt{g^{xs}}}\frac{\partial}{\partial s^m}(\sqrt{g^{xs}}g_{sx}^{mk}), \quad i,j,k,m = 1,\dots,n, \tag{6.59}$$

where $g^{xs} = det(g_{ij}^{xs})$, while g_{sx}^{km} and $g_{\xi x}^{ij}$ are contravariant metric elements of S^{xn} in the coordinates $s^1,\dots,s^n$ and $\xi^1,\dots,\xi^n$, respectively.

Surface grid equations analogous to the domain grid equations (6.33) are also readily obtained for the following monitor metric

$$g^{\mathbf{s}}_{ij} = v(\mathbf{s}) g^{xs}_{ij} \,, \quad i,j = 1,\dots,n \,, \quad v(\mathbf{s}) > 0 \,, \tag{6.60}$$

which is a particular case of the general form (5.58), namely, with $z(\mathbf{s}) = v(\mathbf{s}), F^k_i(\mathbf{s}) = 0$. Since, for the metric (6.60),

$$g^{ij}_{\mathbf{s}} = \frac{1}{v(\mathbf{s})} g^{ij}_{sx} \,, \quad i,j = 1,\dots,n \,,$$

$$g^{\mathbf{s}} = [v(\mathbf{s})]^n g^{xs} \,, \quad g^{xs} = \det(g^{xs}_{ij}) \,,$$

the original grid equations $\Delta_B[\xi^i] = 0$ in (5.4) with respect to $\xi^i(\mathrm{s})$, $i = 1,\dots,n$, in the metric (6.60) are equivalent to

$$\frac{\partial}{\partial s^j}\left\{\sqrt{g^{xs}}[v(\mathbf{s})]^{(n-2)/2} g^{jk}_{xs} \frac{\partial \xi^i}{\partial s^j}\right\} = 0 \,, \ i,j,k = 1,\dots,n \,.$$

These equations for $n = 2$ are equivalent to the Beltrami equations with respect to the metric g^{xs}_{ij} of the physical geometry S^{xn} regardless of $v(\mathbf{s})$. Consequently the inverted Beltrami equations in the metric (6.60) for $n = 2$ are the equations (6.59).

In accordance with (5.3) we find for the metric (6.60)

$$\Delta_B[s^i] = \frac{1}{v(\mathbf{s})} \Delta^x_B[s^i] + \frac{n-2}{2[v(\mathbf{s})]^2} \frac{\partial v}{\partial s^k} g^{ki}_{sx} \,, \quad i,k = 1,\dots,n \,,$$

where Δ^x_B is the operator of Beltrami with respect to the metric g^{xs}_{ij} of the geometry S^{xn} i.e. for a function $b(\mathbf{s})$

$$\Delta^x_B[b] = \frac{1}{\sqrt{g^x}} \frac{\partial}{\partial s^m}\left(\sqrt{g^x} g^{mk}_{sx} \frac{\partial b}{\partial s^k}\right) , \quad k,m = 1,\dots,n \,.$$

Similarly to (6.36) we obtain, using (6.11), the following system of the inverted Beltrami grid equations in the metric (6.60) resolved with respect to $s^i(\boldsymbol{\xi})$

$$\begin{aligned} g^{km}_{\xi x} \frac{\partial^2 s^i}{\partial \xi^k \partial \xi^m} &= \Delta^x_B[s^i] + \frac{1}{f(\mathbf{s})} \frac{\partial f}{\partial s^k} g^{ki}_{sx} \\ &= \Delta^x_B[s^i] + \frac{1}{f(\mathbf{s})} \frac{\partial v}{\partial \xi^k} g^{km}_{\xi x} \frac{\partial s^i}{\partial \xi^m} \,, \ i,k,m = 1,\cdots,n \,, \end{aligned} \tag{6.61}$$

where

$$f(\mathbf{s}) = [v(\mathbf{s})]^{(n-2)/2} \,.$$

In the same way as it was made for (6.37) one can derive a more compact form of the grid equations (6.61)

$$g^{km}_{\xi x} \frac{\partial}{\partial \xi^k}\left(\frac{1}{f(\mathbf{s})} \frac{\partial s^i}{\partial \xi^m}\right) = \frac{1}{f(\mathbf{s})} \Delta^x_B[s^i] \,, \quad i,k,m = 1,\cdots,n \,. \tag{6.62}$$

Inverted Diffusion and Beltrami Equations

It is easily found that the inverted diffusion grid equations (6.12) in the monitor metric (6.60) are of the form

$$\frac{w[\mathbf{s}(\boldsymbol{\xi})]}{v[\mathbf{s}(\boldsymbol{\xi})]} g_{\xi x}^{km} \frac{\partial^2 s^i}{\partial \xi^k \partial \xi^m} = \frac{\partial}{\partial s^j}\left[\frac{w(\mathbf{s})}{v(\mathbf{s})} g_{sx}^{ji}\right], \quad i,j,k,m = 1,\dots,n. \tag{6.63}$$

These equations become the inverted Beltrami equations in the metric (6.60) when $w(\mathbf{s}) = \sqrt{g^{\mathbf{s}}} = [v(\mathbf{s})]^{n/2}\sqrt{g^{xs}}$.

Fluxes-Sources Equations

For the diffusion fluxes-sources surface equations in the monitor metric (6.60) we, similarly to (6.47) and (6.48), readily obtain from (6.19) and (6.23) the following equations

$$J g_{\xi s}^{kk} \frac{\partial}{\partial s^j}\left[\frac{w(\mathbf{s})}{v(\mathbf{s})} g_{sx}^{ji}\right] = \frac{\partial}{\partial \xi^i}\left[\frac{w[\mathbf{s}(\boldsymbol{\xi})]}{v[\mathbf{s}(\boldsymbol{\xi})]} J\Big(g_{\xi x}^{ll}\frac{\partial \xi^j}{\partial s^i} - 2 g_{\xi x}^{jm}\frac{\partial \xi^m}{\partial s^i}\Big)\right], \tag{6.64}$$
$$i,j,k,l,m = 1,\dots,n,$$

and

$$J g_{\xi s}^{kk} \frac{\partial}{\partial s^j}\left[\frac{w(\mathbf{s})}{v(\mathbf{s})} g_{sx}^{ji}\right] = \frac{\partial}{\partial \xi^i}\left[\frac{w[\mathbf{s}(\boldsymbol{\xi})]}{v[\mathbf{s}(\boldsymbol{\xi})]} \Big(2\frac{g_{ki}^{xs}}{J g^{xs}}\frac{\partial s^k}{\partial \xi^i} - g_{\xi x}^{ll}\frac{\partial \xi^j}{\partial s^i}\Big)\right], \tag{6.65}$$
$$i,j,k,l = 1,2,$$

respectively.

Analogously the fluxes-sources surface grid equations (6.28) in the metric (6.60) are as follows:

$$\frac{\partial}{\partial \xi^j}\left(\frac{w(\mathbf{s})}{v(\mathbf{s})} g_{\xi x}^{ij}\right) = 0, \quad i,j = 1,\dots,n. \tag{6.66}$$

The equations (6.64), (6.65), and (6.66) after the substitution $\sqrt{g^{\mathbf{s}}}$ for $w(\mathbf{s})$ become the Beltrami fluxes-sources equations.

6.2.4 Surface Grid Equations with Respect to the Metric of a Monitor Surface

General Inverted Beltrami Equations

If the monitor surface S^{rn} over a surface S^{xn} represented by

$$\mathbf{x}(\mathbf{s}) : S^n \to R^{n+n_1}, \quad \mathbf{x} = (x^1,\dots,x^{n+n_1}), \tag{6.67}$$

is formed by a monitor function $\mathbf{f}(\mathbf{x}) = (f^1(\mathbf{x}), \ldots, f^l(\mathbf{x}))$, i.e. its parametrization in the coordinates $s^1, \ldots, s^n$ is as follows:

$$\mathbf{r}(\mathbf{s}) : S^n \to R^{n+n_1+l}, \quad \mathbf{r}(\mathbf{s}) = \{\mathbf{x}(\mathbf{s}), \mathbf{f}[\mathbf{x}(\mathbf{s})]\}, \tag{6.68}$$

then, in the grid coordinates $\xi^1, \ldots, \xi^n$,

$$\begin{aligned} g^{\boldsymbol{\xi}}_{ij} &= g^{x\xi}_{ij} + \mathbf{f}_{\xi^i} \cdot \mathbf{f}_{\xi^j}, \quad i,j = 1, \ldots, n, \\ g^{x\xi}_{ij} &= \mathbf{x}_{\xi^i} \cdot \mathbf{x}_{\xi^j}, \quad i,j = 1, \ldots, n, \\ \mathbf{r}_{\xi^m \xi^j} &= (\mathbf{x}_{\xi^m \xi^j}, \mathbf{f}_{\xi^m \xi^j}), \quad j,m = 1, \ldots, n, \\ \mathbf{r}_{\xi^i} &= (\mathbf{x}_{\xi^i}, \mathbf{f}_{\xi^i}), \quad i = 1, \ldots, n, \end{aligned} \tag{6.69}$$

where $g^{\boldsymbol{\xi}}_{ij}$ and $g^{x\xi}_{ij}$ are the covariant metric tensors of S^{rn} and S^{xn}, respectively. Consequently the grid equations (6.5) are as follows:

$$g^{mj}_{\boldsymbol{\xi}}(\mathbf{x}_{\xi^m \xi^j} \cdot \mathbf{x}_{\xi^k} + \mathbf{f}_{\xi^m \xi^j} \cdot \mathbf{f}_{\xi^k}) = 0, \quad j,k,m = 1, \ldots, n. \tag{6.70}$$

Since

$$\begin{aligned} \mathbf{x}_{\xi^m \xi^j} \cdot \mathbf{x}_{\xi^k} &= \frac{\partial}{\partial \xi^m}\Big(\frac{\partial \mathbf{x}}{\partial s^p}\frac{\partial s^p}{\partial \xi^j}\Big) \cdot \frac{\partial \mathbf{x}}{\partial s^a}\frac{\partial s^a}{\partial \xi^k} \\ &= \Big(\frac{\partial^2 s^p}{\partial \xi^m \partial \xi^j} g^{xs}_{ap} + \frac{\partial^2 \mathbf{x}}{\partial s^p \partial s^b} \cdot \frac{\partial \mathbf{x}}{\partial s^a}\frac{\partial s^p}{\partial \xi^j}\frac{\partial s^b}{\partial \xi^m}\Big)\frac{\partial s^a}{\partial \xi^k}, \\ &\quad a,b,j,k,m,p - 1, \ldots, n, \end{aligned}$$

we obtain, after multiplying the system (6.70) by $(\partial \xi^k / \partial s^b) g^{bi}_{sx}$,

$$\begin{aligned} &g^{mj}_{\boldsymbol{\xi}}(\mathbf{x}_{\xi^m \xi^j} \cdot \mathbf{x}_{\xi^k} \mid \mathbf{f}_{\xi^m \xi^j} \cdot \mathbf{f}_{\xi^k})\frac{\partial \xi^k}{\partial s^b} g^{bi}_{sx} \\ &= g^{mj}_{\boldsymbol{\xi}}\Big(\frac{\partial^2 s^i}{\partial \xi^m \partial \xi^j} + \mathbf{f}_{\xi^m \xi^j} \cdot \mathbf{f}_{s^b} g^{bi}_{sx}\Big) + g^{pj}_{\mathbf{s}}(\mathbf{x}_{s^p s^j} \cdot \mathbf{x}_{s^b}) g^{bi}_{sx} = 0, \\ &\qquad b,i,j,k,m = 1, \ldots, n. \end{aligned} \tag{6.71}$$

Taking into account that

$$(\mathbf{x}_{s^p s^j} \cdot \mathbf{x}_{s^b}) g^{bi}_{sx} = {}^{\mathbf{x}}\Upsilon^i_{pj}, \quad i,j,p = 1, \ldots, n,$$

where ${}^{\mathbf{x}}\Upsilon^i_{pj}$ is the Christoffel symbol of the second kind of the surface S^{xn} in the coordinates $s^1, \ldots, s^n$, we obtain from (6.71) that the grid system (6.70) resolved with respect to $s^i(\boldsymbol{\xi})$, $i = 1, \ldots, n$, has the following form

$$g^{mj}_{\boldsymbol{\xi}}\Big(\frac{\partial^2 s^i}{\partial \xi^m \partial \xi^j} + \mathbf{f}_{\xi^m \xi^j} \cdot \mathbf{f}_{s^b} g^{bi}_{sx}\Big) = -g^{pj}_{\mathbf{s}}{}^{\mathbf{x}}\Upsilon^i_{pj}, \quad b,i,j,m,p = 1, \ldots, n, \tag{6.72}$$

Analogously to (6.10) we get, in the grid coordinates $\xi^1,\dots,\xi^n$,

$$g^{mj}_{\boldsymbol{\xi}}\frac{\partial^2 \mathbf{f}}{\partial\xi^m\partial\xi^j}=L^{\boldsymbol{\xi}}[\mathbf{f}]=\Delta_B[\mathbf{f}]\,,\quad j,m=1,\dots,n\,,$$

therefore equations (6.72) are also expressed as

$$L^{\boldsymbol{\xi}}[s^i]+\Delta_B[f^k]\frac{\partial f^k}{\partial s^b}g^{bi}_{sx}=-g^{pj}_{\mathbf{s}}{}^{\mathbf{x}}\Upsilon^i_{pj}\,,\qquad b,i,j,p=1,\dots,n\,,\quad k=1,\dots,l\,. \tag{6.73}$$

Simplified Equations

Equations (6.72) and (6.73) are more complicated in comparison with the equations (6.53) prescribed for generating grids in domains. So they may be less malleable for the implementation into numerical codes in the case when the process of grid generation on a surface S^{xn} is coupled with the computation of the points of this surface and the monitor function $\mathbf{f}(\mathbf{x})$ since the quantities $\Delta_B[f^k]$ and ${}^{\mathbf{x}}\Upsilon^i_{pj}$ in (6.72) and (6.73) include the second derivatives with respect to s^i, $i=1,\dots,n$ of the function $\mathbf{f}[\mathbf{x}(\mathbf{s})]$ and the surface parametrization $\mathbf{x}(\mathbf{s})$. However, we can come to the equations of the simpler form (6.53) for generating grids on the surface S^{xn} if we consider as a monitor function over S^{xn} the following function

$$\mathbf{f}_1(\mathbf{s})=\{\mathbf{s},\mathbf{f}[\mathbf{x}(\mathbf{s})]\}\,. \tag{6.74}$$

The monitor surface $S^{r_1 n}$ over S^{xn} with this monitor function $\mathbf{f}_1(\mathbf{s})$ is represented by the parametrization

$$\mathbf{r}_1(\mathbf{s}):S^n\to S^{r_1 n}\,,\quad \mathbf{r}_1(\mathbf{s})=\{\mathbf{x}(\mathbf{s}),\mathbf{s},\mathbf{f}[\mathbf{x}(\mathbf{s})]\}\,. \tag{6.75}$$

Note the monitor surface $S^{r_2 n}$ over S^n with the monitor function $\mathbf{f}_2(\mathbf{s})=\{\mathbf{x}(\mathbf{s}),\mathbf{f}[\mathbf{x}(\mathbf{s})]\}$, and represented correspondently by

$$\mathbf{r}_2(\mathbf{s}):S^n\to S^{r_2 n}\,,\quad \mathbf{r}_2(\mathbf{s})=\{\mathbf{s},\mathbf{x}(\mathbf{s}),\mathbf{f}[\mathbf{x}(\mathbf{s})]\}\,, \tag{6.76}$$

has the same metric tensor as the surface $S^{r_1 n}$. Hence the equations for generating the intermediate transformation $\mathbf{s}(\boldsymbol{\xi}):\Xi^n\to S^n$ with these monitor surfaces are identical and have, in accordance with (6.53), the following form

$$g^{mj}_{\boldsymbol{\xi}}\Big(\frac{\partial^2 s^i}{\partial\xi^m\partial\xi^j}+\mathbf{x}_{\xi^m\xi^j}\cdot\mathbf{x}_{s^i}+\mathbf{f}_{\xi^m\xi^j}\cdot\mathbf{f}_{s^i}\Big)=0\,,\quad i,j,m=1,\dots,m\,, \tag{6.77}$$

where $g^{mj}_{\boldsymbol{\xi}}$ are the contravariant metric elements of the monitor surface $S^{r_2 n}$ in the grid coordinates $\xi^1,\dots,\xi^n$. Note, for the covariant metric tensor of $S^{r_2 n}$ in these coordinates we have

$$g^{\boldsymbol{\xi}}_{ij}=\frac{\partial\mathbf{s}(\boldsymbol{\xi})}{\partial\xi^i}\cdot\frac{\partial\mathbf{s}(\boldsymbol{\xi})}{\partial\xi^j}+\frac{\partial\mathbf{x}[\mathbf{s}(\boldsymbol{\xi})]}{\partial\xi^i}\cdot\frac{\partial\mathbf{x}[\mathbf{s}(\boldsymbol{\xi})]}{\partial\xi^j}+\frac{\partial\mathbf{f}\{\mathbf{x}[\mathbf{s}(\boldsymbol{\xi})]\}}{\partial\xi^i}\cdot\frac{\partial\mathbf{f}\{\mathbf{x}[\mathbf{s}(\boldsymbol{\xi})]\}}{\partial\xi^j}\,,\qquad i,j=1,\dots,n\,.$$

Using the notion of the basic elliptic operator (6.9) the equations (6.77) also have the form

$$L^{\boldsymbol{\xi}}[s^i] + x^p_{s^i} L^{\boldsymbol{\xi}}[x^p] + f^k_{s^i} L^{\boldsymbol{\xi}}[f^k] = 0 \, , \qquad i = 1, \dots, n \, , \quad p = 1, \dots, n + n_1 \, , \quad k = 1, \dots, l \, . \tag{6.78}$$

Equations (6.77) or (6.78) with respect to the components $s^i(\boldsymbol{\xi})$, $i = 1, \dots, n$, of the intermediate transformation $\mathbf{s}(\boldsymbol{\xi})$ include the first derivatives only of the functions $\mathbf{x}(\mathbf{s})$ and $\mathbf{f}[\mathbf{x}(\mathbf{s})]$ in s^i, $i = 1, \dots, n$, , therefore they are more convenient for the implementation into numerical codes in comparison with the equations (6.72) and (6.73). Remind, the grid in S^{xn} is obtained by mapping with $\mathbf{x}(\mathbf{s})$ a grid in S^n generated through the values of the intermediate transformation $\mathbf{s}(\boldsymbol{\xi})$ at the points of a reference grid.

Instead of (6.76) we can also use the parametrization $\mathbf{r}_1(\mathbf{s})$ in the form

$$\mathbf{r}_1(\mathbf{s}) = [N\mathbf{x}(\mathbf{s}), \mathbf{s}, N\mathbf{f}(\mathbf{s})]$$

where N is some positive constant. With respect to this parametrization the equations (6.77) are as follows:

$$g^{mj}_{\boldsymbol{\xi}} \Big(\frac{\partial^2 s^i}{\partial \xi^m \partial \xi^j} + N^2 \mathbf{x}_{\xi^m \xi^j} \cdot \mathbf{x}_{s^i} + N^2 \mathbf{f}_{\xi^m \xi^j} \cdot \mathbf{f}_{s^i} \Big) = 0 \, , \ i, j, m = 1, \dots, m \, . \tag{6.79}$$

It is obvious that the difference between the solutions of (6.79) and (6.72) is lessened when N is increased.

Note that, similarly to (6.41), the equations (6.79) also have the following form

$$L^{\boldsymbol{\xi}}[s^i] + N^2 L^{\boldsymbol{\xi}}[x^j] \frac{\partial x^j}{\partial s^i} + N^2 L^{\boldsymbol{\xi}}[f^k] \frac{\partial f^k}{\partial s^i} = 0 \, , \qquad i = 1, \dots, n \, , \quad j = 1, \dots, n + n_1 \, , \quad k = 1, \dots, l \, . \tag{6.80}$$

Inverted Diffusion Equations

In general case when $w(\mathbf{s}) \neq \sqrt{g^{\mathbf{s}}}$ the inverted diffusion grid equations (6.12) in the metric of the monitor surface S^{rn} do not have the form (6.53). Their expressions are computed by substituting in (6.12) the formulas (5.67) for the contravariant elements of the monitor metric of S^{rn}.

Fluxes-Sources Equation

Similarly to the inverted diffusion grid equations, the fluxes-sources grid equations in the metric of the monitor surface S^{rn} over the physical surface S^{xn} are obtained from (6.19), (6.23), and (6.28) where the contravariant metric elements $g^{ij}_{\mathbf{s}}$ and Jacobian $g^{\mathbf{s}} = det(g^{\mathbf{s}}_{ij})$ are computed by the formulas (5.67) and (5.69), respectively.

6.3 Role of the Mean Curvature

This section shows how the mean curvature, which is one of the abstract geometric characteristics reviewed in Chap. 4, is involved in grid generation technologies.

6.3.1 Mean Curvature and Inverted Beltrami Grid Equations

General Formulas

In the case of a regular surface $S^{rn} \subset R^{n+1}$ defined by the parametrization

$$\mathbf{r}(\mathbf{s}) : S^n \to R^{n+1}, \quad \mathbf{r}(\mathbf{s}) = [r^1(\mathbf{s}), \dots, r^{n+1}(\mathbf{s})] ,$$

with its natural metric elements $g^{\mathbf{s}}_{ij}$ specified in the coordinates $s^1, \dots, s^n$ as

$$g^{\mathbf{s}}_{ij} = \mathbf{r}_{s^i} \cdot \mathbf{r}_{s^j} , \quad i, j = 1, \dots, n ,$$

we have, from (6.2), (4.22), and (4.24),

$$\Delta_B[s^i] = -g^{kj}_{\mathbf{s}} \Upsilon^i_{kj} = -g^{kj}_{\mathbf{s}} g^{im}_{\mathbf{s}} (\mathbf{r}_{s^k s^j} \cdot \mathbf{r}_{s^m}) , \quad i, j, k, m = 1, \dots, n . \tag{6.81}$$

Now remind that the quantity

$$K_m = \frac{1}{n} g^{kj}_{\mathbf{s}} \mathbf{r}_{s^k s^j} \cdot \mathbf{n} , \quad j, k = 1, \dots, n , \tag{6.82}$$

where $\mathbf{n}$ is an $(n+1)$-dimensional unit normal vector to S^{rn} in R^{n+1}, is the mean curvature of the monitor surface S^{rn} with respect to this normal $\mathbf{n}$.

Note the mean curvature with respect to the same normal is an invariant of parametrizations of S^{rn}.

It appears that the mean curvature of S^{rn} is connected with the parametrization $\mathbf{r}(\mathbf{s}) : S^n \to S^{rn} \subset R^{n+1}$ by the following relation

$$\Delta_B[\mathbf{r}] = n K_m \mathbf{n} , \tag{6.83}$$

where $\mathbf{n}$ is the same vector used in (6.82). Indeed we have

$$\Delta_B[\mathbf{r}] = \frac{1}{\sqrt{g^{\mathbf{s}}}} \frac{\partial}{\partial s^j} (\sqrt{g^{\mathbf{s}}} g^{gi}_{\mathbf{s}} \mathbf{r}_{s^i}) = g^{ji}_{sr} \mathbf{r}_{s^j s^i} + \Delta_B[s^k] \mathbf{r}_{s^k} , \quad i, j, k = 1, \cdots, n , \tag{6.84}$$

and applying the relation (6.81) to the last item of (6.84) we obtain

$$\Delta_B[\mathbf{r}] = g^{ji}_{sr} [\mathbf{r}_{s^j s^i} - g^{mk}_{sr} (\mathbf{r}_{s^j s^i} \cdot \mathbf{r}_{s^m}) \mathbf{r}_{s^k}] , \quad i, j, k, m = 1, \cdots, n . \tag{6.85}$$

Now taking into account equation (2.6), yielding that the expansion of the vector $\mathbf{r}_{s^i s^j}$ in the basis $(\mathbf{r}_{s^1}, \cdots, \mathbf{r}_{s^n}, \mathbf{n})$ is expressed as

$$\mathbf{r}_{s^i s^j} = g_{sr}^{mk}(\mathbf{r}_{s^i s^j} \cdot \mathbf{r}_{s^m})\mathbf{r}_{s^k} + (\mathbf{r}_{s^i s^j} \cdot \mathbf{n})\mathbf{n} ,$$

we find, applying this expansion to (6.85) and using (6.82),

$$\Delta_B[\mathbf{r}] = g_{sr}^{ji}(\mathbf{r}_{s^j s^i} \cdot \mathbf{n})\mathbf{n} = nK_m\mathbf{n} , \quad i,j = 1,\cdots,n ,$$

i.e. equation (6.83) is valid. Consequently from the relation (6.84) we obtain

$$g_{sr}^{ji}\mathbf{r}_{s^j s^i} + \Delta_B[s^k]\mathbf{r}_{s^k} = nK_m\mathbf{n} , \quad i,j,k = 1,\cdots,n . \tag{6.86}$$

This formula is valid in arbitrary coordinates, in particular, in the grid coordinates $\xi^1,\cdots,\xi^n$ satisfying the Beltrami equations in (5.4) it results in (6.8).

Application to a Monitor Surface over a Domain

Let the parametric transformation $\mathbf{r}(\mathbf{s})$ for S^{rn} be specified as $\mathbf{r}(\mathbf{s}) = [\mathbf{s}, f(\mathbf{s})]$ with a scalar-valued monitor function $f(\mathbf{s})$. Then $S^{rn} \subset R^{n+1}$ is a monitor surface over S^n, whose covariant metric tensor in the coordinates $s^1,\ldots,s^n$ is computed as

$$g_{ij}^{\mathbf{s}} = \mathbf{r}_{s^i} \cdot \mathbf{r}_{s^j} = \delta_j^i + f_{s^i} f_{s^j} , \quad i,j = 1,\cdots,n . \tag{6.87}$$

Taking advantage of (5.68) for $l = 1$ we find

$$g_{\mathbf{s}}^{im} f_{s^m} = \Big(\delta_m^i - \frac{1}{g^{\mathbf{s}}} f_{s^i} f_{s^m}\Big) f_{s^m} = \Big(1 - \frac{1}{g^{\mathbf{s}}} f_{s^m} f_{s^m}\Big) f_{s^i} = \frac{1}{g^{\mathbf{s}}} f_{s^i} , \qquad i,m = 1,\ldots,n . \tag{6.88}$$

since in accordance with (5.69)

$$g^{\mathbf{s}} = 1 + f_{s^m} f_{s^m}, \quad m = 1,\ldots,n.$$

So, using (6.2), we conclude that in the metric (6.87)

$$\Delta_B[s^i] = -g_{\mathbf{s}}^{kj} g_{\mathbf{s}}^{im} f_{s^k s^j} f_{s^m} = -\frac{1}{g^{\mathbf{s}}} g_{\mathbf{s}}^{kj} f_{s^k s^j} f_{s^i} , \ i,j,k,m = 1,\ldots,n . \tag{6.89}$$

For the parametrization $\mathbf{r}(\mathbf{s}) = [\mathbf{s}, f(\mathbf{s})]$ we find

$$\mathbf{r}_{s^i} = (\underbrace{0,\ldots,0}_{i-1},1,\underbrace{0,\ldots,0}_{n-i}, f_{s^i}) , \quad i = 1,\ldots,n ,$$

so it is obvious that for the unit normal $\mathbf{n}$ we can take

$$\mathbf{n} = \frac{1}{\sqrt{g^{\mathbf{s}}}}(-f_{s^1},\ldots,-f_{s^n},1) . \tag{6.90}$$

For the above expressions for $\mathbf{r}(\mathbf{s})$ and $\mathbf{n}$

$$\mathbf{r}_{s^k s^j} \cdot \mathbf{n} = \frac{1}{\sqrt{g^{\mathrm{s}}}} f_{s^k s^j} , \quad j,k = 1,\dots,n ,$$

so, in accordance with (6.82), the mean curvature of this monitor surface S^{rn} with respect to the normal (6.90) is computed by the following formula

$$K_m = \frac{1}{n\sqrt{g^{\mathrm{s}}}} g_{\mathrm{s}}^{kj} f_{s^k s^j} , \quad j,k = 1,\dots,n . \tag{6.91}$$

Thus in the metric (6.87) equations (6.89), as well as (6.2), have the following form

$$\Delta_B[s^i] = -\frac{n}{\sqrt{g^{\mathrm{s}}}} K_m f_{s^i} , \quad i = 1,\dots,n . \tag{6.92}$$

Consequently the inverted Beltrami grid equations (6.11) in application to a domain $X^n = S^n$ with a scalar monitor function $f(\mathbf{x})$ are expressed through the mean curvature of S^{rn} with respect to the normal (6.90) as follows:

$$g_{\boldsymbol{\xi}}^{ij} \frac{\partial^2 s^k}{\partial \xi^i \partial \xi^j} = -\frac{n}{\sqrt{g^{\mathrm{s}}}} K_m f_{s^k} , \quad i,j,k = 1,\dots,n . \tag{6.93}$$

So if $K_m = 0$, i.e. the monitor surface S^{rn} is a minimal surface, then equations (6.93) have the simple form of equations (6.31) (with the zero right-hand part). Some results related to the construction of minimal n-dimensional surfaces, providing such equations, can be found in the monograph by Fomenko and Thi (1991).

Notice, another form for the expression of the mean curvature of the monitor surface with the monitor metric (6.87) can be computed using the $(n+1)$th component of (6.83). Thus we have, with respect to the normal (6.90),

$$K_m = \frac{1}{n}\sqrt{g^{\mathrm{s}}}\,\Delta_B[f] . \tag{6.94}$$

Availing us of (6.88) we find, in the case of the parametrization $\mathbf{r}(\mathbf{s}) = [\mathbf{s}, f(\mathbf{s})]$,

$$\Delta_B[f] = \frac{1}{\sqrt{g^{\mathrm{s}}}} \frac{\partial}{\partial s^j}\Big(\sqrt{g^{\mathrm{s}}}\, g_{\mathrm{s}}^{jk} \frac{\partial f}{\partial s^k}\Big) = \frac{1}{\sqrt{g^{\mathrm{s}}}} \frac{\partial}{\partial s^j}\Big(\frac{1}{\sqrt{g^{\mathrm{s}}}} \frac{\partial f}{\partial s^j}\Big) , \quad j,k = 1,\dots,n .$$

Thus from (6.94)

$$K_m = \frac{1}{n} \frac{\partial}{\partial s^j}\Big(\frac{1}{\sqrt{g^{\mathrm{s}}}} \frac{\partial f}{\partial s^j}\Big) , \quad j = 1,\dots,n , \tag{6.95}$$

with respect to the normal (6.90). Since the points of the monitor surface $S^{rn} \subset R^{n+1}$ in the coordinates $s^1,\dots,s^n,s^{n+1}$ can be found from the solution of the equation

$$s^{n+1} - f(s^1,\dots,s^n) = 0 ,$$

the formula (6.95) is a particular case of (4.111) for

$$\varphi(s^1, \dots, s^n, s^{n+1}) \equiv s^{n+1} - f(s^1, \dots, s^n)$$

and

$$g_{ij}^{rs} = \delta_j^i, \quad i, j = 1, \dots, n+1 .$$

6.3.2 Mean Curvature and Control of Grid Clustering

Fundamental Formula

It is well-known that when S^{xn} is a domain with the Euclidean metric tensor then the operator of Beltrami in this metric is the Laplace operator and the spacing between $(n-1)$-dimensional grid hypersurfaces $\xi^i = \text{const}$ in S^{xn} related to the solution of the Laplace equations, for both $n = 2$ and $n = 3$, increases near a boundary convex segment and, conversely, the spacing decreases when the boundary segment in concave (see Fig. 6.1). It is shown further that similar facts are also valid for the grid hypersurfaces related to the solution of the Beltrami equations in arbitrary n-dimensional regular surfaces S^{xn}.

Throughout this subsection we assume i_0, $1 \le i_0 \le n$ as a fixed index, i.e. the summation in the formulas with i_0 repeated is not carried out over this index.

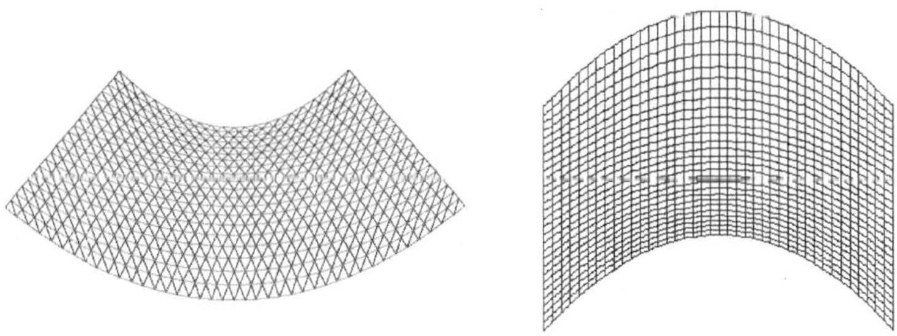

Fig. 6.1. Grid rarefaction (clustering) near convex (concave) boundary segments

Relative Spacing Between Coordinate Surfaces

Let $s^1, \dots, s^n$ be an arbitrary local coordinate system of an n-dimensional regular surface S^{xn} represented by (5.1). We first consider in the surface S^{xn} a family of the coordinate $(n-1)$-dimensional hypersurfaces $s^{i_0} = const.$ Analogously to the specification by (4.12) we can readily find that the vector

$\mathbf{n}_{i_0}$ lying in the tangent n-dimensional plane to S^{xn} and which is expressed in the form

$$\mathbf{n}_{i_0} = \frac{1}{\sqrt{g_{sx}^{i_0 i_0}}} g_{sx}^{i_0 j} \mathbf{x}_{s^j} , \quad j = 1, \dots , n , \tag{6.96}$$

is a unit normal to the coordinate hypersurface $s^{i_0} = c_0$. Here g_{sx}^{ij} is the (ij)th element of the contravariant metric tensor of S^{xn} in the coordinates $s^1, \dots , s^n$ and the values of g_{sx}^{ij} and $\mathbf{x}_{s^j}$ are considered at the points of S^n for which $s^{i_0} = c_0$. Indeed

$$\mathbf{n}_{i_0} \cdot \mathbf{x}_{s^k} = \frac{1}{\sqrt{g_{sx}^{i_0 i_0}}} g_{sx}^{i_0 j} \mathbf{x}_{s^j} \cdot \mathbf{x}_{s^k} = \frac{1}{\sqrt{g_{sx}^{i_0 i_0}}} g_{sx}^{i_0 j} g_{jk}^{xs} = \frac{1}{\sqrt{g_{sx}^{i_0 i_0}}} \delta_k^{i_0} ,$$
$$j, k = 1, \dots , n ,$$

i.e. $\mathbf{n}_{i_0}$ is orthogonal to the coordinate surface $s^{i_0} = c_0$ in S^{xn}. Further

$$\begin{aligned} \mathbf{n}_{i_0} \cdot \mathbf{n}_{i_0} &= \frac{1}{\sqrt{g_{sx}^{i_0 i_0}}} g_{sx}^{i_0 j} \mathbf{x}_{s^j} \cdot \frac{1}{\sqrt{g_{sx}^{i_0 i_0}}} g_{sx}^{i_0 k} \mathbf{x}_{s^k} \\ &= \frac{1}{g_{sx}^{i_0 i_0}} g_{sx}^{i_0 j} g_{sx}^{i_0 k} g_{jk}^{xs} = 1 , \quad j, k = 1, \dots , n , \end{aligned}$$

i.e. $\mathbf{n}_{i_0}$ is a unit vector.

Let us denote by l_h the distance between a point on the hypersurface $s^{i_0} = c_0$ and the nearest point on the hypersurface $s^{i_0} = c_0 + h$ in S^{xn}. We have

$$\begin{aligned} l_h &= (\mathbf{n}_{i_0} \cdot \mathbf{x}_{s^{i_0}})h + O(h^2) = h \frac{1}{\sqrt{g_{sx}^{i_0 i_0}}} g_{sx}^{i_0 j} \mathbf{x}_{s^j} \cdot \mathbf{x}_{s^{i_0}} + O(h^2) \\ &= h \frac{1}{\sqrt{g_{sx}^{i_0 i_0}}} g_{sx}^{i_0 j} g_{j i_0}^{xs} + O(h^2) = h \frac{1}{\sqrt{g_{sx}^{i_0 i_0}}} + O(h^2) , \quad j = 1, \dots , n . \end{aligned}$$

So the quantity $1/\sqrt{g_{sx}^{i_0 i_0}}$ with i_0 fixed reflects the relative spacing between the corresponding points on the coordinate hypersurfaces $s^{i_0} = c_0 + h$ and $s^i = c_0$ on S^{xn} (see Fig. 6.2 for $n = 2$).

Rate of Change of the Relative Spacing

The vector $\mathbf{n}_{i_0}$ is orthogonal to the coordinate hypersurface $s^{i_0} = c_0$ in S^{xn} and therefore the derivative of $1/\sqrt{g_{sx}^{i_0 i_0}}$ in the $\mathbf{n}_{i_0}$ direction is the rate of change of the relative spacing between the coordinate hypersurfaces $s^{i_0} =$ *const*. Using (6.96) we obtain

$$\begin{aligned} \frac{\mathrm{d}}{\mathrm{d}\mathbf{n}_{i_0}} \Big(\frac{1}{\sqrt{g_{sx}^{i_0 i_0}}} \Big) &= \frac{1}{\sqrt{g_{sx}^{i_0 i_0}}} g_{sx}^{i_0 j} \frac{\partial}{\partial s^j} \Big(\frac{1}{\sqrt{g_{sx}^{i_0 i_0}}} \Big) \\ &= \frac{1}{\sqrt{g_{sx}^{i_0 i_0}}} \nabla \Big(s^{i_0}, \frac{1}{\sqrt{g_{sx}^{i_0 i_0}}} \Big) , \quad j = 1, \dots , n , \end{aligned} \tag{6.97}$$

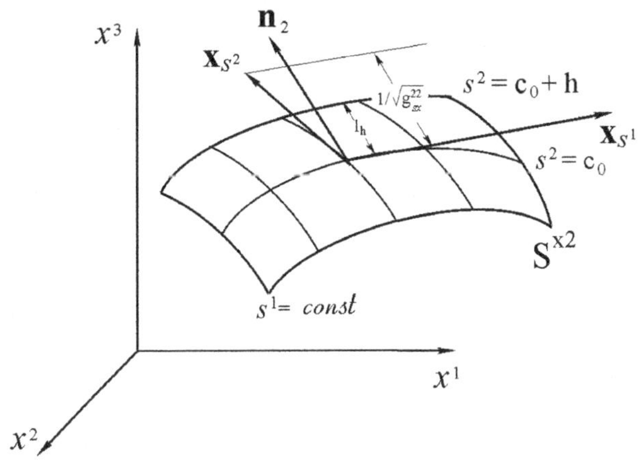

Fig. 6.2. Spacing of the coordinate lines $s^2 = const$ on the regular surface

where $\nabla(\, , \,)$ is the Beltrami's mixed differential parameter. On the other hand

$$\begin{aligned}\frac{\mathrm{d}}{\mathrm{d}\mathbf{n}_{i_0}}\left(\frac{1}{\sqrt{g_{sx}^{i_0 i_0}}}\right) &= -\frac{1}{2\sqrt{(g_{sx}^{i_0 i_0})^3}}\frac{\mathrm{d}}{\mathrm{d}\mathbf{n}_{i_0}} g_{sx}^{i_0 i_0} \\ &= -\frac{1}{2(g_{sx}^{i_0 i_0})^2} g_{sx}^{i_0 j}\frac{\partial}{\partial s^j} g_{sx}^{i_0 i_0} , \quad j = 1, \dots , n .\end{aligned} \tag{6.98}$$

Availing us of (4.25) in this equation yields

$$\frac{\mathrm{d}}{\mathrm{d}\mathbf{n}_{i_0}}\left(\frac{1}{\sqrt{g_{sx}^{i_0 i_0}}}\right) = \frac{1}{(g_{sx}^{i_0 i_0})^2} g_{sx}^{i_0 l} g_{sx}^{i_0 j} \Upsilon_{lj}^{i_0} , \quad j, l = 1, \dots , n . \tag{6.99}$$

Note this equation is valid for an arbitrary coordinate system $s^1, \dots , s^n$.

Relations to the Mean Curvature

In order to connect the rate of change of the relative spacing of the coordinate hypersurfaces with geometrical characteristics we need to consider the following general situation in the theory of matrices. Let (a_{ij}), $i, j = 1, \dots , n$, $n \geq 2$, be a nondegenerate symmetric matrix of rank n and (a^{ij}), $i, j = 1, \dots , n$, be its inverse matrix. Let $a^{i_0 i_0} \neq 0$ for some fixed index i_0, $1 \leq i_0 \leq n$. Define a matrix (b^{ij}) where

$$b^{ij} = \frac{1}{a^{i_0 i_0}}(a^{i_0 i_0} a^{ij} - a^{i_0 i} a^{i_0 j}) , \quad i, j = 1, \dots , n . \tag{6.100}$$

Let $(a_{ij}^{i_0})$ and $(b_{i_0}^{ij})$ be the $(n-1) \times (n-1)$ matrices obtained by deleting the i_0th row and i_0th column of the matrices (a_{ij}) and (b^{ij}), respectively.

Theorem 5. *The matrix* $(b^{ij}_{i_0})$ *is the inverse of* $(a^{i_0}_{ij})$.

Proof. It is sufficient to show that

$$a^{i_0}_{ij} b^{jl}_{i_0} = \delta^i_l \,, \quad i,j,l = 1,\dots,n \,, \text{ and } i,j,l \neq i_0 \,. \tag{6.101}$$

Here and further the entries $i = 1,\dots,n$, and $i \neq k$ mean $i = 1,\dots,k-1,k+1,\dots,n$.

From (6.100) we readily see

$$b^{i_0 k} = b^{k i_0} = 0 \,, \quad k = 1,\dots,n \,,$$

therefore

$$a^{i_0}_{ij} b^{jl}_{i_0} = a_{im} b^{ml} \,, \quad i,j,l,m = 1,\dots,n \,, \text{ and } i,j,l \neq i_0 \,.$$

Since (6.100)

$$\begin{aligned} a_{im} b^{ml} &= a_{im} a^{ml} - \frac{a_{im}}{a^{i_0 i_0}} a^{i_0 m} a^{i_0 l} \\ &= \delta^i_l - \frac{\delta^i_{i_0}}{a^{i_0 i_0}} a^{i_0 l} = \delta^i_l \,, \quad i,l,m = 1,\dots,n \,, \text{ and } i,l \neq i_0 \,, \end{aligned}$$

i.e. (6.101) is valid. This proves the theorem. □

Now we apply the matrix consideration given above to the matrix (g^{xs}_{ij}) which is the covariant metric tensor of the regular surface S^{xn} in the coordinates $s^1,\dots,s^n$.

Designate by $(g^{i_0}_{ij})$ the matrix obtained by deleting the i_0th row and i_0th column of (g^{xs}_{ij}). It is clear that $(g^{i_0}_{ij})$ is the covariant metric tensor of the coordinate hypersurface $s^{i_0} = c_0$ in the coordinates $s^1,\dots,s^{i_0-1},s^{i_0+1},\dots,s^n$. The matrix which is inverse to $(g^{i_0}_{ij})$ is the contravariant metric tensor of this coordinate hypersurface $s^{i_0} = c_0$ in the same coordinates $s^1,\dots,s^{i_0-1}$, $s^{i_0+1},\dots,s^n$. Let this matrix be designated by $(g^{ij}_{i_0})$. Since

$$g^{i_0 i_0}_{sx} = \det(g^{i_0}_{ij})/g^{xs} \neq 0 \,,$$

where $g^{xs} = \det(g^{xs}_{ij})$, we find, availing us of theorem 5,

$$g^{ij}_{i_0} = \frac{1}{g^{i_0 i_0}_{sx}} (g^{i_0 i_0}_{sx} g^{ij}_{sx} - g^{i_0 i}_{sx} g^{i_0 j}_{sx}) \,, \quad i,j = 1,\dots,n \,, \text{ and } i,j \neq i_0 \,.$$

Therefore

$$g^{i_0 i}_{sx} g^{i_0 j}_{sx} = g^{i_0 i_0}_{sx} (g^{ij}_{sx} - g^{ij}_{i_0}) \,, \quad i,j = 1,\dots,n \,, \text{ and } i,j \neq i_0 \,. \tag{6.102}$$

Since

$$g^{i_0 l}_{sx} g^{i_0 j}_{sx} \Upsilon^{i_0}_{lj} = g^{i_0 k}_{sx} g^{i_0 p}_{sx} \Upsilon^{i_0}_{kp} + 2 g^{i_0 i_0}_{sx} g^{i_0 j}_{sx} \Upsilon^{i_0}_{i_0 j} - g^{i_0 i_0}_{sx} g^{i_0 i_0}_{sx} \Upsilon^{i_0}_{i_0 i_0} ,$$
$$j, k, l, p = 1, \dots, n , \text{ and } k, p \neq i_0 ,$$

we obtain, using (6.102),

$$g^{i_0 l}_{sx} g^{i_0 j}_{sx} \Upsilon^{i_0}_{lj} = g^{i_0 i_0}_{sx} (g^{kp}_{sx} - g^{kp}_{i_0}) \Upsilon^{i_0}_{kp} + 2 g^{i_0 i_0}_{sx} g^{i_0 j}_{sx} \Upsilon^{i_0}_{i_0 j} - g^{i_0 i_0}_{sx} g^{i_0 i_0}_{sx} \Upsilon^{i_0}_{i_0 i_0}$$
$$= g^{i_0 i_0}_{sx} g^{lj}_{sx} \Upsilon^{i_0}_{lj} - g^{i_0 i_0}_{sx} g^{kp}_{i_0} \Upsilon^{i_0}_{kp} , \qquad (6.103)$$
$$j, k, l, p = 1, \dots, n , \text{ and } k, p \neq i_0 .$$

Now we note that from (4.28)

$$\Upsilon^{i_0}_{kp} = g^{i_0 l}_{sx} [\mathbf{x}_{s^k s^p} \cdot \mathbf{x}_{s^l}] , \quad k, l, p = 1, \dots, n ,$$

so, since (6.96),

$$\Upsilon^{i_0}_{kp} = \sqrt{g^{i_0 i_0}_{sx}} \mathbf{x}_{s^k s^p} \cdot \mathbf{n}_{i^0} , \quad k, p = 1, \dots, n ,$$

where $\mathbf{n}_{i^0}$ is the unit normal to the coordinate hypersurface $s^{i_0} = c_0$ in S^{xn}. Thus, in the designation (4.79),

$$\Upsilon^{i_0}_{kp} = \sqrt{g^{i_0 i_0}_{sx}} b_{kp} , \quad k, p = 1, \dots, n , \text{ and } k, p \neq i_0 , \qquad (6.104)$$

where

$$b_{kp} = \mathbf{x}_{s^k s^p} \cdot \mathbf{n}_{i^0} , \quad k, p = 1, \dots, n , \text{ and } k, p \neq i_0 ,$$

is the (kp)th element of the second fundamental form of the coordinate hypersurface $s^{i_0} = c_0$ in S^{xn}. Therefore, using (4.81) and (6.104), we find

$$g^{kp}_{i_0} \Upsilon^{i_0}_{kp} = \sqrt{g^{i_0 i_0}_{sx}} g^{kp}_{i_0} b_{kp} = (n-1) \sqrt{g^{i_0 i_0}_{sx}} K_m(s^{i_0}) ,$$
$$k, p = 1, \dots, n , \text{ and } k, p \neq i_0 , \qquad (6.105)$$

where $K_m(s^{i_0})$ is the mean curvature of the coordinate hypersurface $s^{i_0} = c_0$ in S^{xn} with respect to the normal $\mathbf{n}_{i_0}$. Further, from (6.2), we obtain

$$g^{lj}_{sx} \Upsilon^{i_0}_{lj} = -\Delta^x_B [s^{i_0}] , \quad j, l = 1, \dots, n ,$$

where Δ^x_B is the operator of Beltrami in the metric of S^{xn}. Substituting this equation and (6.105) in (6.103) gives

$$g^{i_0 l}_{sx} g^{i_0 j}_{sx} \Upsilon^{i_0}_{lj} = -g^{i_0 i_0}_{sx} \Delta^x_B [s^{i_0}] - (n-1)(g^{i_0 i_0}_{sx})^{3/2} K_m(s^{i_0}) , \quad j, l = 1, \dots, n .$$

Therefore, using (6.99), we conclude that the rate of change of the relative spacing of the coordinate hypersurfaces $s^{i_0} = \mathit{const}$ is expressed through the mean curvature and the Beltrami's second differential parameter as follows:

$$\frac{\mathrm{d}}{\mathrm{d}\mathbf{n}_{i_0}}\left(\frac{1}{\sqrt{g_{sx}^{i_0 i_0}}}\right) = -\frac{1}{g_{sx}^{i_0 i_0}}\Delta_B^x[s^{i_0}] - \frac{n-1}{\sqrt{g_{sx}^{i_0 i_0}}}K_m(s^{i_0}) . \tag{6.106}$$

The application of (6.97) to this equation gives the following relation between the mean curvature of the coordinate hypersurface $s^{i_0} = c_0$ and the Beltrami's mixed and second differential parameters:

$$(n-1)K_m(s^{i_0}) + \nabla\left(s^{i_0}, \frac{1}{\sqrt{g_{sx}^{i_0 i_0}}}\right) + \frac{\Delta_B^x[s^{i_0}]}{\sqrt{g_{sx}^{i_0 i_0}}} = 0 . \tag{6.107}$$

Note the equations (6.106) and (6.107) are readily rewritten for the case of the monitor surface S^{rn}, namely, it suffices to substitute $\mathbf{s}$ for sr in these equations. In particular, when S^{rn} is the monitor surface over S^n with a scalar-valued monitor function $f(\mathbf{s})$, i.e. S^{rn} is represented by

$$\mathbf{r}(\mathbf{s}) : S^n \to R^{n+1} , \quad \mathbf{r}(\mathbf{s}) = [\mathbf{s}, f(\mathbf{s})] ,$$

then, availing us of (6.92), we obtain that the equation (6.106) has the following specific form

$$\frac{\mathrm{d}}{\mathrm{d}\mathbf{n}_{i_0}}\left(\frac{1}{\sqrt{g_{\mathbf{s}}^{i_0 i_0}}}\right) = \frac{n}{\sqrt{g^{\mathbf{s}} g_{\mathbf{s}}^{i_0 i_0}}}K_m(S^{rn}) f_{s^{i_0}} - \frac{n-1}{\sqrt{g_{\mathbf{s}}^{i_0 i_0}}}K_m(s^{i_0}) \tag{6.108}$$

where $g^{\mathbf{s}} = 1 + f_{s^i} f_{s^i}$, $i = 1, \dots, n$, $K_m(S^{rn})$ is the mean curvature of S^{rn} in R^{n+1} with respect to the unit normal (6.90).

Two-Dimensional Case

Let S^{x2} be an arbitrary two-dimensional regular surface. Since for the coordinate line $s^{i_0} = c_0$ in S^{x2}

$$K_m(s^{i_0}) = \sigma_{i_0} , \tag{6.109}$$

where σ_{i_0} is the geodesic curvature of the curve $s^{i_0} = c_0$ in S^{xn}, the equations derived from (6.106) in the two-dimensional case have the form

$$\begin{aligned} \frac{\mathrm{d}}{\mathrm{d}\mathbf{n}_1}\left(\frac{1}{\sqrt{g_{sx}^{11}}}\right) &= -\frac{1}{g_{sx}^{11}}\Delta_B^x[s^1] - \frac{\sigma_1}{\sqrt{g_{sx}^{11}}} , \\ \frac{\mathrm{d}}{\mathrm{d}\mathbf{n}_2}\left(\frac{1}{\sqrt{g_{sx}^{22}}}\right) &= -\frac{1}{g_{sx}^{22}}\Delta_B^x[s^2] - \frac{\sigma_2}{\sqrt{g_{sx}^{22}}} . \end{aligned} \tag{6.110}$$

While equation (6.107) in the two-dimensional case yields

$$\begin{aligned} \sigma_1 + \nabla(s^1, 1/\sqrt{g_{sx}^{11}}) + \Delta_B^x[s^1]/\sqrt{g_{sx}^{11}} = 0 , \\ \sigma_2 + \nabla(s^2, 1/\sqrt{g_{sx}^{22}}) + \Delta_B^x[s^2]/\sqrt{g_{sx}^{22}} = 0 . \end{aligned} \tag{6.111}$$

Basic Relation to Grid Coordinates

Let us apply formulas (6.106) and (6.110) to the grid coordinates $\xi^1, \dots, \xi^n$ in S^{xn} obtained by the composition of the parametrization (5.1) and intermediate transformation (5.2).

Basic Theorem

We designate by v_p the rate of change of the relative spacing between the grid hypersurfaces $\xi^p = const$ in S^{xn}, i.e., analogously to (4.63),

$$v_p = \frac{\mathrm{d}}{\mathrm{d}\mathbf{n}_p}\Big(\frac{1}{\sqrt{g^{pp}_{\xi x}}}\Big) = \frac{1}{(g^{pp}_{\xi x})^2} g^{pl}_{\xi x} g^{pj}_{\xi x} \Upsilon^p_{lj} \, , \; j, l, p = 1, \dots, n \, , \; p \text{ fixed} \, , \qquad (6.112)$$

here Υ^p_{lj} is the Christoffel symbol of the second rank of S^{xn} in the grid coordinates $\xi^1, \dots, \xi^n$, $\mathbf{n}_p$ is the normal to the grid hypersurface $\xi^p = const$, namely, similar to (6.96),

$$\mathbf{n}_p = \frac{1}{\sqrt{g^{pp}_{\xi x}}} g^{pj}_{\xi x} \frac{\partial}{\partial \xi^j} \mathbf{x}[\mathbf{s}(\boldsymbol{\xi})] \, , \quad j, p = 1, \dots, n \, , \quad p \text{ fixed} \, . \qquad (6.113)$$

We see that if $v_p < 0 (v_p > 0)$ then the nodes of the coordinate grid cluster (rarefy) in the $\mathbf{n}_p$ direction, i.e. v_p is a measure of change of grid spacing. We also call it a measure of grid clustering. The formulas (6.106) and (6.110), rewritten in the grid coordinates, result in the following:

Theorem 6. *Let $\mathbf{x}(\mathbf{s})$ in (5.1) and $\mathbf{s}(\boldsymbol{\xi})$ in (5.2) be nondegenerate transformations of the class $C^2[S^n]$ and $C^2[\Xi^n]$, respectively. Then*

$$v_p = -\frac{1}{g^{pp}_{\xi x}} \Delta^x_B[\xi^p] - \frac{n-1}{\sqrt{g^{pp}_{\xi x}}} K_m(\xi^p) \, , \quad p = 1, \dots, n \, , \quad p \textit{ fixed} \, , \qquad (6.114)$$

where Δ^x_B is the operator of Beltrami defined by (5.3) in the metric of S^{xn}; the function $\xi^p(\mathbf{s})$ is the pth component of the transformation $\boldsymbol{\xi}(\mathbf{s}) : S^n \to \Xi^n$ inverse to (5.2); $K_m(\xi^p)$ is the geodesic curvature of the curve $\xi^p = c_0$ in S^{x2} when $n = 2$, while $K_m(\xi^p)$, when $n > 2$, is the mean curvature of the hypersurface $\xi^p = c_0$ in S^{xn}.

Remarks

We assume here that the logical domain Ξ^n in the boundary value problem (5.4) formulated for generating boundary conforming grids is a rectangular n-dimensional parallelepiped $0 \le \xi^i \le l_i$, $i = 1, \dots, n$, and the $(n-1)$-dimensional boundary plane $\xi^i = 0$ or $\xi^i = l_i$ is mapped onto some $(n-1)$-dimensional segment of the boundary of S^{xn}.

Formula (6.114) demonstrates how the measure v_p of grid clustering near a boundary hypersurface $\xi^p = c_0$ in S^{xn} depends on its mean curvature. In particular, when the grid coordinate function $\xi^p(\mathbf{s})$, $p = 1, \dots, n$, is subject to the corresponding pth equation of the grid system

$$\Delta^x_B[\xi^i] = 0 \,, \quad i = 1, \dots, n \,, \tag{6.115}$$

in the original metric of the physical geometry S^{xn} then (6.114) yields

$$v_p = -\frac{n-1}{\sqrt{g^{pp}_{\xi x}}} K_m(\xi^p) \,, \quad p = 1, \dots, n \,, \quad p \text{ fixed} \,. \tag{6.116}$$

So the sign of v_p is determined by the sign of $K_m(\xi^p)$. Note the equations (6.115) proposed for two-dimensional domains by Winslow (1967) and for surfaces by Warsi (1981) are the most popular for the generation of fixed grids in domains and on two-dimensional surfaces.

If S^{xn} is a domain S^n with the Euclidean metric then the equations (6.115) are the Laplace equations

$$\nabla^2[\xi^i] \equiv \frac{\partial}{\partial s^j}\Big(\frac{\partial \xi^i}{\partial s^j}\Big) = 0 \,, \quad i, j = 1 \dots, n \,. \tag{6.117}$$

In the monographs of Thompson J.F., Warsi Z.U.A., and Mastin C.W. (1985) and Liseikin V.D. (1999) there was proved that the nodes of the coordinate grid obtained in S^n by the solution of the equations inverse to (6.117) on a rectangular computational domain $\Xi^n : 0 \le \xi^i \le l_i$, $i = 1, \dots, n$, cluster near concave boundary segments of S^n and rarefy near its convex boundary segments (see Fig. 6.1). However the formula (6.116) yields more strong conclusions for n-dimensional domains when $n \ge 3$. In order to formulate these results we first note that the unit normal $\mathbf{n}_p$, specified by (6.113), in this case is as follows

$$\mathbf{n}_p = \frac{1}{\sqrt{g^{pp}_{\xi s}}} g^{pj}_{\xi s} \mathbf{s}_{\xi^j} \,, \quad j, p = 1, \dots, n \,, \quad p \text{ fixed} \,,$$

where

$$g^{pj}_{\xi s} = \frac{\partial \boldsymbol{\xi}}{\partial s^p} \cdot \frac{\partial \boldsymbol{\xi}}{\partial s^j} \,, \quad j, p = 1, \dots, n \,.$$

Since

$$\mathbf{n}_p \cdot \mathbf{s}_{\xi^p} = \frac{1}{\sqrt{g^{pp}_{\xi s}}} > 0 \,, \quad p = 1, \dots, n \,, \quad p \text{ fixed} \,,$$

the unit normal $\mathbf{n}_p$ is directed to the interior of S^n at the points of the boundary hypersurface $\xi^p = 0$. Contrary, at the points of the hypersurface $\xi^p = l_p$ it is directed to the exterior of S^n. Therefore the inequality $v_p > 0$ ($v_p < 0$) at the points of the boundary hypersurface $\xi^p = 0$ means that the grid nodes cluster (rarefy) near it. Contrary for the hypersurface $\xi^p = l_p$ in S^n the inequality $v_p > 0$ ($v_p < 0$) means rarefaction (clustering) of grid nodes near it. Thus the nodes of the grid produced by the equations inverted to (6.117) will also cluster (rarefy) near the boundary $\xi^p = l_p$ if this segment is not concave (convex) but has the negative (positive) mean curvature, for example, it is a saddle surface.

Formula (6.116) yields also a new result in the theory of surface grid generation. Namely, the nodes of the coordinate grid on the surface S^{x2} generated through the equations inverse to (6.115) cluster (rarefy) near concave (convex) segments of the boundary of S^{x2}. Figure 6.3 of the surface grid generated through the solution of equations (6.115) for $n = 2$ demonstrates node clustering near its concave boundary segments. The right-hand part of Fig. 6.3 illustrates the grid in a parametric domain S^2.

Fig. 6.3. A surface grid with clustering near concave boundary segments.

Application of Theorem to Beltrami Grid Equations

Formula (6.114) gives one an opportunity for finding explicitly the necessary values of control functions in grid equations to provide the generation of grids whose nodes cluster or rarefy, if it is required, near specified boundary segments of an arbitrary physical geometry S^{xn}.

We assume that the logical domain Ξ^n in the boundary value problem (5.4) formulated for generating grids is the standard unit cube, i.e. $0 \le \xi^i \le 1$, $i = 1, \ldots, n$. Besides this, let the $(n-1)$-dimensional boundary plane $\xi^i = 0$ or $\xi^i = 1$ of Ξ^n for some i, $1 \le i \le n$, be mapped onto some segment of the boundary of S^{rn}, i.e. the segment is an $(n-1)$-dimensional coordinate hypersurface. For the sake of definitiveness we consider further the boundary segments $\xi^p = 0$ whose normal (6.113) is directed to the interior of the physical geometry S^{xn}. So the condition $v_p > 0 (v_p < 0)$ at the points of such a boundary segment means grid clustering (rarefaction) near it. For the segments with opposite direction of the normal (6.113) all results are readily reformulated.

We establish in this section a relation between the Beltrami's second differential parameter $\Delta_B[\xi^i]$ with respect to the monitor surface S^{rn} over S^n and the magnitude $\nabla^2[\xi^i]$ which, in fact, is also the Beltrami's second differential parameter of $\xi^i(\mathbf{s})$ with respect to the Euclidean metric of S^n.

General Multidimensional Case

Let a grid in S^n be controlled by a monitor surface S^{rn} over S^n represented by a scalar-valued monitor function $f(\mathbf{s})$, i.e.

$$\mathbf{r}(\mathbf{s}) : S^n \to R^{n+1} , \quad \mathbf{r}(s^1, \dots , s^n) = [s^1, \dots , s^n, f(s^1, \dots , s^n)] , \tag{6.118}$$

and whose metric is borrowed from R^{n+1}, namely,

$$g^{\mathbf{s}}_{ij} = \mathbf{r}_{s^i} \cdot \mathbf{r}_{s^j} = \delta^i_j + f_{s^i} f_{s^j} , \quad i, j = 1, \dots , n . \tag{6.119}$$

This situation occurs when adaptive grids are generated by the numerical solution of the inverted Beltrami equations in an n-dimensional domain X^n which is identified with S^n.

Let some functions $\xi^i(\mathbf{s})$, $i = 1, \dots , n$, comprise a new coordinate system. We establish now a relation between $\nabla^2[\xi^i]$ and $\Delta_B[\xi^i]$, where Δ_B is the operator of Beltrami (5.3) with respect to the metric of the monitor surface S^{rn}. Expanding the differentiation in $\Delta_B[\xi^i]$ gives

$$\begin{aligned} \Delta_B[\xi^i] &= \frac{1}{\sqrt{g^{\mathbf{s}}}} \frac{\partial}{\partial s^j} \Big(\sqrt{g^{\mathbf{s}}} g^{jk}_{\mathbf{s}} \frac{\partial \xi^i}{\partial s^k} \Big) \\ &= g^{jk}_{\mathbf{s}} \frac{\partial^2 \xi^i}{\partial s^j \partial s^k} + \Delta_B[s^k] \frac{\partial \xi^i}{\partial s^k} , \quad i, j, k = 1, \dots , n . \end{aligned} \tag{6.120}$$

As the covariant metric elements $g^{\mathbf{s}}_{ij}$, $i, j = 1, \dots , n$, are specified by (6.119) so, in accordance with (5.68) and (5.69) for $v(\mathbf{s}) \equiv 1$,

$$\begin{aligned} g^{\mathbf{s}} &= \det(g^{\mathbf{s}}_{ij}) = 1 + \frac{\partial f}{\partial s^j} \frac{\partial f}{\partial s^j}, \quad i, j = 1, \dots , n, \\ g^{jk}_{\mathbf{s}} &= \delta^j_k - \frac{1}{g^{\mathbf{s}}} \frac{\partial f}{\partial s^j} \frac{\partial f}{\partial s^k} , \quad j, k = 1, \dots , n . \end{aligned} \tag{6.121}$$

Therefore equations (6.120) with respect to the metric (6.119) have the following form

$$\begin{aligned} \Delta_B[\xi^i] &= \Big(\delta^j_k - \frac{1}{g^{\mathbf{s}}} \frac{\partial f}{\partial s^j} \frac{\partial f}{\partial s^k} \Big) \frac{\partial^2 \xi^i}{\partial s^j \partial s^k} + \Delta_B[s^k] \frac{\partial \xi^i}{\partial s^k} \\ &= \nabla^2[\xi^i] - \frac{1}{g^{\mathbf{s}}} \frac{\partial f}{\partial s^j} \frac{\partial f}{\partial s^k} \frac{\partial^2 \xi^i}{\partial s^j \partial s^k} + \Delta_B[s^k] \frac{\partial \xi^i}{\partial s^k} , \quad i, j, k = 1, \dots , n , \end{aligned}$$

where $\nabla^2 = \frac{\partial}{\partial s^j}\frac{\partial}{\partial s^j}$, $j = 1,\dots,n$. These equations give the following connection between $\nabla^2[\xi^i]$ and $\Delta_B[\xi^i]$:

$$\nabla^2[\xi^i] = \Delta_B[\xi^i] - \Delta_B[s^k]\frac{\partial \xi^i}{\partial s^k} + \frac{1}{g^{\mathrm{s}}}\frac{\partial f}{\partial s^j}\frac{\partial f}{\partial s^k}\frac{\partial^2 \xi^i}{\partial s^j \partial s^k}, \qquad (6.122)$$
$$i,j,k = 1,\dots,n\,. \square$$

Note the Laplace operator ∇^2 is the Beltrami operator with respect to the Euclidean metric in S^n therefore, from (6.2)

$$\nabla^2[\xi^n] = -g^{kl}_{\xi s}\Gamma^n_{lk}\,, \quad l,k = 1,\dots,n\,,$$

where the metric elements and Christoffel symbols are in the coordinates $\xi^1,\dots,\xi^n$ and defined by the formulas

$$g^{s\xi}_{ij} = \mathbf{s}_{\xi^i}\cdot\mathbf{s}_{\xi^j} = \frac{\partial s^k}{\partial \xi^i}\frac{\partial s^k}{\partial \xi^j}\,, \qquad i,j,k = 1,\dots,n\,,$$
$$g^{ij}_{\xi s} = \nabla\xi^i\cdot\nabla\xi^j = \frac{\partial \xi^i}{\partial s^k}\frac{\partial \xi^j}{\partial s^k}\,, \qquad i,j,k = 1,\dots,n\,, \qquad (6.123)$$
$$\Gamma^m_{ij} = \mathbf{s}_{\xi^i\xi^j}\cdot\nabla\xi^m = \frac{\partial^2 s^l}{\partial\xi^i\partial\xi^j}\frac{\partial \xi^m}{\partial s^l}\,, \; i,j,l,m = 1,\dots,n\,,$$

here Γ^m_{ij}, $i,j,m = 1,\dots,n$, are the Christoffel symbols of S^n in the coordinates $\xi^1,\dots,\xi^n$,

$$\nabla\xi^i = \left(\frac{\partial \xi^i}{\partial s^1},\dots,\frac{\partial \xi^i}{\partial s^n}\right), \quad i = 1,\dots,n\,.$$

Now, applying (6.2) to $\Delta_B[\xi^i]$ and $\Delta_B[s^k]$, we find from (6.122)

$$g^{lj}_{\boldsymbol{\xi}}\,{}^{\boldsymbol{\xi}}\Upsilon^i_{lj} - g^{lj}_{\mathrm{s}}\,{}^{\mathrm{s}}\Upsilon^k_{lj}\frac{\partial \xi^i}{\partial s^k} - g^{lj}_{\xi s}\Gamma^i_{lj} - \frac{1}{g^{\mathrm{s}}}\frac{\partial f}{\partial s^j}\frac{\partial f}{\partial s^k}\frac{\partial^2 \xi^i}{\partial s^j\partial s^k} = 0\,, \qquad (6.124)$$
$$i,j,k,l = 1,\dots,n\,,$$

where ${}^{\boldsymbol{\xi}}\Upsilon^i_{lj}$, ${}^{\mathrm{s}}\Upsilon^k_{lj}$ are the Christoffel symbols of the second kind of the monitor surface S^{rn} in the coordinates $\xi^1,\dots,\xi^n$ and $s^1,\dots,s^n$, respectively, while Γ^i_{lj} are the Christoffel symbols of the second kind of the parametric domain S^n in the coordinates $\xi^1,\dots,\xi^n$. Note the equations (6.120), (6.122), and (6.124) are valid for an arbitrary coordinate system $\xi^1,\dots,\xi^n$.

Let now the coordinates $\xi^1,\dots,\xi^n$ be obtained by the solution of the problem (5.4) with respect to the metric (6.119). Then applying (6.122) with $i = n$ and the condition $\Delta_B[\xi^n] = 0$ to (6.106) we obtain

$$\frac{\mathrm{d}}{\mathrm{d}\mathbf{n}_n^{s\xi}}\left(\frac{1}{\sqrt{g_{\xi s}^{nn}}}\right) = -\frac{1}{g_{\xi s}^{nn}}\nabla^2[\xi^n] - \frac{n-1}{\sqrt{g_{\xi s}^{nn}}}K_m^{s\xi}(\xi^n)$$

$$= -\frac{1}{g_{\xi s}^{nn} g^s}\frac{\partial f}{\partial s^j}\frac{\partial f}{\partial s^k}\frac{\partial^2 \xi^n}{\partial s^j \partial s^k} + \frac{1}{g_{\xi s}^{nn}}\Delta_B[s^k]\frac{\partial \xi^n}{\partial s^k} - \frac{n-1}{\sqrt{g_{\xi s}^{nn}}}K_m^{s\xi}(\xi^n)\,, \qquad (6.125)$$

$$j,k = 1,\dots,n\,.$$

This formula establishes a connection between the rate of change of the relative spacing of the grid hypersurfaces $\xi^n = const$ in S^n near a hypersurface $\xi^n = c_0$ and the monitor function and the mean curvature of this hypersurface.

Analogous formula, for the coordinate surfaces $\xi^i = const$, is obtained from (6.125) by substituting the fixed index i for n.

Now we establish a relation between the Christoffel symbols of the second kind of the monitor surface S^{rn} and domain S^n in the coordinates $\xi^1,\dots,\xi^n$ designated as ${}^{\boldsymbol{\xi}}\Upsilon^i_{jk}$ and Γ^i_{jk}, respectively. Using (4.48) with the identification $\xi^i = s^i$ we find

$$ {}^{\boldsymbol{\xi}}\Upsilon^i_{jk} = \Gamma^i_{jk} + \frac{1}{g^s}\frac{\partial f}{\partial s^p}\frac{\partial \xi^i}{\partial s^p}\nabla^{\boldsymbol{\xi}}_{jk}(f)\,, \quad i,j,k,l,m = 1,\dots,n\,, \qquad (6.126)$$

where

$$\nabla^{\boldsymbol{\xi}}_{jk}(f) = (f_{\xi^j\xi^k} - f_{\xi^l}\Gamma^l_{jk}) = \left(f_{\xi^j\xi^k} - f_{\xi^l}\frac{\partial^2 s^m}{\partial \xi^j \partial \xi^k}\frac{\partial \xi^l}{\partial s^m}\right)$$

$$= \left(f_{\xi^j\xi^k} - \frac{\partial f}{\partial s^l}\frac{\partial^2 s^l}{\partial \xi^j \partial \xi^k}\right), \quad i,k,l,m = 1,\dots,n\,, \qquad (6.127)$$

is the mixed derivative of $f[\mathbf{s}(\boldsymbol{\xi})]$ in the metric $g^{s\xi}_{ij}$. Since $\nabla^{\boldsymbol{\xi}}_{jk}(f)$ is a covariant tensor of the second rank we obtain

$$\nabla^{\boldsymbol{\xi}}_{jk}(f) = f_{s^l} f_{s^p}\frac{\partial s^l}{\partial \xi^j}\frac{\partial s^p}{\partial \xi^k}\,, \quad j,k,l,p = 1,\dots,n\,.$$

Further for $g^{jk}_{\boldsymbol{\xi}}$ we have from (6.121) and (4.15)

$$g^{jk}_{\boldsymbol{\xi}} = g^{jk}_{\xi s} - \frac{1}{g^s}\frac{\partial f}{\partial s^m}\frac{\partial \xi^j}{\partial s^m}\frac{\partial f}{\partial s^l}\frac{\partial \xi^k}{\partial s^l}\,, \quad j,k,l,m = 1,\dots,n\,,$$

therefore

$$\begin{aligned} g_{\boldsymbol{\xi}}^{jk}{}^{\boldsymbol{\xi}}\Upsilon_{jk}^{i} &= \Big(g_{\xi s}^{jk} - \frac{1}{g^{\mathrm{s}}}\frac{\partial f}{\partial s^m}\frac{\partial \xi^j}{\partial s^m}\frac{\partial f}{\partial s^l}\frac{\partial \xi^k}{\partial s^l}\Big) \\ &\quad\times\Big(\Gamma_{jk}^{i} + \frac{1}{g^{\mathrm{s}}}\frac{\partial f}{\partial s^p}\frac{\partial \xi^i}{\partial s^p}\nabla_{jk}^{\boldsymbol{\xi}}(f)\Big) \\ &= g_{\xi s}^{jk}\Gamma_{jk}^{i} - \frac{1}{g^{\mathrm{s}}}\frac{\partial f}{\partial s^m}\frac{\partial \xi^j}{\partial s^m}\frac{\partial f}{\partial s^l}\frac{\partial \xi^k}{\partial s^l}\Gamma_{jk}^{i} \\ &\quad+\frac{1}{g^{\mathrm{s}}}\frac{\partial f}{\partial s^p}\frac{\partial \xi^i}{\partial s^p}\frac{\partial^2 f}{\partial s^m \partial s^m} \\ &\quad-\frac{1}{(g^{\mathrm{s}})^2}\frac{\partial f}{\partial s^m}\frac{\partial f}{\partial s^l}\frac{\partial^2 f}{\partial s^m \partial s^l}\frac{\partial f}{\partial s^p}\frac{\partial \xi^i}{\partial s^p}\,, \\ &\quad i,j,k,l,m,p = 1,\dots,n\,. \end{aligned} \tag{6.128}$$

In accordance with (4.28)

$${}^{\mathrm{s}}\Upsilon_{ij}^{k} = -\frac{1}{g^{\mathrm{s}}}\frac{\partial^2 f}{\partial s^i \partial s^j}\frac{\partial f}{\partial s^k}\,, \qquad i,j,k = 1,\dots,n\,,$$

so using (6.2) and (6.121) gives

$$\begin{aligned} -\Delta_B[s^p] &= \frac{1}{g^{\mathrm{s}}} g_{\mathrm{s}}^{lm}\frac{\partial^2 f}{\partial s^l \partial s^m}\frac{\partial f}{\partial s^p} \\ &= \frac{1}{g^{\mathrm{s}}}\Big(\delta_m^l - \frac{1}{g^{\mathrm{s}}}\frac{\partial f}{\partial s^l}\frac{\partial f}{\partial s^m}\Big)\frac{\partial^2 f}{\partial s^l \partial s^m}\frac{\partial f}{\partial s^p} \\ &= \frac{1}{g^{\mathrm{s}}}\Big(\frac{\partial^2 f}{\partial s^m \partial s^m} - \frac{1}{g^{\mathrm{s}}}\frac{\partial f}{\partial s^m}\frac{\partial f}{\partial s^l}\frac{\partial^2 f}{\partial s^m \partial s^l}\Big)\frac{\partial f}{\partial s^p}\,, \quad l,m,p = 1,\dots,n\,, \end{aligned}$$

and consequently equations (6.128) also have the form

$$\begin{aligned} g_{\boldsymbol{\xi}}^{jk}\Upsilon_{jk}^{i} = g_{\xi s}^{jk}\Gamma_{jk}^{i} - \Delta_B[s^p]\frac{\partial \xi^i}{\partial s^p} - \frac{1}{g^{\mathrm{s}}}\frac{\partial f}{\partial s^m}\frac{\partial \xi^j}{\partial s^m}\frac{\partial f}{\partial s^l}\frac{\partial \xi^k}{\partial s^l}\Gamma_{jk}^{i}\,, \\ i,j,k,l,m,p = 1,\dots,n\,, \end{aligned}$$

which is equivalent to (6.122) and (6.124). Hence, in the case of the grid coordinates, satisfying (5.4), we find

$$\begin{aligned} \nabla^2[\xi^i] = -\Delta_B[s^p]\frac{\partial \xi^i}{\partial s^p} - \frac{1}{g^{\mathrm{s}}}\frac{\partial f}{\partial s^m}\frac{\partial \xi^j}{\partial s^m}\frac{\partial f}{\partial s^l}\frac{\partial \xi^k}{\partial s^l}\Gamma_{jk}^{i}\,, \\ i,j,k,l,m,p = 1,\dots,n\,. \;\square \end{aligned} \tag{6.129}$$

Availing us of (6.129) we also obtain another form of (6.125), assuming $n = i$,

$$\begin{aligned}
\frac{\mathrm{d}}{\mathrm{d}\mathbf{n}_i^{s\xi}}\Bigl(\frac{1}{\sqrt{g^{ii}_{\xi s}}}\Bigr) &= \frac{1}{g^{ii}_{\xi s} g^{\mathrm{s}}} \frac{\partial f}{\partial s^p}\frac{\partial \xi^j}{\partial s^p}\frac{\partial f}{\partial s^l}\frac{\partial \xi^k}{\partial s^l}\Gamma^i_{jk} \\
&\quad + \frac{1}{g^{ii}_{\xi s}}\Delta_B[s^k]\frac{\partial \xi^i}{\partial s^k} - \frac{n-1}{\sqrt{g^{ii}_{\xi s}}}K^{s\xi}_m(\xi^i)\,, \\
&\quad i,j,k,l,p=1,\dots,n\,, \qquad i \text{ fixed}\,.
\end{aligned} \tag{6.130}$$

Using (6.92) and the relation

$$\mathrm{grad}\xi^i = \sqrt{g^{ii}_{\xi s}}\mathbf{n}_i^{s\xi}\,, \qquad i=1,\dots,n\,, \qquad i \text{ fixed}\,,$$

we conclude that

$$\Delta_B[s^k]\frac{\partial \xi^i}{\partial s^k} = -n\frac{\sqrt{g^{ii}_{\xi s}}}{\sqrt{g^{\mathrm{s}}}}\mathrm{grad} f\cdot \mathbf{n}_i^{s\xi}K_m(S^{rn})\,, \qquad i=1,\dots,n\,, \qquad i \text{ fixed}\,,$$

where the quantity

$$K_m(S^{rn}) = \frac{1}{n\sqrt{g^{\mathrm{s}}}}g^{kj}_{\mathrm{s}} f_{s^k s^j}\,, \qquad j,k=1,\dots,n\,,$$

is, in accordance with (6.91), the mean curvature of the monitor surface S^{rn} in R^{n+1} with respect to the unit normal (6.90). Therefore (6.130) also has the following form

$$\begin{aligned}
\frac{\mathrm{d}}{\mathrm{d}\mathbf{n}_i^{s\xi}}\Bigl(\frac{1}{\sqrt{g^{ii}_{\xi s}}}\Bigr) &= \frac{1}{g^{ii}_{\xi s} g^{\mathrm{s}}} \frac{\partial f}{\partial s^p}\frac{\partial \xi^j}{\partial s^p}\frac{\partial f}{\partial s^l}\frac{\partial \xi^k}{\partial s^l}\Gamma^i_{jk} \\
&- \frac{1}{\sqrt{g^{ii}_{\xi s}}}\Bigl[\frac{nK_m(S^{rn})}{\sqrt{g^{\mathrm{s}}}}\mathrm{grad} f\cdot \mathbf{n}_i^{s\xi} + (n-1)K^{s\xi}_m(\xi^i)\Bigr]\,, \\
& i,j,k,l,p=1,\dots,n\,, \qquad i \text{ fixed}\,.
\end{aligned} \tag{6.131}$$

Now we establish a relation between the mean curvature of the grid hypersurfaces $\xi^i = c_0$ in S^{rn} and $\xi^i = c_0$ in S^n. Let the mean curvature of this hypersurface in S^{rn} be designated by $K^{r\xi}_m(\xi^i)$. For the sake of simplicity, we assume $i=n$.

We designate by g^{ij}_{sn} and g^{ij}_{rn} the contravariant metric elements in the coordinates $\xi^1,\dots,\xi^{n-1}$ of the grid hypersurface $\xi^n = c_0$ in S^n and on S^{rn}, respectively. Then from (6.100) we obtain

$$g^{ij}_{sn} = \frac{1}{g^{nn}_{\xi s}}(g^{nn}_{\xi s}g^{ij}_{\xi s} - g^{in}_{\xi s}g^{jn}_{\xi s})\,, \quad i,j=1,\dots,n-1\,,$$

$$g^{ij}_{rn} = \frac{1}{g^{nn}_{\boldsymbol{\xi}}}(g^{nn}_{\boldsymbol{\xi}}g^{ij}_{\boldsymbol{\xi}} - g^{in}_{\boldsymbol{\xi}}g^{jn}_{\boldsymbol{\xi}})\,, \quad i,j=1,\dots,n-1\,.$$

Since the elements of the covariant metric tensor of the grid hypersurface $\xi^n = c_0$ in S^{rn} in the coordinates $\xi^1,\dots,\xi^n$ are defined by

$$g^{\boldsymbol{\xi}}_{ij} = \mathbf{r}_{\xi^i} \cdot \mathbf{r}_{\xi^j} = \mathbf{s}_{\xi^i} \cdot \mathbf{s}_{\xi^j} + \frac{\partial f}{\partial \xi^i}\frac{\partial f}{\partial \xi^j} = g^{s\xi}_{ij} + \frac{\partial f}{\partial \xi^i}\frac{\partial f}{\partial \xi^j}\,, \quad i,j = 1,\dots,n-1\,,$$

we also find from (5.67) and (4.15)

$$g^{ij}_{rn} = g^{ij}_{sn} - \frac{1}{1+\nabla^n(f)} g^{il}_{sn} g^{jp}_{sn} \frac{\partial f}{\partial \xi^l}\frac{\partial f}{\partial \xi^p}\,, \quad i,j,k,l,m,p = 1,\dots,n-1\,,$$

where

$$\nabla^n(f) = g^{kl}_{sn} \frac{\partial f}{\partial \xi^k}\frac{\partial f}{\partial \xi^l}\,, \quad k,l = 1,\dots,n-1\,.$$

Therefore, availing us of (6.105) and (6.126), we conclude that

$$\begin{aligned}
K^{r\xi}_m(\xi^n) &= \frac{1}{(n-1)\sqrt{g^{nn}_{\boldsymbol{\xi}}}} g^{ij}_{rn}\, {}^{\boldsymbol{\xi}}\Upsilon^n_{ij} \\
&= \frac{1}{(n-1)\sqrt{g^{nn}_{\boldsymbol{\xi}}}} \left(g^{ij}_{sn} - \frac{1}{1+\nabla^n(f)} g^{il}_{sn} g^{jp}_{sn} \frac{\partial f}{\partial \xi^l}\frac{\partial f}{\partial \xi^p} \right) \\
&\times \left(\Gamma^n_{ij} + \frac{1}{g^s} \frac{\partial f}{\partial s^k}\frac{\partial \xi^n}{\partial s^k} \nabla_{\xi^i\xi^j}(f) \right) \\
&= \frac{\sqrt{g^{nn}_{s\xi}}}{\sqrt{g^{nn}_{\boldsymbol{\xi}}}} K^{s\xi}_m(\xi^n) + \frac{1}{(n-1)\sqrt{g^{nn}_{\boldsymbol{\xi}}}} \left[\frac{1}{g^s} \frac{\partial f}{\partial s^k}\frac{\partial \xi^n}{\partial s^k} \nabla_{\xi^i\xi^j}(f) g^{ij}_{rn} \right. \\
&\left. - \frac{1}{1+\nabla^n(f)} g^{il}_{sn} g^{jp}_{sn} \frac{\partial f}{\partial \xi^l}\frac{\partial f}{\partial \xi^p} \Gamma^n_{ij} \right], \\
&i,j,l,p = 1,\dots,n-1\,, \quad k = 1,\dots,n\,,
\end{aligned} \tag{6.132}$$

where $K^{s\xi}_m(\xi^n)$ is the mean curvature of the coordinate hypersurface $\xi^n = c_0$ in S^n. □

Control of Grid Clustering near Boundary Segments

Let now the monitor function be specified as

$$f(\mathbf{s}) = g[\varphi(\mathbf{s})]\,, \quad \mathbf{s} \in S^n\,, \tag{6.133}$$

where $\varphi(\mathbf{s})$ is such scalar-valued function that the $(n-1)$-dimensional surface defined from the equation $\varphi(\mathbf{s}) = 0$ coincides with the coordinate surface $\xi^i = c_0$ for some c_0. We designate this $(n-1)$-dimensional surface by $S^{n-1}_{c_0}$. We also assume that all expressions throughout this paragraph are considered at the points of the surface $S^{n-1}_{c_0}$ and

$$\begin{aligned}
\text{grad}\,\varphi(\mathbf{s}) \cdot \boldsymbol{\nabla}\xi^i(\mathbf{s}) = \varphi_{s^k}(\mathbf{s}) \frac{\partial \xi^i}{\partial s^k}(\mathbf{s}) > 0\,, \quad \mathbf{s} \in S^{n-1}_{c_0}\,, \\
i,k = 1,\dots,n\,, \quad i \text{ fixed}\,.
\end{aligned} \tag{6.134}$$

With these assumptions it is readily found that

$$\mathbf{n}_i^{s\xi} = \text{grad}\ \varphi / |\text{grad}\ \varphi| \ ,$$
$$\text{grad} f = g' \text{grad}\ \varphi \ ,$$

and, consequently,

$$\text{grad}\ f \cdot \mathbf{n}_i^{s\xi} = g' |\text{grad}\ \varphi| \ .$$

Further, availing us of (4.106), we have

$$K_m(S_{c_0}^{n-1}) = \frac{1}{(n-1)|\text{grad}\ \varphi|^3} (\varphi_{s^i} \varphi_{s^j} \varphi_{s^i s^j} - |\text{grad}\ \varphi|^2 \varphi_{s^l s^l}) \ , \qquad i,j,l = 1, \dots, n \ , \tag{6.135}$$

where $K_m(S_{c_0}^{n-1})$ is the mean curvature of the surface $S_{c_0}^{n-1}$ in S^n.

Analogously we can compute the mean curvature $K_m(S^{rn})$ of the monitor surface S^{rn} in R^{n+1} with respect to the normal (4.77), using the formula (4.83). For this purpose we find, taking advantage of (6.133),

$$\begin{aligned} f_{s^i} &= g' \varphi_{s^i} \ , \quad i = 1, \dots, n \ , \\ g^{\mathrm{s}} &= 1 + \nabla(f) = 1 + (g')^2 |\text{grad}\ \varphi|^2 \ . \end{aligned} \tag{6.136}$$

Similarly

$$f_{s^i s^j} = g' \varphi_{s^i s^j} + g'' \varphi_{s^i} \varphi_{s^j} \ , \quad i,j = 1, \dots, n \ ,$$

so

$$\begin{aligned} \nabla^2[f] &= g' \nabla^2[\varphi] + g'' |\text{grad}\ \varphi|^2 \ , \\ f_{s^i} f_{s^j} f_{s^i s^j} &= (g')^2 \varphi_{s^i} \varphi_{s^j} (g' \varphi_{s^i s^j} + g'' \varphi_{s^i} \varphi_{s^j}) \\ &= (g')^3 \varphi_{s^i} \varphi_{s^j} \varphi_{s^i s^j} + (g')^2 g'' |\text{grad}\ \varphi|^4 \ , \quad i,j = 1, \dots, n \ . \end{aligned}$$

Availing us of these relations and (6.135), we obtain

$$\begin{aligned} &\nabla^2[f] - \frac{1}{g^{\mathrm{s}}} f_{s^i} f_{s^j} f_{s^i s^j} \\ &= \frac{1}{g^{\mathrm{s}}} \{[1 + (g')^2 |\text{grad}\ \varphi|^2][g' \nabla^2[\varphi] + g'' |\text{grad}\ \varphi|^2] \\ &- (g')^3 \varphi_{s^i} \varphi_{s^j} \varphi_{s^i s^j} - (g')^2 g'' |\text{grad}\ \varphi|^4\} \\ &= \frac{1}{g^{\mathrm{s}}} \{g' \nabla^2[\varphi] - (n-1)(g')^3 |\text{grad}\ \varphi|^3 K_m(S_{c_0}^{n-1}) + g'' |\text{grad}\ \varphi|^2\} \ , \\ &i,j = 1, \dots, n \ . \end{aligned}$$

Hence (4.105) gives

$$K_m(S^{rn}) = \frac{1}{n(g^{\mathrm{s}})^{3/2}}\{g'\nabla^2[\varphi] \\ -(n-1)(g')^3|\mathrm{grad}\ \varphi|^3 K_m(S^{n-1}_{c_0}) + g''|\mathrm{grad}\ \varphi|^2\}\ , \tag{6.137}$$

Now using this equation we conclude that

$$\begin{aligned} &\frac{n}{\sqrt{g^{\mathrm{s}}}}\mathrm{grad}\ f\cdot \mathbf{n}_i^{s\xi} K_m(S^{rn}) + (n-1)K_m(S^{n-1}_{c_0}) \\ &= \frac{ng'|\mathrm{grad}\ \varphi|}{\sqrt{g^{\mathrm{s}}}} K_m(S^{rn}) + (n-1)K_m(S^{n-1}_{c_0}) \\ &= \frac{g'|\mathrm{grad}\ \varphi|}{(g^{\mathrm{s}})^2}\{g'\nabla^2[\varphi] + g''|\mathrm{grad}\ \varphi|^2\} \\ &+(n-1)\Big(1 - \frac{(g')^4|\mathrm{grad}\ \varphi|^4}{(g^{\mathrm{s}})^2}\Big)K_m(S^{n-1}_{c_0})\ , \quad i=1,\dots,n\ . \end{aligned} \tag{6.138}$$

Further we note that in the case of (6.133) we obtain

$$\begin{aligned} &\frac{\partial f}{\partial s^p}\frac{\partial \xi^j}{\partial s^p}\frac{\partial f}{\partial s^l}\frac{\partial \xi^k}{\partial s^l}\Gamma^i_{jk} = \frac{\partial f}{\partial \xi^p} g^{pj}_{\xi s}\frac{\partial f}{\partial \xi^l} g^{lk}_{\xi s}\Gamma^i_{jk} \\ &= \Big(\frac{\partial f}{\partial \xi^i}\Big)^2 g^{ij}_{\xi s} g^{ik}_{\xi s}\Gamma^i_{jk} = (g^{ii}_{\xi s})^2\Big(\frac{\partial f}{\partial \xi^i}\Big)^2 \frac{\mathrm{d}}{\mathrm{d}\mathbf{n}_i^{s\xi}}\Big(\frac{1}{\sqrt{g^{ii}_{\xi s}}}\Big) \\ &= g^{ii}_{\xi s}\nabla(f)\frac{\mathrm{d}}{\mathrm{d}\mathbf{n}_i^{s\xi}}\Big(\frac{1}{\sqrt{g^{ii}_{\xi s}}}\Big)\ , \quad i,j,k,l,p = 1,\dots,n\ ,\ i \text{ fixed}\ , \end{aligned} \tag{6.139}$$

since (6.112) with $p = i$. Now, substituting (6.138) and (6.139) in (6.131), we find

$$\frac{\mathrm{d}}{\mathrm{d}\mathbf{n}_i^{s\xi}}\Big(\frac{1}{\sqrt{g^{ii}_{\xi s}}}\Big) = -\frac{1}{\sqrt{g^{ii}_{\xi s} g^{\mathrm{s}}}} b[\mathbf{s}, g'(0), g''(0)]\ ,\ i=1,\dots,n\ ,\ i \text{ fixed}\ , \tag{6.140}$$

where

$$b[\mathbf{s}, g'(0), g''(0)] = g'(0)|\mathrm{grad}\ \varphi|\{g'(0)\nabla^2[\varphi] \\ +g''(0)|\mathrm{grad}\ \varphi|^2\} + (n-1)[1 + 2(g'(0))^2|\mathrm{grad}\ \varphi|^2]K_m(S^{n-1}_{c_0})\ .$$

Thus the positive (negative) sign of the expression $b[\mathbf{s}, g'(0), g''(0)]$ indicates the grid clustering (rarefaction) near the coordinate surface $\varphi(\mathbf{s}) = 0$. If this coordinate surface is a boundary surface of the domain S^n then the quantity $K_m(S^{n-1}_{c_0})$ is known and, by a suitable specification of the constants $a = g'(0)$ and $b = g''(0)$, we can realize the necessary sign of this expression. Consequently the monitor function $f = g(\varphi)$ in (6.133) can be chosen, for example, by

$$g(\varphi) = a\varphi + \frac{1}{2}b\varphi^2\ . \tag{6.141}$$

Another example of the monitor function gives the following formula

$$g(\varphi) = \frac{a^2}{b}\exp\Big(\frac{b}{a}\varphi\Big) .$$

Let $g'(0)$ be specified such that $g'(0) \neq 0$ then

$$b[s, g'(0), g''(0)] > 0$$

if

$$g''(0) > \max_{s \in S^{n-1}_{c_0}} -\frac{1}{g'(0)|\text{grad } \varphi|^3} D_1 , \tag{6.142}$$

where

$$D_1 = [g'(0)]^2 |\text{grad } \varphi| \nabla^2[\varphi] + (n-1)[1 + 2[g'(0)]^2 |\text{grad } \varphi|^2] K_m(S^{n-1}_{c_0}) ,$$

while

$$b[s, g'(0), g''(0)] < 0$$

if

$$g''(0) < \min_{s \in S^{n-1}_{c_0}} -\frac{1}{g'(0)|\text{grad } \varphi|^3} D_1 . \tag{6.143}$$

For example, assuming $g'(0) = 1$, we readily find that

$$b[\mathbf{s}, 1, g''(0)] > 0$$

if

$$g''(0) > \max_{\mathbf{s} \in S^{n-1}_{c_0}} -\frac{1}{|\text{grad } \varphi|^3} D_2 , \tag{6.144}$$

where

$$D_2 = |\text{grad } \varphi| \nabla^2[\varphi] + (n-1)(1 + 2|\text{grad } \varphi|^2) K_m(S^{n-1}_{c_0}) .$$

Analogously

$$b[\mathbf{s}, 1, g''(0)] < 0$$

if

$$g''(0) < \min_{\mathbf{s} \in S^{n-1}_{c_0}} -\frac{1}{|\text{grad } \varphi|^3} D_2 . \tag{6.145}$$

Since the terms in the formula (6.140) are computed locally, we can choose a single function of the form (6.133) to realize the necessary requirements of grid clustering near opposite segments of the boundary of S^n.

For example, let $\xi^i = c_0$ and $\xi^i = c_1$ be two opposite boundary segments of S^n found from the equations $\varphi_0(\mathbf{s}) = 0$ and $\varphi_1(\mathbf{s}) = 0$, respectively. Then assuming in (6.133) and (6.141)

$$\varphi(\mathbf{s}) = \varphi_0(\mathbf{s})\varphi_1(\mathbf{s}) ,$$

we can, analogously to (6.142) and (6.143), find $g'(0)$ and $g''(0)$ to provide the necessary grid behavior near these boundary segments.

Two-Dimensional Case

In the two-dimensional case the mean curvature in (6.130) is replaced by the curvature of the corresponding curvilinear line, in accordance with the relation,

$$K_m^{s\xi}(\zeta^i) = k_i \ , \quad i = 1,2 \ ,$$

where k_i is the curvature of the curve $\xi^i = c_0$ in S^2. We find from (6.132), in the two-dimensional case,

$$\sigma_2 = \frac{{}^{\boldsymbol{\xi}}\Upsilon^2_{11}}{\sqrt{g^{22}_{\boldsymbol{\xi}} g^{\boldsymbol{\xi}}_{11}}} = \frac{\sqrt{g^{22}_{\xi s} g^{s\xi}_{11}}}{\sqrt{g^{22}_{\boldsymbol{\xi}} g^{\boldsymbol{\xi}}_{11}}}(k_2 + b_2) \ , \tag{6.146}$$

where σ_2 is the geodesic curvature of the curve $\xi^2 = c_0$ on S^{r2}, k_2 is the curvature of the curve $\xi^2 = c_0$ in S^2, while

$$b_2 = \frac{1}{g^{\mathrm{s}}\sqrt{g^{22}_{\xi s} g^{s\xi}_{11}}} \frac{\partial f}{\partial s_k} \frac{\partial \xi_2}{\partial s_k} \frac{\partial^2 f}{\partial s_l \partial s_m} \frac{\partial s_m}{\partial \xi_1} \frac{\partial s_l}{\partial \xi_1} \ , \qquad k,l,m = 1,2 \ .$$

Thus

$$k_2 = \frac{\sqrt{g^{22}_{\boldsymbol{\xi}}}}{\sqrt{g^{22}_{\xi s} g^{s\xi}_{11}}} \sigma_2 - b_2 \ , \tag{6.147}$$

where

$$k_2 = \frac{1}{\sqrt{g^{22}_{\xi s} g^{s\xi}_{11}}} \frac{\partial^2 s^l}{\partial \xi^1 \partial \xi^1} \frac{\partial \xi^2}{\partial s^l} = \frac{1}{\sqrt{g^{22}_{\xi s} g^{s\xi}_{11}}} \Upsilon^2_{11} \ , \tag{6.148}$$

is the curvature of the coordinate line $\xi^2 = \text{const}$ in the parametric domain S^2. Substituting this expression in (6.110) gives

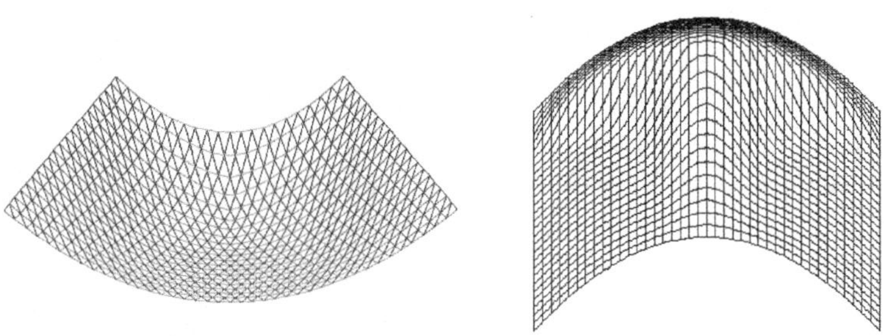

Fig. 6.4. Grid clustering (rarefaction) near a convex (concave) boundary segment

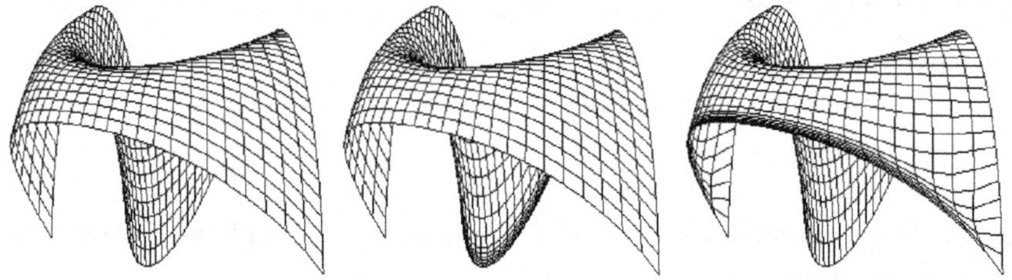

Fig. 6.5. Examples of surface grids with node clustering near boundary segments.

$$\frac{\mathrm{d}}{\mathrm{d}\mathbf{n}_2^{s\xi}}\left(\frac{1}{\sqrt{g^{22}_{\xi s}}}\right) = \frac{1}{g^{22}_{\xi s}}\Delta_B[s_k]\frac{\partial \xi^2}{\partial s^k} - \frac{\sqrt{g^{22}_{\boldsymbol{\xi}}}}{g^{22}_{\xi s}g^{s\xi}_{11}}\sigma_2 + \frac{b_2}{\sqrt{g^{22}_{\xi s}}}$$
$$-\frac{1}{g^{22}_{\xi s}g^{\mathrm{s}}}\frac{\partial f}{\partial s^j}\frac{\partial f}{\partial s^k}\frac{\partial^2 \xi^2}{\partial s^j \partial s^k}, \quad k = 1, 2 . \tag{6.149}$$

Since

$$\frac{b_2}{\sqrt{g^{22}_{\xi s}}} - \frac{1}{g^{22}_{\xi s}g^{\mathrm{s}}}\frac{\partial f}{\partial s^j}\frac{\partial f}{\partial s^k}\frac{\partial^2 \xi^2}{\partial s^j \partial s^k} = \frac{1}{g^{\mathrm{s}}g^{22}_{\xi s}g^{s\xi}_{11}}\Big(\frac{\partial f}{\partial s^k}\frac{\partial \xi^2}{\partial s^k}\frac{\partial^2 f}{\partial s^l \partial s^m}\frac{\partial s^m}{\partial \xi^1}\frac{\partial s^l}{\partial \xi^1}$$
$$-g^{s\xi}_{11}\frac{\partial f}{\partial s^j}\frac{\partial f}{\partial s^k}\frac{\partial^2 \xi^2}{\partial s^j \partial s^k}\Big), \quad j, k, l, m = 1, 2 ,$$

so in contrast to (6.116) a sign of the change rate of the grid spacing near a boundary segment in S^2 is not determined by the curvature of this segment only; it depends as well on the derivatives of the monitor function $f(\mathbf{s})$. Therefore one can control the grid spacing near the boundary of S^2 with a proper choice of this function.

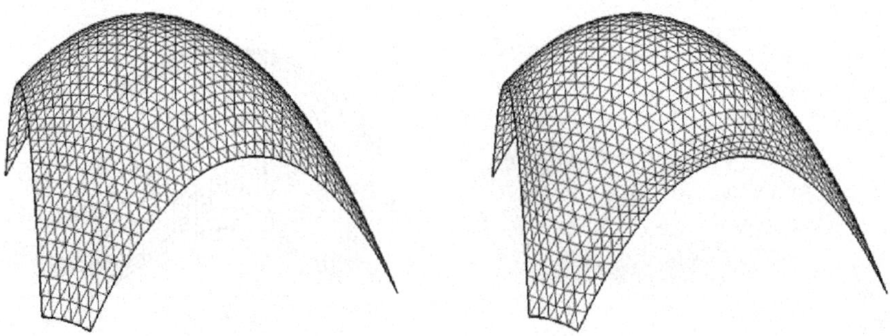

Fig. 6.6. Triangular surface grids.

Thus using (6.149) with the choice of the monitor function $f(s^1, s^2)$ and the boundary transformation $\mathbf{s}(\xi^1, \text{const})$ we can, in principle, provide grid nodes clustering near the boundary curve $\xi^2 = 0$ on S^{r2} or in S^2, respectively, regardless of its geometry in S^2. Analogous results are obtained for the grid lines near the boundaries $\xi^2 = 1$ and $\xi^1 = 1$ and $\xi^1 = 0$.

Figure 6.4 demonstrates domain grids with the clustering (rarefaction) provided near the convex (concave) segments of domains by monitor functions of the form (6.133).

In the same way there is provided grid clustering (rarefaction) near boundaries of physical surfaces. Figures 6.5 and 6.6 demonstrate quadrilateral and triangular surface grids generated without node clustering (left-hand) and with node clustering (right-hand) realized by a function of the type (6.133) near boundary segments.

Application to Diffusion Grid Equations

The second popular approach for generating grids is based on the solution of diffusion equations

$$\frac{\partial}{\partial s^j}\left[w(\mathbf{s})\frac{\partial \xi^i}{\partial s^j}\right] = 0\,, \quad i,j = 1,\ldots n\,, \tag{6.150}$$

where $w(\mathbf{s}) > 0$ is a diffusion function specified by the user to provide grid control. Equations (6.150) were proposed for $n = 2$ by Danaev N.T., Liseikin V.D., and Yanenko N.N. (1980) and Winslow A.M. (1981) for the generation of adaptive grids in domains.

The transformed equations, for finding the intermediate map (5.2), in which the dependent and independent variables are mutually altered have, in accordance with (6.37), the following form

$$g^{kj}_{\xi s}\frac{\partial}{\partial \xi^j}\left(\frac{1}{w(\mathbf{s})}\frac{\partial s^l}{\partial \xi^k}\right) = 0\,, \quad j,k,l = 1,\ldots n\,, \tag{6.151}$$

where

$$g^{kj}_{\xi s} = \frac{\partial \xi^k}{\partial s^i}\frac{\partial \xi^j}{\partial s^i}\,, \quad i,j,k = 1,\ldots n\,.$$

If S^{xn} is a domain S^n with the Euclidean metric then the operator of Beltrami Δ_B coincides with the Laplace operator ∇^2 and besides this

$$\text{grad}\ \xi^p = \left(\frac{\partial \xi^p}{\partial s^1},\ldots,\frac{\partial \xi^p}{\partial s^n}\right) = \sqrt{g^{pp}_{\xi x}}\mathbf{n}_p\,, \quad p = 1,\ldots,n\,, \quad p \text{ fixed}\,.$$

Therefore we obtain, if (6.150) is held,

$$\nabla^2[\xi^p] = -\frac{\sqrt{g^{pp}_{\xi x}}}{w}\text{grad}\ w\cdot\mathbf{n}_p\,, \quad p = 1,\ldots,n\,, \quad p \text{ fixed}\,,$$

and consequently formulas (6.114) and (6.150) yield

$$v_p = \frac{1}{\sqrt{g^{pp}_{\xi x}}}\left(\frac{1}{w}\text{grad } w \cdot \mathbf{n}_p - (n-1)K_m(\xi^p)\right), \qquad p = 1, \ldots, n, \quad p \text{ fixed}. \tag{6.152}$$

The normal $\mathbf{n}_p$ and the mean curvature of the boundary of the domain S^n is known beforehand so the characteristic v_p of the rate of change of the grid spacing near a boundary segment is defined explicitly by the diffusion function $w(\mathbf{s})$ in the vicinity of the segment. Thus the grid points cluster (rarefy) near the boundary segment $\xi^p = 0$ if $w(\mathbf{s})$ is such that $v_p > 0$ ($v_p < 0$) in (6.152) at its points.

The specification of the diffusion function $w(\mathbf{s})$ is facilitated if it is sought in the form

$$w(\mathbf{s}) = g[\varphi(\mathbf{s})]$$

where $\varphi(\mathbf{s})$ is a scalar-valued function such that the equation $\varphi(\mathbf{s}) = 0$ determines the boundary hypersurface $\xi^p = 0$. The vector

$$\text{grad } \varphi = (\varphi_{s^1}, \ldots, \varphi_{s^n})$$

is orthogonal to this hypersurface and consequently parallel to $\mathbf{n}_p$. We assume that $\varphi(\mathbf{s})$ is such that the both vectors gradφ and $\mathbf{n}_p$ are of the same direction, i.e.

$$\mathbf{n}_p = \frac{1}{|\text{grad } \varphi|}\text{grad } \varphi .$$

Since

$$\text{grad } w = g'\text{grad } \varphi,$$

we readily obtain from (6.152)

$$v_p = \frac{1}{\sqrt{g^{pp}_{\xi x}}}\left(\frac{g'}{g}|\text{grad } \varphi| - (n-1)K_m(\xi^p)\right), \quad p = 1, \ldots, n, \quad p \text{ fixed}.$$

Thus $v_p > 0$ (for grid clustering) if $g(\varphi) \neq 0$ and

$$\frac{g'(0)}{g(0)} > \max_{\mathbf{s}|_{\varphi(\mathbf{s})=0}} \frac{n-1}{|\text{grad } \varphi|}K_m(\xi^p),$$

while $g(\varphi)$ we can, for example, specify as

$$g(\varphi) = g(0) + g'(0)\varphi$$

or

$$g(\varphi) = g(0)\exp\left[\frac{g'(0)}{g(0)}\varphi\right].$$

Note for the hypersurface $\varphi(\mathbf{s}) = 0$ in S^n the invariant $K_m(\xi^p)$ can, in accordance with (4.111), be computed by the following formula

$$K_m(\xi^p) = -\frac{1}{n-1}\frac{\partial}{\partial s^j}\Big(\frac{1}{\sqrt{\nabla(\varphi)}}\varphi_{s^j}\Big)\,, \quad j = 1, \dots, n\,,$$

where

$$\nabla(\varphi) = \varphi_{s^k}\varphi_{s^k}\,, \quad k = 1, \dots, n\,.$$

Analogously, with the help of a function $w(\mathbf{s}) = g[\varphi(\mathbf{s})]$ in diffusion equations (5.16) there is controlled the sign of the measure of grid clustering near boundary segments of a regular surface S^{xn}. Figure 6.7 illustrates surface grids obtained by the numerical solution of equations (5.26) (left-hand) and (5.16) (right-hand) with node rarefaction (clustering) near the boundary segment "a".

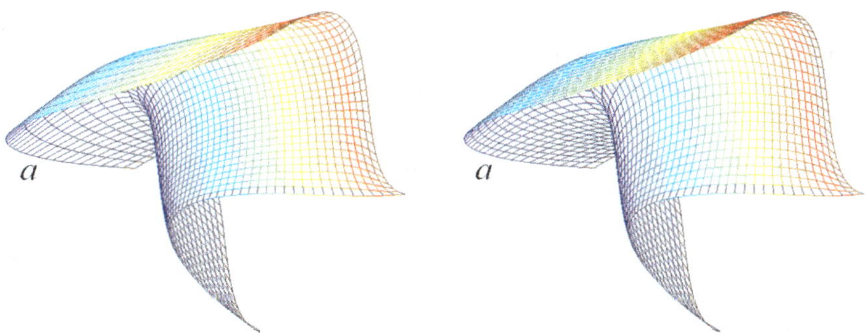

Fig. 6.7. Examples of surface grids generated through diffusion equations.

The approach for controlling grid behavior near boundary segments is readily applied to the generation of meshes by the Poisson system (5.93). A detailed description of how the source functions P^i in (5.93) are specified for this purpose is in the monograph by Liseikin (2004).

6.4 Practical Grid Equations

This section discusses in detail one-, two-, and three-dimensional grid equations adjusted to generating grids on the boundary curves and surfaces of three-dimensional blocks and in the interior of the blocks. The equations are considered with respect to both monitor surfaces and monitor metrics.

6.4.1 Equations for Generating Grids on Curves

General Equation

Let a curve S^{x1} in the k-dimensional space R^k be represented as

$$\mathbf{x}(s) : [0,1] \to R^k , \quad \mathbf{x} = (x^1, \dots , x^k) . \tag{6.153}$$

For a general monitor metric g^s over S^{x1}, in particular (5.58), one-dimensional inverted diffusion grid equation is obtained from (6.22) as

$$\frac{d}{d\xi}\left[\frac{g^s}{w(s)}\frac{ds}{d\xi}\right] = 0. \tag{6.154}$$

For the inverted Beltrami grid equation we have, assuming in (6.154) $w(s) = \sqrt{g^s}$,

$$\frac{d}{d\xi}\left[\sqrt{g^s}\frac{ds}{d\xi}\right] = 0. \tag{6.155}$$

Of course the equations (6.153) and (6.155) are also the corresponding fluxes-sources one-dimensional grid equations.

Equation with Respect to the Metric of a Monitor Curve

Let $\mathbf{f}(\mathbf{x}) = [f^1(\mathbf{x}), \dots , f^l(\mathbf{x})]$ be a monitor function, determining the monitor curve S^{r1} over S^{x1}. Then S^{r1} is parametrized by the following transformation

$$\mathbf{r}(s) : [0,1] \to R^{l+k} , \quad \mathbf{r}(s) = \{\mathbf{x}(s), \mathbf{f}[\mathbf{x}(s)]\} , \tag{6.156}$$

consequently the covariant g^s and contravariant g_s metric tensor of S^{r1} in the coordinate s is

$$g^s = \mathbf{r}_s \cdot \mathbf{r}_s = g^{xs} + \mathbf{f}_s \cdot \mathbf{f}_s \quad \text{and} \quad g_s = 1/g^s , \tag{6.157}$$

respectively, where

$$g^{xs} = \mathbf{x}_s \cdot \mathbf{x}_s , \tag{6.158}$$

is the covariant metric tensor of S^{x1}, while $\mathbf{f}_s = d\mathbf{f}[\mathbf{x}(s)]/ds$. So the intermediate transformation $s(\xi)$ for generating a grid on S^{x1} is subject, in accordance with (6.11), to the following equation

$$g_\xi \frac{\mathrm{d}^2 s}{\mathrm{d}\xi^2} = \frac{1}{\sqrt{g^s}}\frac{\mathrm{d}}{\mathrm{d}s}(\sqrt{g^s}g_s) = \frac{1}{\sqrt{g^s}}\frac{\mathrm{d}}{\mathrm{d}s}\left(\frac{1}{\sqrt{g^s}}\right) , \tag{6.159}$$

where g_ξ is the contravariant tensor of S^{r1} in the grid coordinate ξ, i.e.

$$g_\xi = 1/g^\xi , \quad g^\xi = g^s(\mathrm{d}s/\mathrm{d}\xi)^2 .$$

Equation (6.159) is readily converted to

$$\frac{d^2s}{d\xi^2} = \sqrt{g^\xi}\frac{d}{d\xi}\left(\frac{1}{\sqrt{g^s}}\right) = -\frac{1}{2g^s}\frac{ds}{d\xi}\frac{d}{d\xi}g^s , \tag{6.160}$$

and consequently to the following divergent form

$$\frac{d}{d\xi}\left(\sqrt{g^s}\frac{ds}{d\xi}\right) = 0 , \quad \xi \in [0,1] , \tag{6.161}$$

i.e. to the equation (6.155).

Simplified Equation

Note if $k = 1$ in (6.153), i.e. S^{x1} is an interval $[a, b]$, then $\mathbf{x}(s)$ in (6.153) is a scalar function $x(s)$ mapping the interval [0,1] onto $[a, b]$ and consequently the metric tensor g^s in the grid equations (6.159 – 6.161) has the form

$$g^s = (x_s)^2 + \mathbf{f}_s \cdot \mathbf{f}_s .$$

In particular, when $x(s) \equiv s$, then

$$g^s = 1 + \mathbf{f}_s \cdot \mathbf{f}_s ,$$

while the grid equation in this particular case is also, in accordance with (6.5),

$$\frac{d^2s}{d\xi^2} + \frac{d^2\mathbf{f}[s(\xi)]}{d\xi^2} \cdot \frac{d\mathbf{f}(s)}{ds} = 0 . \tag{6.162}$$

If the monitor function $\mathbf{f}(s)$ is a scalar-valued function $f(s)$ then applying (3.10) with the identification $x = s$ and $u = f$ we find

$$\frac{d^2f[s(\xi)]}{d\xi^2} = \frac{d^2f}{ds^2}\left(\frac{ds}{d\xi}\right)^2 + f_s\frac{d^2s}{d\xi^2} = k[1 + (f_s)^2]^{3/2}\left(\frac{ds}{d\xi}\right)^2 + f_s\frac{d^2s}{d\xi^2} ,$$

where k is the curvature of the monitor surface S^{r1}. Substituting this expression in (6.162) gives the following one-dimension grid equation

$$\frac{d^2s}{d\xi^2} + k\sqrt{1 + (f_s)^2}f_s\left(\frac{ds}{d\xi}\right)^2 = 0 , \tag{6.163}$$

for a scalar-valued monitor function.

A form of a curve grid equation similar to (6.162) can be obtained for an arbitrary curve S^{x1} parametrized by (6.153) if we, analogously to (6.74), specify a monitor function $\mathbf{f}_1(s)$ over S^{x1} as

$$\mathbf{f}_1(s) = \{s, \mathbf{f}[\mathbf{x}(s)]\} .$$

Then, similarly to (6.77), the curve grid equation is as follows:

$$\frac{d^2s}{d\xi^2} + \frac{d^2\mathbf{x}[s(\xi)]}{d\xi^2} \cdot \frac{d\mathbf{x}(s)}{ds} + \frac{d^2\mathbf{f}\{\mathbf{x}[s(\xi)]\}}{d\xi^2} \cdot \frac{d\mathbf{f}(s)}{ds} = 0 . \tag{6.164}$$

6.4.2 Equations for Generating Grids on Two-Dimensional Surfaces

Surface Grid Equations in General Monitor Metric

Inverted Diffusion Equations.

Let a two-dimensional surface S^{x2} in R^k be represented by a parametrization

$$\mathbf{x}(\mathbf{s}) : S^2 \to R^k , \quad \mathbf{x} = (x^1, \dots, x^k) . \tag{6.165}$$

The most general diffusion form of the two-dimensional surface grid equations in an arbitrary monitor metric $g^{\mathbf{s}}_{ij}$ over S^{x2}, in particular (5.58), follows from (6.12) as

$$w[\mathbf{s}(\boldsymbol{\xi})] g^{km}_{\boldsymbol{\xi}} \frac{\partial^2 s^i}{\partial \xi^k \partial \xi^m} = \frac{\partial}{\partial s^j} [w(\mathbf{s}) g^{ji}_{\mathbf{s}}], \quad i, j, k, m = 1, 2, \tag{6.166}$$

where

$$g^{ij}_{\boldsymbol{\xi}} = g^{mp}_{\mathbf{s}} \frac{\partial \xi^i}{\partial s^m} \frac{\partial \xi^j}{\partial s^p}, \quad i, j, m, p = 1, 2.$$

The grid on S^{x2} generated through the equations (6.166) (this system is equivalent to the diffusive system (5.16) with $n = 2$) is obtained by mapping a reference grid in Ξ^2 on S^{x2} by the function $\mathbf{x}[\mathbf{s}(\boldsymbol{\xi})]$ where $\mathbf{s}(\boldsymbol{\xi}) : \Xi^2 \to S^2$ is the intermediate transformation whose components $s^i(\boldsymbol{\xi})$, $i = 1, 2$, satisfy these equations. In the same way, the grid on S^{x2} is obtained by mapping with $\mathbf{x}(\mathbf{s})$ the grid in S^2 found by the numerical solution of (6.166) on the reference grid in Ξ^2.

The coefficients $g^{ij}_{\boldsymbol{\xi}}$ in (6.166) can be found by two ways:

1. by expressing these elements through $g^{\boldsymbol{\xi}}_{mp}$,
2. by expressing the elements $\dfrac{\partial \xi^i}{\partial s^j}$ in the formulas for $g^{km}_{\boldsymbol{\xi}}$ through $\dfrac{\partial s^m}{\partial \xi^p}$.

In the first way the quantities $g^{\boldsymbol{\xi}}_{ij}$ are computed by the formula

$$g^{\boldsymbol{\xi}}_{ij} = g^{\mathbf{s}}_{mp} \frac{\partial s^m}{\partial \xi^i} \frac{\partial s^p}{\partial \xi^j}, \quad i, j, m, p = 1, \dots, n.$$

Note that, similarly to (2.21), in the two-dimensional case the elements $g^{ij}_{\boldsymbol{\xi}}$ of the contravariant monitor metric tensor in the grid coordinates ξ^1, ξ^2 are computed through the elements $g^{\boldsymbol{\xi}}_{ij}$ of the covariant monitor metric tensor $(g^{\boldsymbol{\xi}}_{km})$ in the same coordinates ξ^1, ξ^2 by the formula

$$g^{ij}_{\boldsymbol{\xi}} = \frac{(-1)^{i+j}}{g^{\boldsymbol{\xi}}} g^{\boldsymbol{\xi}}_{3-i 3-j}, \quad i, j = 1, 2 ; \quad i, j \text{ fixed} , \tag{6.167}$$

where

$$g^{\boldsymbol{\xi}} = \det(g^{\boldsymbol{\xi}}_{ij}) = g^{\mathbf{s}} J^2, \quad J = \det\Big(\frac{\partial s^i}{\partial \xi^j}\Big).$$

Now introducing the operator L^2 by the formula

$$L^2[v] = g^{\boldsymbol{\xi}}_{22} \frac{\partial^2 v}{\partial \xi^1 \partial \xi^1} - 2 g^{\boldsymbol{\xi}}_{12} \frac{\partial^2 v}{\partial \xi^1 \partial \xi^2} + g^{\boldsymbol{\xi}}_{11} \frac{\partial^2 v}{\partial \xi^2 \partial \xi^2} \tag{6.168}$$

and taking into account that

$$L^2[v] = g^{\mathbf{s}} J^2 L^{\boldsymbol{\xi}}[v] \,, \tag{6.169}$$

where $L^{\boldsymbol{\xi}}$ is the basic operator defined by (6.9) for $n = 2$, we find from (6.166) the following surface grid system

$$w[\mathbf{s}(\boldsymbol{\xi})] L^2[s^i] = g^{\mathbf{s}} J^2 \frac{\partial}{\partial s^j} [w(\mathbf{s}) g^{ji}_{\mathbf{s}}] \,, \quad i,j = 1,2 \,. \tag{6.170}$$

In the second way we get, availing us of the identities (2.1)

$$g^{km}_{\boldsymbol{\xi}} = \frac{1}{J^2} b^{km}, \quad k,m = 1,2,$$

where

$$b^{km} = (-1)^{k+l+m+p} g^{lp}_{\mathbf{s}} \frac{\partial s^{3-l}}{\partial \xi^{3-k}} \frac{\partial s^{3-p}}{\partial \xi^{3-m}}, \quad k,l,m,p = 1,2, \quad k,m \ \text{fixed}.$$

Thus from the equations (6.166) we also obtain the following equivalent diffusion grid system

$$w[\mathbf{s}(\boldsymbol{\xi})] b^{km} \frac{\partial^2 s^i}{\partial \xi^k \partial \xi^m} = J^2 \frac{\partial}{\partial s^j} [w(\mathbf{s}) g^{ji}_{\mathbf{s}}], \quad i,j,k,m = 1,2, \tag{6.171}$$

The transformed equations (6.170) and (6.171) are in fact the equations (6.166) multiplied by $g^{\boldsymbol{\xi}}$ and J^2, respectively. Though the equations (6.170) and (6.171) are equivalent to the equations (6.166) if J does not vanish at any point of Ξ^n, they are more convenient for solving by iterative numerical methods because they do not include J as a denominator. Thus the numerical solution of these equations can be fulfilled if the Jacobian vanishes in the process of iterations.

Note the functionals of grid smoothness (5.32) and diffusion (5.52) as well as the fluxes-sources equations (6.19), (6.23), and (6.28) include J as a denominator. Therefore the process of finding the grid nodes through the minimization of the functionals or through the numerical solution of the fluxes-sources equations rules out any opportunity for the Jacobian to vanish even at one point.

Inverted Beltrami Equations.

The inverted two-dimensional Beltrami surface grid equations are yielded from (6.166), (6.170), and (6.171) by assuming in these equations $w(\mathbf{s}) = \sqrt{g^{\mathbf{s}}}$.

Fluxes-Sources Equations.

Since (2.1)

$$\frac{\partial \xi^p}{\partial s^k}\frac{\partial \xi^p}{\partial s^m}=\frac{1}{J^2}a_{km}, \quad k,m,p=1,2,$$

where

$$a_{km}=(-1)^{k+m}\frac{\partial s^{3-k}}{\partial \xi^j}\frac{\partial s^{3-m}}{\partial \xi^j}, \quad j,k,m=1,2, \quad k,m \text{ fixed},$$

so we get from the fluxes-sources diffusion grid system (6.23) the following equivalent system for generating grids on two-dimensional surfaces

$$a_{km}\frac{\partial}{\partial s^i}[w(\mathbf{s})g_{\mathbf{s}}^{km}]=J\frac{\partial}{\partial \xi^j}\Big\{w[\mathbf{s}(\boldsymbol{\xi})]\Big(2\frac{g_{pi}^{\mathbf{s}}}{Jg^{\mathbf{s}}}\frac{\partial s^p}{\partial \xi^j}-(-1)^{i+j}\frac{1}{J^2}b^{ll}\frac{\partial s^{3-i}}{\partial \xi^{3-j}}\Big)\Big\},$$
$$i,j,k,l,m,p=1,2, \quad i \text{ fixed}. \tag{6.172}$$

This system of equations is in fact the system (6.23) multiplied by J.

Another form of the fluxes-sources two-dimensional grid equations is presented by equations (6.29).

Inverted Beltrami Equations with Respect to the Metric of a Monitor Surface

Let us consider here a monitor surface S^{r2} formed by the values of a vector-valued function $\mathbf{f}(\mathbf{x})=[f^1(\mathbf{x}),\dots,f^l(\mathbf{x})]$ over a two-dimensional surface S^{x2} represented by (6.165). The monitor surface S^{r2} is represented in the coordinates s^1,s^2 by the following parametrization

$$\mathbf{r}(\mathbf{s}):S^2\to R^{l+k}, \quad \mathbf{r}(\mathbf{s})=\{\mathbf{x}(\mathbf{s}),\mathbf{f}[\mathbf{x}(\mathbf{s})]\}, \quad \mathbf{s}=(s^1,s^2). \tag{6.173}$$

Consequently for the elements g_{ij}^{xs} and $g_{ij}^{\mathbf{s}}$ in the coordinates s^1,s^2 of the covariant metric tensor of S^{x2} and S^{r2}, respectively, we find

$$\begin{aligned} g_{ij}^{xs}&=\mathbf{x}_{s^i}\cdot\mathbf{x}_{s^j}, \quad i,j=1,2,\\ g_{ij}^{\mathbf{s}}&=\mathbf{r}_{s^i}\cdot\mathbf{r}_{s^j}=g_{ij}^{xs}+\mathbf{f}_{s^i}\cdot\mathbf{f}_{s^j}, \quad i,j=1,2. \end{aligned} \tag{6.174}$$

Analogously in the grid coordinates ξ^1,ξ^2

$$\begin{aligned} g_{ij}^{\boldsymbol{\xi}}&=g_{ij}^{x\xi}+\frac{\partial\mathbf{f}\{\mathbf{x}[\mathbf{s}(\boldsymbol{\xi})]\}}{\partial\xi^i}\cdot\frac{\partial\mathbf{f}\{\mathbf{x}[\mathbf{s}(\boldsymbol{\xi})]\}}{\partial\xi^j}, \quad i,j=1,2,\\ g_{ij}^{x\xi}&=\frac{\partial\mathbf{x}[\mathbf{s}(\boldsymbol{\xi})]}{\partial\xi^i}\cdot\frac{\partial\mathbf{x}[\mathbf{s}(\boldsymbol{\xi})]}{\partial\xi^j}, \quad i,j=1,2. \end{aligned} \tag{6.175}$$

Using the description (6.168) of the operator L^2 we find from the grid systems (6.11) and (6.73) with $n=2$, suitable to generate grids on two-dimensional surfaces, the following systems

$$L^2[s^i] = g^{\mathbf{s}} J^2 \Delta_B[s^i] \;, \quad i = 1, 2 \;, \tag{6.176}$$

and

$$\begin{gathered} L^2[s^i] + g^{\mathbf{s}} J^2 \Big[\Delta_B[f^a] \frac{\partial f^a}{\partial s^b} g^{bi}_{sx} + g^{pj\mathbf{x}}_{\mathbf{s}} \Upsilon^i_{pj} \Big] = 0 \;, \\ b, i, j, p = 1, 2 \;, \quad a = 1, \dots, l \;, \end{gathered} \tag{6.177}$$

respectively. Contrary to (6.11) and (6.73) these systems are nor divided by J.

Similarly to (6.171) there are obtained the equations by multiplying (6.11) and (6.73) with J^2.

Simplified Equations

The equations (6.177) are simplified if the monitor function over S^{x2} is chosen in the form

$$\mathbf{f}[\mathbf{x}(\mathbf{s})] = [\mathbf{s}, \mathbf{v}(\mathbf{s})] \;, \quad \mathbf{v}(\mathbf{s}) = [v^1(\mathbf{s}), \dots, v^l(\mathbf{s})] \;. \tag{6.178}$$

The metric tensor of this monitor surface S^{r2} over S^{x2}, formed by the monitor function (6.178), coincides with the metric tensor of the monitor surface over S^2 formed by the monitor function $\mathbf{f}_1[\mathbf{s}] = [\mathbf{x}(\mathbf{s}), \mathbf{v}(\mathbf{s})]$ so, in accordance with (6.78), the grid equations with respect to $s^i(\boldsymbol{\xi})$ are in this case as follows:

$$\begin{gathered} L^2[s^i] + \frac{\partial x^m}{\partial s^i} L^2[x^m] + \frac{\partial v^p}{\partial s^i} L^2[v^p] = 0 \;, \\ i = 1, 2 \;, \quad m = 1, \dots, k \;, \quad p = 1, \dots, l \;. \end{gathered} \tag{6.179}$$

Note the metric element $g^{\boldsymbol{\xi}}_{ij}$ in (6.179) for the operator L^2 are computed by the formula

$$\begin{gathered} g^{\boldsymbol{\xi}}_{ij} = \frac{\partial s^a(\boldsymbol{\xi})}{\partial \xi^i} \frac{\partial s^a(\boldsymbol{\xi})}{\partial \xi^j} + \frac{\partial x^m[\mathbf{s}(\boldsymbol{\xi})]}{\partial \xi^i} \frac{\partial x^m[\mathbf{s}(\boldsymbol{\xi})]}{\partial \xi^j} + \frac{\partial v^p[\mathbf{s}(\boldsymbol{\xi})]}{\partial \xi^i} \frac{\partial v^p[\mathbf{s}(\boldsymbol{\xi})]}{\partial \xi^j} \;, \\ i, j, a = 1, 2 \;, \quad m = 1, \dots, k \;, \quad p = 1, \dots, l \;. \end{gathered}$$

If the surface S^{x2} lies in R^3 and is represented by a parametrization

$$\mathbf{x}(\mathbf{s}) = (\mathbf{s}, x^3(\mathbf{s})) = (s^1, s^2, x^3(s^1, s^2)) \tag{6.180}$$

then the simplified equations can be obtained for an arbitrary monitor function $\mathbf{f}(\mathbf{x})$ since the parametrization (6.173) of the monitor surface S^{r2} has the following form

$$\mathbf{r}(\mathbf{s}) = \{s^1, s^2, x^3(s^1, s^2), \mathbf{f}[\mathbf{x}(s^1, s^2)]\} \;, \tag{6.181}$$

i.e. S^{r2} is also the monitor surface over the domain S^2, formed by the values of the vector-valued function

$$\mathbf{g}(\mathbf{s}) = \{x^3(\mathbf{s}), \mathbf{f}[\mathbf{x}(\mathbf{s})]\} .$$

Note a nondegenerate, smooth surface $S^{x2} \subset R^3$ is always represented locally as the graph of a bivariate function, i.e. in the form similar to (6.180). In the case of the parametrization (6.181) of the monitor surface S^{r2} over S^2 we can also generate grids on S^{x2} with the use of equations (6.78), for $n = 2$, which analogously to (6.179) are converted to

$$\begin{aligned} &L^2[s^i] + \frac{\partial x^3}{\partial s^i} L^2[x^3] + \frac{\partial f^p}{\partial s^i} L^2[f^p] = 0 , \\ &i = 1,2 , \quad p = 1, \dots , l , \end{aligned} \tag{6.182}$$

where

$$\frac{\partial x^3}{\partial s^i} = \frac{\partial x^3[\mathbf{s}]}{\partial s^i} , \qquad \frac{\partial f^p}{\partial s^i} = \frac{\partial f[\mathbf{x}(\mathbf{s})]}{\partial s^i} , \qquad i = 1,2 ,$$

while the metric elements $g^{\boldsymbol{\xi}}_{ij}$ in the description (6.168) of the operator L^2 are as follows:

$$g^{\boldsymbol{\xi}}_{ij} = \frac{\partial \mathbf{s}(\boldsymbol{\xi})}{\partial \xi^i} \cdot \frac{\partial \mathbf{s}(\boldsymbol{\xi})}{\partial \xi^j} + \frac{\partial x^3[\mathbf{s}(\boldsymbol{\xi})]}{\partial \xi^i} \frac{\partial x^3[\mathbf{s}(\boldsymbol{\xi})]}{\partial \xi^j} + \frac{\partial \mathbf{f}[\mathbf{x}(\mathbf{s}(\boldsymbol{\xi}))]}{\partial \xi^i} \cdot \frac{\partial \mathbf{f}[\mathbf{x}(\mathbf{s}(\boldsymbol{\xi}))]}{\partial \xi^j} , \quad i,j = 1,2 .$$

The form (6.180) of the parametrization of S^{x2} allows one to include the mean curvature K_m of this surface into the grid equations with respect to the metric of this surface, i.e. without a monitor function. Indeed the parametrization (6.180) also represents the parametrization of the monitor surface S^{r2} over S^2 with a scalar-valued monitor function $f(\mathbf{s}) = x^3(\mathbf{s})$. Therefore, availing us of (6.93) for $n = 2$ and (6.169), we obtain the following grid equations for generating a fixed grid on S^{x2}

$$L^2[s^i] + 2\sqrt{g^{\mathbf{s}}} J^2 K_m \frac{\partial x^3}{\partial s^i} = 0 , \quad i = 1,2 . \tag{6.183}$$

6.4.3 Equations for Generating Grids in Domains

Two-Dimensional Domains

There are two approaches for controlling the grid behavior in a two-dimensional domain $X^2 \subset R^2$: with the use of monitor surfaces and monitor manifolds.

General Monitor Metric.

Two-dimensional domain grid equations in an arbitrary monitor metric $g^{\mathbf{s}}_{ij}$ over the domain S^2 are naturally obtained from the equations (6.170), (6.171), and (6.172) for two-dimensional surfaces, assuming in the corresponding formulas $g^{x\mathbf{s}}_{ij} = \delta^i_j, \ \ i,j = 1,2$.

Equations with Respect to a Monitor Surface.

Let

$$\mathbf{f}(\mathbf{x}) : X^2 \to R^l \ , \quad \mathbf{x} = (x^1, x^2) \ , \quad \mathbf{f} = (f^1, \dots, f^l) \ ,$$

be a monitor function over X^2. Then the monitor surface S^{r2} over X^2 is parametrized as follows:

$$\mathbf{r}(\mathbf{x}) : X^2 \to R^{2+l} \ , \quad \mathbf{r}(\mathbf{x}) = [x^1, x^2, \mathbf{f}(\mathbf{x})] \ .$$

Particular forms of the grid equations in the metric of the monitor surface S^{r2} are obtained from (6.53) or (6.55) with $n = 2$ by the assumption $x^i = s^i$, $i = 1, 2$. Since (6.167), (6.169) the equations are as follows:

$$L^2[x^i] + \frac{\partial f^k}{\partial x^i} L^2[f^k] = 0 \ , \quad i = 1, 2 \ , \quad k = 1, \dots, l \ . \tag{6.184}$$

Here the elements $g^{\boldsymbol{\xi}}_{ij}$ of the metric tensor of S^{r2} in the definition (6.168) of the operator L^2 are computed by the formula

$$g^{\boldsymbol{\xi}}_{ij} = \frac{\partial x^m}{\partial \xi^i} \frac{\partial x^m}{\partial \xi^j} + \frac{\partial f^k[\mathbf{x}(\boldsymbol{\xi})]}{\partial \xi^i} \frac{\partial f^k[\mathbf{x}(\boldsymbol{\xi})]}{\partial \xi^j} \ , \quad i, j, m = 1, 2 \ , \quad k = 1, \dots, l \ .$$

Analogously to (6.171) the equations (6.184) are also equivalent to

$$b^{jm} \frac{\partial^2 x^i}{\partial \xi^j \partial \xi^m} + \frac{\partial f^k}{\partial x^i} b^{jm} \frac{\partial^2 f^k[\mathbf{x}(\boldsymbol{\xi})]}{\partial \xi^j \partial \xi^m}, \quad i, j, m = 1, 2, \quad k = 1, \dots, l. \tag{6.185}$$

In the case $\mathbf{f}(\mathbf{x})$ is a scalar-valued monitor function $f(\mathbf{x})$ over X^2, the grid equations are also represented by (6.93) with $n = 2$, $s^i = x^i$, $i = 1, 2$. So the equations (6.184) have one more equivalent form

$$L^2[x^i] + 2 g^{\boldsymbol{\xi}} \sqrt{g^{\mathbf{x}}} J^2 K_m f_{x^i} = 0 \ , \quad i = 1, 2 \ , \tag{6.186}$$

where

$$g^{\boldsymbol{\xi}}_{ij} = \frac{\partial x^m}{\partial \xi^i} \frac{\partial x^m}{\partial \xi^j} + \frac{\partial f[\mathbf{x}(\boldsymbol{\xi})]}{\partial \xi^i} \frac{\partial f[\mathbf{x}(\boldsymbol{\xi})]}{\partial \xi^j} \ , \quad i, j, m = 1, 2 \ ,$$

$$g^{\boldsymbol{\xi}} = \det(g^{\boldsymbol{\xi}}_{ij}) \ , \quad g^{\mathbf{x}} = 1 + (f_{x^1})^2 + (f_{x^2})^2 \ ,$$

$$K_m = \frac{1}{2\sqrt{g^{\mathbf{x}}}} g^{kj}_{\mathbf{x}} f_{x^k x^j} \ , \quad j, k = 1, 2 \ ,$$

$g^{kj}_{\mathbf{x}}$, $j, k = 1, 2$, are the elements of the contravariant metric tensor of S^{r2} in the coordinates x^1, x^2. Since the elements $g^{\mathbf{x}}_{ij}$, $i, j = 1, 2$, of the covariant metric tensor of S^{r2} in the coordinates x^1, x^2, are expressed by

$$g^{\mathbf{x}}_{ij} = \delta^i_j + f_{x^i} f_{x^j} \ , \quad i, j = 1, 2 \ ,$$

and analogously to (6.167),

$$g_{\mathbf{x}}^{ij} = (-1)^{i+j} g_{3-i3-j}^{\mathbf{x}} / g^{\mathbf{x}} , \quad i,j = 1,2 , \quad i,j \text{ fixed} ,$$

we find

$$K_m = \frac{1}{2(g^{\mathbf{x}})^{3/2}} B , \tag{6.187}$$

where

$$B = (1 + f_{x^2} f_{x^2}) f_{x^1 x^1} - 2 f_{x^1} f_{x^2} f_{x^1 x^2} + (1 + f_{x^1} f_{x^1}) f_{x^2 x^2} .$$

Thus equations (6.186) can be written in the following form

$$L^2[x^i] + \frac{J^2 B}{1 + (f_{x^1})^2 + (f_{x^2})^2} f_{x^i} = 0 , \quad i = 1,2 . \tag{6.188}$$

Three-Dimensional Domains

Analogously to the two-dimensional case discussed above we consider two types of three-dimensional grid equations for generating grids in a three-dimensional domain $X^3 \subset R^3$.

Equation in General Monitor Metric

Inverted Diffusion Equations.

Three-dimensional inverted equations with respect to an arbitrary monitor metric $g_{ij}^{\mathbf{s}}$, $i,j = 1,2,3$, over the domain S^3 and weight function $w(\mathbf{s}) > 0$ are described by the formula (6.12) with $n = 3$. However for the purpose of practical applications we, analogously to the two-dimensional cases considered above, change in two ways the coefficients $g_{\boldsymbol{\xi}}^{ij}$ in the description of the operator $L^{\boldsymbol{\xi}}$ by multiplying them with $g^{\boldsymbol{\xi}}$ and J^2, in order the Jacobian J was not included in the equations as a denominator. Such equations can be solved by iterative methods if the Jacobian vanishes during some iterations at the points of Ξ^n.

In the first way the elements of the inverse matrix $(g_{\boldsymbol{\xi}}^{ij})$, $i,j = 1,2,3$, can be found from the elements of $(g_{ij}^{\boldsymbol{\xi}})$ by the following general formula

$$\begin{aligned} g_{\boldsymbol{\xi}}^{ij} &= \frac{1}{g^{\boldsymbol{\xi}}} (g_{i+1j+1}^{\boldsymbol{\xi}} g_{i+2j+2}^{\boldsymbol{\xi}} - g_{i+1j+2}^{\boldsymbol{\xi}} g_{i+2j+1}^{\boldsymbol{\xi}}) , \quad i,j = 1, \dots , n , \\ g^{\boldsymbol{\xi}} &= \det(g_{ij}^{\boldsymbol{\xi}}) , \end{aligned} \tag{6.189}$$

in which any index, say i, is identified with $i \pm 3$, so, for instance, $g_{45}^{\boldsymbol{\xi}} = g_{12}^{\boldsymbol{\xi}}$.

So the grid equations (6.12) with $n = 3$ are as follows

$$w[\mathbf{s}(\boldsymbol{\xi})]L^3[s^i] = g^{\mathbf{s}} J^2 \frac{\partial}{\partial s^j}(w[\mathbf{s}]g_{\mathbf{s}}^{ji}), \quad i,j = 1,2,3, \tag{6.190}$$

where

$$\begin{aligned} L^3[v] = {} & [g_{22}^{\boldsymbol{\xi}} g_{33}^{\boldsymbol{\xi}} - (g_{23}^{\boldsymbol{\xi}})^2] \frac{\partial^2 v}{\partial \xi^1 \partial \xi^1} + 2[g_{23}^{\boldsymbol{\xi}} g_{13}^{\boldsymbol{\xi}} - g_{12}^{\boldsymbol{\xi}} g_{33}^{\boldsymbol{\xi}}] \frac{\partial^2 v}{\partial \xi^1 \partial \xi^2} \\ & + 2[g_{12}^{\boldsymbol{\xi}} g_{23}^{\boldsymbol{\xi}} - g_{22}^{\boldsymbol{\xi}} g_{13}^{\boldsymbol{\xi}}] \frac{\partial^2 v}{\partial \xi^1 \partial \xi^3} + [g_{11}^{\boldsymbol{\xi}} g_{33}^{\boldsymbol{\xi}} - (g_{13}^{\boldsymbol{\xi}})^2] \frac{\partial^2 v}{\partial \xi^2 \partial \xi^2} \\ & + 2[g_{12}^{\boldsymbol{\xi}} g_{13}^{\boldsymbol{\xi}} - g_{11}^{\boldsymbol{\xi}} g_{23}^{\boldsymbol{\xi}}] \frac{\partial^2 v}{\partial \xi^2 \partial \xi^3} + [g_{11}^{\boldsymbol{\xi}} g_{22}^{\boldsymbol{\xi}} - (g_{12}^{\boldsymbol{\xi}})^2] \frac{\partial^2 v}{\partial \xi^3 \partial \xi^3} . \end{aligned}$$

Analogously, we obtain equivalent equations, in the second way, expressing in the contravariant elements

$$g_{\boldsymbol{\xi}}^{ij} = g_{\mathbf{s}}^{km} \frac{\partial \xi^i}{\partial s^k} \frac{\partial \xi^j}{\partial s^m}, \quad i,j,k,m = 1,2,3,$$

the terms $\partial \xi^i / \partial s^j$ by $\partial s^k / \partial \xi^m$, in accordance with formulas (2.2).

Inverted Beltrami Equations.

Three-dimensional inverted Beltrami Equations are obtained from the inverted diffusion equation (6.190) by substituting in these formulas $\sqrt{g^{\mathbf{s}}}$ for $w(\mathbf{s})$.

Fluxes-Sources Equations.

Three-dimensional fluxes-sources equations for the general monitor metric are of the form (6.19) with $n = 3$. Equivalent forms of these equations are obtained by changing the terms $\partial \xi^i / \partial s^j$ and $g_{\boldsymbol{\xi}}^{ij}$ with the elements of the matrix $(\partial s^i / \partial \xi^j)$ and $(g_{ij}^{\boldsymbol{\xi}})$, respectively.

Equations with Respect to a Monitor Surface.

Let a monitor function

$$\mathbf{f}(\mathbf{x}) : X^3 \to R^l , \quad \mathbf{x} = (x^1, x^2, x^3) , \quad \mathbf{f} = (f^1, \dots, f^l) ,$$

over X^3 have been specified. Then the monitor surface S^{r3} over X^3 has the following parametrization

$$\mathbf{r}(\mathbf{x}) : X^3 \to R^{3+l} , \quad \mathbf{r}(\mathbf{x}) = [x^1, x^2, x^3, \mathbf{f}(\mathbf{x})] .$$

The grid equations in X^3 are, for example, the equations (6.53) in which $n = 3$, $x^i = s^i$, $i = 1,2,3$, while the contravariant metric tensor $(g_{\boldsymbol{\xi}}^{ij})$, $i,j = 1,2,3$, of the surface S^{r3} in the grid coordinates ξ^1, ξ^2, ξ^3, is the matrix inverse to the covariant metric tensor $(g^{\boldsymbol{\xi}}_{ij})$, $i, j = 1, 2, 3$, where

$$g^{\boldsymbol{\xi}}_{ij} = \frac{\partial x^m(\boldsymbol{\xi})}{\partial \xi^i}\frac{\partial x^m(\boldsymbol{\xi})}{\partial \xi^j} + \frac{\partial f^k[\mathbf{x}(\boldsymbol{\xi})]}{\partial \xi^i}\frac{\partial f^k[\mathbf{x}(\boldsymbol{\xi})]}{\partial \xi^j}\,, \qquad i, j, m = 1, 2, 3\,, \quad k = 1, \ldots, l\,. \tag{6.191}$$

Thus the equations for gridding the domain X^3 with the monitor function $\mathbf{f}(\mathbf{x}) : X^3 \to R^l$ are transformed from (6.53) to

$$L^3[x^i] + L^3[f^k]\frac{\partial f^k}{\partial x^i} = 0\,, \quad i = 1, 2, 3\,, \quad k = 1, \ldots, l\,. \tag{6.192}$$

If $\mathbf{f}(\mathbf{x})$ is a scalar-valued monitor function then, analogously to the two-dimensional case considered above, the grid equations are represented by (6.93) with $n = 3$, $s^i = x^i$, $i = 1, 2, 3$. Similarly to (6.186) these equations also have the form, following from (6.93) for $n = 3$,

$$L^3[x^i] + 3g^{\boldsymbol{\xi}}\sqrt{g^{\mathbf{x}}}J^2 K_m f_{x^i} = 0\,, \quad i = 1, 2, 3\,, \tag{6.193}$$

where, in accordance with (5.69) and (6.91),

$$g^{\mathbf{x}} = 1 + (f_{x^1})^2 + (f_{x^2})^2 + (f_{x^3})^2\,,$$

$$K_m = \frac{1}{3\sqrt{g^{\mathbf{x}}}} g^{kj}_{\mathbf{x}} f_{x^k x^j}\,, \quad j, k = 1, 2, 3\,,$$

$$J = \det\left(\frac{\partial x^i}{\partial \xi^j}\right).$$

Here $g^{kj}_{\mathbf{x}}$, $j, k = 1, 2, 3$, are the elements of the contravariant metric tensor of S^{r3} in the coordinates x^1, x^2, x^3. These elements are computed, for example, by the formula (5.68) with $n = 3$ in which $\mathbf{s}$ is identified with $\mathbf{x}$, i.e.

$$g^{ij}_{\mathbf{x}} = \delta^j_i - \frac{1}{g^{\mathbf{x}}} f_{x^j} f_{x^i}\,, \quad j, i = 1, 2, 3\,.$$

7 Numerical Implementation of Grid Generators

The systems of grid equations (6.11), (6.12), and their modifications described in Chap. 6 allow one to generate grids in domains or on surfaces in a unified manner, regardless of their dimension. In particular, these systems can be applied to produce grids in spatial blocks by means of the successive generation of grids on curvilinear edges, faces, and parallelepipeds, using the solution at a step $i < n$ as the Dirichlet boundary condition for the following step $i + 1 \leq n$. Thus both the interior and the boundary grid points of a domain or surface can be calculated by the similar elliptic solver.

This chapter reviews some finite-difference numerical algorithms for grid generation based on the equations and functionals discussed in Chaps. 5 and 6.

In the chapter the following rule is observed: the indices related to the numeration of cells and/or cell vertices are considered as fixed, i.e. the summation in formulas over such indices is not carried out.

7.1 Method of Fractional Steps

In this section we describe one version of the algorithm of fractional steps proposed by Yanenko (1971). Other versions of this algorithm that can be readily implemented for solving the resulting multidimensional grid equations, in particular, the popular ADI (alternating direction implicit) method are reviewed by Kovenya, Tarnavskii, and Chernyi (1990), Fletcher (1997), and Langtangen (2003).

7.1.1 One-Dimensional Equation

Here we consider a curve S^{x1} specified by a parametrization

$$\mathbf{x}(s) : [0, 1] \to R^k , \quad \mathbf{x} = (x^1, \dots , x^k) .$$

For generating a grid on the curve S^{x1} we use the inverted equation (6.155), in which g^s is a monitor covariant metric over the curve, in particular, specified in the form (5.58) for $n = 1$, i.e.

$$g^s = z(s)g^{xs} + F^m(s)F^m(s)\ , \quad m = 1, \dots, l\ ,$$

$$g^{xs} = \frac{\partial \mathbf{x}}{\partial s} \cdot \frac{\partial \mathbf{x}}{\partial s}.$$

The metric g^s can also be the metric of a monitor curve S^{r1} prescribed by a monitor function for controlling grid properties

$$\mathbf{f}(\mathbf{x}) : G^k \to R^l\ , \quad \mathbf{f} = (f^1, \dots, f^l)\ ,$$

where G^k is a domain in R^k containing S^{x1}. As a result the monitor curve S^{r1} over S^{x1} is parametrized by

$$\mathbf{r}(s) : [0,1] \to R^{l+k}\ , \quad \mathbf{r}(s) = (\mathbf{x}(s), \mathbf{f}[\mathbf{x}(s)])\ ,$$

and consequently

$$g^s = g^{rs} = \mathbf{x}_s \cdot \mathbf{x}_s + \mathbf{f}_s \cdot \mathbf{f}_s = \frac{\mathrm{d}\mathbf{x}}{\mathrm{d}s} \cdot \frac{\mathrm{d}\mathbf{x}}{\mathrm{d}s} + \frac{\mathrm{d}\mathbf{f}[\mathbf{x}(s)]}{\mathrm{d}s} \cdot \frac{\mathrm{d}\mathbf{f}[\mathbf{x}(s)]}{\mathrm{d}s}\ .$$

Numerical Algorithm

The numerical grid on S^{x1} is computed after solving the Dirichlet boundary value problem with respect to $s(\xi)$ for the equation (6.155), i.e.

$$\begin{aligned} &\frac{\mathrm{d}}{\mathrm{d}\xi}\left(\sqrt{g^s}\frac{\mathrm{d}s}{\mathrm{d}\xi}\right) = 0\ , \quad 0 < \xi < 1\ , \\ &s(0) = 0\ , \quad s(1) = 1\ . \end{aligned} \tag{7.1}$$

Namely, the grid nodes $\mathbf{x}_j$, $j = 0, 1, \dots, N$, on S^{x1} are defined by the relation

$$\mathbf{x}_j = \mathbf{x}(s(jh))\ , \quad j = 0, 1, \dots, N\ , \quad h = 1/N\ ,$$

or by

$$\mathbf{x}_j = \mathbf{x}(s_j)\ , \quad j = 0, 1, \dots, N\ , \quad h = 1/N\ ,$$

here s_j, $j = 0, 1, \dots, N$, is a difference function obtained by the numerical solution on a uniform grid $\xi_j = jh$, $j = 0, 1, \dots, N$, of the Dirichlet problem (7.1).

Iterative Scheme

The nonlinear problem (7.1) is solved by an iterative process which is engendered by the numerical solution of the following parabolic problem with respect to a function $s(\xi, t)$

$$\begin{aligned} &\frac{\partial s}{\partial t} - \frac{\partial}{\partial \xi}\left(\sqrt{g^s}\frac{\partial s}{\partial \xi}\right) = 0\ , \quad 0 \le \xi \le 1\ , \quad 0 \le t \le T\ , \\ &s(0,t) = 0\ , \quad s(1,t) = 1\ , \quad s(\xi, 0) = s_0(\xi)\ . \end{aligned} \tag{7.2}$$

The problem (7.2) is approximated on the uniform grid $(ih, n\tau)$ with respect to s_i^n, $i = 0, 1, \dots, N$, $n = 0, 1, \dots$, by the following natural stencil

$$\begin{aligned} &\frac{s_i^{n+1} - s_i^n}{\tau} = \frac{1}{h^2}[v_{i+1/2}^n(s_{i+1}^{n+1} - s_i^{n+1}) - v_{i-1/2}^n(s_i^{n+1} - s_{i-1}^{n+1})] , \\ &s_0^n = 0 , \quad s_N^n = 1 , \quad s_i^0 = s_0(ih) , \quad h = 1/N , \end{aligned} \tag{7.3}$$

where

$$v_{i+1/2}^n = \frac{1}{2}\left(\sqrt{g^s(s_i^n)} + \sqrt{g^s(s_{i+1}^n)}\right) , \quad i = 0, 1, \dots, N-1. \tag{7.4}$$

The scheme (7.3) is implicit. Its solution is obtained from the algorithm which is expounded by the application to the following well-known difference reference problem

$$\begin{aligned} &A_i^{n+1}s_{i-1}^{n+1} - C_i^{n+1}s_i^{n+1} + B_i^{n+1}s_{i+1}^{n+1} = -F_i^n , \quad i = 1, 2, \dots, N-1 , \\ &s_0^{n+1} = a , \quad s_N^{n+1} = b . \end{aligned} \tag{7.5}$$

The solution of (7.5) is found through the following recursive formulas

$$s_i^{n+1} = \alpha_{i+1}^{n+1}s_{i+1}^{n+1} + \beta_{i+1}^{n+1} , \quad i = 1, \dots, N-1 , \quad s_N^{n+1} = b , \tag{7.6}$$

where

$$\begin{aligned} \alpha_{i+1}^{n+1} &= \frac{B_i^{n+1}}{C_i^{n+1} - \alpha_i^{n+1}A_i^{n+1}} , \quad i = 1, \dots, N-1 , \quad \alpha_1^{n+1} = 0 , \\ \beta_{i+1}^{n+1} &= \frac{A_i^{n+1}\beta_i^{n+1} + F_i^n}{C_i^{n+1} - \alpha_i^{n+1}A_i^{n+1}} , \quad i = 1, \dots, N-1 , \quad \beta_1^{n+1} = a . \end{aligned} \tag{7.7}$$

Thus assuming in (7.5) $a = 0$, $b = 1$, and

$$\begin{aligned} &A_i^{n+1} = v_{i-1/2}^n , \quad B_i^{n+1} = v_{i+1/2}^n , \quad C_i^{n+1} = v_{i-1/2}^n + v_{i+1/2}^n + \theta , \\ &F_i^n = \theta s_i^n , \quad \theta = h^2/\tau , \quad i = 1, \dots, N-1 , \end{aligned} \tag{7.8}$$

we obtain a solution of (7.3) at a step $n+1$ if it is known at the previous step n. Note the values of the initial function s_i^0, $i = 0, 1, \dots, N$, are specified by the user. Naturally it may be assumed that

$$s_i^0 = ih , \quad i = 0, \dots, N , \quad h = 1/N .$$

As an approximate numerical solution of (7.1) there is taken the solution s_i^n, $i = 0, 1, \dots, N$, of (7.3) at a step number n if

$$\max_{0 \le i \le N} \frac{|s_i^{n+1} - s_i^n|}{\tau} \le \varepsilon , \tag{7.9}$$

for some sufficiently small $\varepsilon > 0$.

Step-by-Step Algorithm

The algorithm described above is presented here in a step-by-step manner.
Step 1.
Define an initial grid distribution of the parametric interval [0,1] by introducing a monotone difference function s^0_i, $i = 0, \dots, N$, such that $s^0_0 = 0$, $s^0_N = 1$.
Step 2.
Compute the difference function $v^0_{i+1/2}$, $i = 0, \dots, N-1$, by formula (7.4).
Step 3.
Compute the difference functions A^1_i, B^1_i, C^1_i, F^0_i, $i = 1, \dots, N-1$, by formulas in (7.8).
Step 4.
Compute the coefficients α^1_i and β^1_i, $i = 1, \dots, N$, by formulas in (7.7) with $a = 0$.
Step 5.
Compute the difference solution s^1_i, $i = 0, \dots, N$, of the first step through the formula (7.6) taking into account $s^1_0 = 0$, $s^1_N = b = 1$.
Step 6.
Return to step 2 assuming $s^0_0 = s^1_i$, $i = 0, \dots, N$, where s^1_i is the solution obtained at the step 5.

Continue until the tolerance requirement (7.9) is observed.

The algorithm described is readily reformulated for the numerical solution of the inverted diffusion equation (6.154), namely, by substituting in (7.2) and (7.4) $g^{\mathbf{s}}/w(\mathbf{s})$ for $\sqrt{g^{\mathbf{s}}}$.

7.1.2 Two-Dimensional Equations

In this section a finite-difference numerical algorithm for generating grids in two-dimensional domains and surfaces is described.

Algorithms for Generating Grids in Two-Dimensional Domains

Boundary Value Problem

Let us first discuss the grid algorithm for a two-dimensional domain S^2. We shell use for the logical domain Ξ^2 the unit square: $\Xi^2 = \{0 \le \xi^1, \xi^2 \le 1\}$. Let the transformation $\mathbf{s}(\boldsymbol{\xi})$ for generating a grid in S^2 be specified on the boundary of Ξ^2, i.e. there is a map

$$\boldsymbol{\varphi}(\boldsymbol{\xi}) : \partial\Xi^2 \to \partial S^2 , \quad \boldsymbol{\varphi} = (\varphi^1, \varphi^2) \tag{7.10}$$

which is continuous on $\partial\Xi^2$. Note the one-dimensional transformation on any segment of $\partial\Xi^2$ can be computed by the algorithm described in Sect. 7.1.1. The generation of a grid in S^2 is based on the numerical solution of the

Dirichlet problem for the most general systems of inverted diffusion equations (6.170) or (6.171). In particular, the boundary value problem for the system (6.170) has the following form

$$
\begin{aligned}
&w[\mathbf{s}(\boldsymbol{\xi})]L^2[s^i] = g^{\mathbf{s}}J^2\frac{\partial}{\partial s^j}\left[w(\mathbf{s})g_{\mathbf{s}}^{ij}\right], \quad i,j=1,2\,, \\
&s^i(\boldsymbol{\xi})\,\Big|_{\partial\Xi^2} = \varphi^i(\boldsymbol{\xi}), \quad i=1,2\,,
\end{aligned}
\tag{7.11}
$$

where

$$
L^2[v] = g_{22}^{\boldsymbol{\xi}}\frac{\partial^2 v}{\partial\xi^1\partial\xi^1} - 2g_{12}^{\boldsymbol{\xi}}\frac{\partial^2 v}{\partial\xi^1\partial\xi^2} + g_{11}^{\boldsymbol{\xi}}\frac{\partial^2 v}{\partial\xi^2\partial\xi^2}\,. \tag{7.12}
$$

As another equivalent system of the inverted diffusion grid equations in (7.11) there can be the system (6.171).

Assuming in (7.11) $w(\mathbf{s}) = \sqrt{g^{\mathbf{s}}}$ yields also the boundary value problem for finding grids by the inverted Beltrami equations.

Parabolic Equations

The nonlinear boundary–value problem (7.11) is solved by an iterative process. For this purpose (7.11) is replaced by the following boundary-value parabolic problem with respect to the functions $s^i(\xi^1,\xi^2,t)$, $i=1,2:$

$$
\begin{aligned}
&\frac{\partial s^i}{\partial t} = w[\mathbf{s}(\boldsymbol{\xi})]L^2[s^i] - g^{\mathbf{s}}J^2\frac{\partial}{\partial s^j}[w(\mathbf{s})g_{\mathbf{s}}^{ij}]\,, \quad i,j=1,2\,, \\
&s^i(\boldsymbol{\xi},t) = \varphi^i(\boldsymbol{\xi})\,, \quad \boldsymbol{\xi}\in\partial\Xi^2\,, \quad t\geq 0\,, \\
&s^i(\boldsymbol{\xi},0) = s_0^i(\boldsymbol{\xi})\,, \quad \boldsymbol{\xi}\in\Xi^2\,,
\end{aligned}
\tag{7.13}
$$

where $s_0^i(\boldsymbol{\xi})$ is the i-th component of the initial transformation

$$
\mathbf{s}_0(\boldsymbol{\xi}) : \Xi^2 \to S^2, \quad \mathbf{s}_0(\boldsymbol{\xi}) = [s_0^1(\boldsymbol{\xi}), s_0^2(\boldsymbol{\xi})],
$$

specified by the user.

The solution $\mathbf{s}(\boldsymbol{\xi},t)$ satisfying (7.13) aspires to the solution of (7.11) when $t\to\infty$. Therefore an approximate solution of (7.11) is obtained from the solution of (7.13) computed for some sufficiently large value $t=T_0$.

Initial Transformation

The initial transformation

$$
\mathbf{s}(\boldsymbol{\xi},0) = \mathbf{s}_0(\boldsymbol{\xi}) : \Xi^2 \to S^2\,.
$$

can be found by propagating the values of $\boldsymbol{\varphi}(\boldsymbol{\xi}) = [\varphi^1(\boldsymbol{\xi}), \varphi^2(\boldsymbol{\xi})]$ from the boundary points into the interior of the domain Ξ^2, for example, if Ξ^2 is a unit cube through the formula of the Lagrange two-dimensional transfinite interpolation. This formula has the following recursive form for the components $s^i(\boldsymbol{\xi},0)$ of the mapping $\mathbf{s}_0(\boldsymbol{\xi})$:

$$\begin{aligned} F^i(\xi^1,\xi^2) &= \alpha^i_{01}(\xi^1)\varphi^i(0,\xi^2)+\alpha^i_{11}(\xi^1)\varphi^i(1,\xi^2)\,, \\ s^i(\xi^1,\xi^2,0) &= F^i(\xi^1,\xi^2)+\alpha^i_{02}(\xi^2)[\varphi^i(\xi^1,0)-F^i(\xi^1,0)] \\ &\quad +\alpha^i_{12}(\xi^2)[\varphi^i(\xi^1,1)-F^i(\xi^1,1)]\,, \quad i=1,2, \quad i \text{ fixed}\,, \end{aligned} \tag{7.14}$$

where the functions $\alpha^i_{kj}(s)$, $0 \le s \le 1$, (referred to as blending functions) are subject to the following restrictions

$$\alpha^i_{0j}(0)=\alpha^i_{1j}(1)=1\,, \quad \alpha^i_{0j}(1)=\alpha^i_{1j}(0)=0\,. \tag{7.15}$$

In particular for the simplest expressions of the blending functions

$$\alpha^i_{0j}(s)=1-s\,, \quad \alpha^i_{1j}(s)=s\,,$$

satisfying (7.15) we find from (7.14)

$$\begin{aligned} F^i(\xi^1,\xi^2) &= (1-\xi^1)\varphi^i(0,\xi^2)+\xi^1\varphi^i(1,\xi^2)\,, \\ s^i(\xi^1,\xi^2,0) &= F^i(\xi^1,\xi^2)+(1-\xi^2)[\varphi^i(\xi^1,0)-F^i(\xi^1,0)] \\ &\quad +\xi^2[\varphi^i(\xi^1,1)-F^i(\xi^1,1)]\,, \quad i=1,2\,. \end{aligned} \tag{7.16}$$

Iterative Algorithms for Generating Quadrilateral Grids

Let us introduce, for convenience, new dependent variables $s(\boldsymbol{\xi},t)=s^1(\boldsymbol{\xi},t)$, $v(\boldsymbol{\xi},t)=s^2(\boldsymbol{\xi},t)$, $g_{km}=g^{\boldsymbol{\xi}}_{km}$, $k,m=1,2$. With respect to the corresponding discrete vector-valued function $\mathbf{s}^n=(s^n_{ij},v^n_{ij})$, $0\le i\le N_1$, $0\le j\le N_2$, $0\le n$, the problem (7.13) is approximated on the rectangular grid $(ih_1,jh_2,n\tau)$, $h_1=1/N_1$, $h_2=1/N_2$, in the logical domain $\Xi^2\times[0,T]$, where Ξ^2 is a square (Fig. 7.1), by the scheme

$$\begin{aligned} \frac{s^{n+1/2}_{ij}-s^n_{ij}}{\tau} &= w(\mathbf{s}^n)_{ij}[g_{22}(\mathbf{s}^n)_{ij}\mathbf{L}_{ij}(s^{n+1/2})-2g_{12}(\mathbf{s}^n)_{ij}\mathbf{M}_{ij}(s^n) \\ &\quad +g_{11}(\mathbf{s}^n)_{ij}\mathbf{S}_{ij}(s^n)]+\mathbf{P}_{ij}(\mathbf{s}^n)\,, \\ &\quad 1\le i\le N_1-1\,, \quad 1\le j\le N_2-1\,, \quad n\ge 0\,, \end{aligned} \tag{7.17}$$

$$\begin{aligned} \frac{v^{n+1/2}_{ij}-v^n_{ij}}{\tau} &= w(\mathbf{s}^n)_{ij}[g_{22}(\mathbf{s}^n)_{ij}\mathbf{L}_{ij}(v^{n+1/2})-2g_{12}(\mathbf{s}^n)_{ij}\mathbf{M}_{ij}(v^n) \\ &\quad +g_{11}(\mathbf{s}^n)_{ij}\mathbf{S}_{ij}(v^n)]+\mathbf{Q}_{ij}(\mathbf{s}^n)\,, \\ &\quad 1\le i\le N_1-1\,, \quad 1\le j\le N_2-1\,, \quad n\ge 0\,, \end{aligned} \tag{7.18}$$

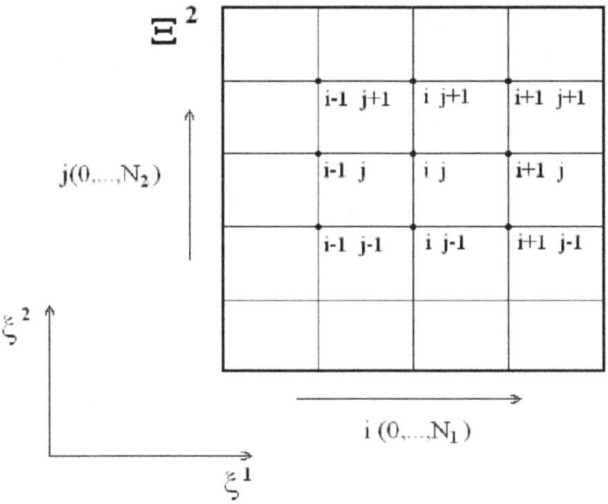

Fig. 7.1. Two-dimensional quadrilateral stencil for finite differences

$$\frac{s_{ij}^{n+1}-s_{ij}^{n+1/2}}{\tau}=g_{11}(\mathbf{s}^n)_{ij}\mathbf{S}_{ij}(s^{n+1}-s^n)\,,\qquad 1\le i\le N_1-1\,,\quad 1\le j\le N_2-1\,,\quad n\ge 0\,, \tag{7.19}$$

$$\frac{v_{ij}^{n+1}-v_{ij}^{n+1/2}}{\tau}=g_{11}(\mathbf{s}^n)_{ij}\mathbf{S}_{ij}(v^{n+1}-v^n)\,,\qquad 1\le i\le N_1-1\,,\quad 1\le j\le N_2-1\,,\quad n\ge 0\,, \tag{7.20}$$

where

$$\mathbf{L}_{ij}(z)=\frac{z_{i+1j}-2z_{ij}+z_{i-1j}}{(h_1)^2}\,,$$

$$\mathbf{M}_{ij}(z)=\frac{z_{i+1j+1}-z_{i-1j+1}-z_{i+1j-1}+z_{i-1j-1}}{4h_1h_2}\,,$$

$$\mathbf{S}_{ij}(z) = \frac{z_{ij+1} - 2z_{ij} + z_{ij-1}}{(h_2)^2},$$

$$\mathbf{P}_{ij}(\mathbf{s}^n) = \left\{g^{\mathbf{s}} J^2 \frac{\partial}{\partial s^k}[w(\mathbf{s}^n) g_{\mathbf{s}}^{k1}(\mathbf{s}^n)]\right\}_{ij}$$

$$= \left\{g^{\mathbf{s}} J(-1)^{k+l} \frac{\partial s^{3-k}}{\partial \xi^{3-l}} \frac{\partial}{\partial \xi^l}\left[w(\mathbf{s}^n) g_{\mathbf{s}}^{k1}(\mathbf{s}^n)\right]\right\}_{ij}$$

$$= (g^{\mathbf{s}} J)_{ij} \Big[\frac{v_{ij+1}^n - v_{ij-1}^n}{2h_2} \frac{w(\mathbf{s}^n)_{i+1j} g_{\mathbf{s}}^{11}(\mathbf{s}^n)_{i+1j} - w(\mathbf{s}^n)_{i-1j} g_{\mathbf{s}}^{11}(\mathbf{s}^n)_{i-1j}}{2h_1}$$

$$- \frac{v_{i+1j}^n - v_{i-1j}^n}{2h_1} \frac{w(\mathbf{s}^n)_{ij+1} g_{\mathbf{s}}^{11}(\mathbf{s}^n)_{ij+1} - w(\mathbf{s}^n)_{ij-1} g_{\mathbf{s}}^{11}(\mathbf{s}^n)_{ij-1}}{2h_2}$$

$$- \frac{s_{ij+1}^n - s_{ij-1}^n}{2h_2} \frac{w(\mathbf{s}^n)_{i+1j} g_{\mathbf{s}}^{12}(\mathbf{s}^n)_{i+1j} - w(\mathbf{s}^n)_{i-1j} g_{\mathbf{s}}^{12}(\mathbf{s}^n)_{i-1j}}{2h_1}$$

$$+ \frac{s_{i+1j}^n - s_{i-1j}^n}{2h_1} \frac{w(\mathbf{s}^n)_{ij+1} g_{\mathbf{s}}^{12}(\mathbf{s}^n)_{ij+1} - w(\mathbf{s}^n)_{ij-1} g_{\mathbf{s}}^{12}(\mathbf{s}^n)_{ij-1}}{2h_2}\Big],$$

$$\mathbf{Q}_{ij}(\mathbf{s}^n) = \left\{g^{\mathbf{s}} J^2 \frac{\partial}{\partial s^k}[w(\mathbf{s}^n) g_{\mathbf{s}}^{k1}(\mathbf{s}^n)]\right\}_{ij}$$

$$= \left\{g^{\mathbf{s}} J(-1)^{k+l} \frac{\partial s^{3-k}}{\partial \xi^{3-l}} \frac{\partial}{\partial \xi^l}[w(\mathbf{s}^n) g_{\mathbf{s}}^{k2}(\mathbf{s}^n)]\right\}_{ij}$$

$$= (g^{\mathbf{s}} J)_{ij} \Big[\frac{v_{ij+1}^n - v_{ij-1}^n}{2h_2} \frac{w(\mathbf{s}^n)_{i+1j} g_{\mathbf{s}}^{12}(\mathbf{s}^n)_{i+1j} - w(\mathbf{s}^n)_{i-1j} g_{\mathbf{s}}^{12}(\mathbf{s}^n)_{i-1j}}{2h_1}$$

$$- \frac{v_{i+1j}^n - v_{i-1j}^n}{2h_1} \frac{w(\mathbf{s}^n)_{ij+1} g_{\mathbf{s}}^{12}(\mathbf{s}^n)_{ij+1} - w(\mathbf{s}^n)_{ij-1} g_{\mathbf{s}}^{12}(\mathbf{s}^n)_{ij-1}}{2h_2}$$

$$- \frac{s_{ij+1}^n - s_{ij-1}^n}{2h_2} \frac{w(\mathbf{s}^n)_{i+1j} g_{\mathbf{s}}^{22}(\mathbf{s}^n)_{i+1j} - w(\mathbf{s}^n)_{i-1j} g_{\mathbf{s}}^{22}(\mathbf{s}^n)_{i-1j}}{2h_1}$$

$$+ \frac{s_{i+1j}^n - s_{i-1j}^n}{2h_1} \frac{w(\mathbf{s}^n)_{ij+1} g_{\mathbf{s}}^{22}(\mathbf{s}^n)_{ij+1} - w(\mathbf{s}^n)_{ij-1} g_{\mathbf{s}}^{22}(\mathbf{s}^n)_{ij-1}}{2h_2}\Big],$$

$$
\begin{aligned}
g_{11}(\mathbf{s}^n)_{ij} = g^{\mathbf{s}}_{11}(\mathbf{s}^n)_{ij}\left(\frac{s^n_{i+1j} - s^n_{i-1j}}{2h_1}\right)^2 \\
+2g^{\mathbf{s}}_{12}(\mathbf{s}^n)_{ij}\frac{s^n_{i+1j} - s^n_{i-1j}}{2h_1}\frac{v^n_{i+1j} - v^n_{i-1j}}{2h_1} \\
+g^{\mathbf{s}}_{22}(\mathbf{s}^n)_{ij}\left(\frac{v^n_{i+1j} \quad v^n_{i-1j}}{2h_1}\right)^2 ,
\end{aligned}
$$

$$
\begin{aligned}
g_{12}[\mathbf{s}^n]_{ij} &= g^{\mathbf{s}}_{11}(\mathbf{s}^n)_{ij}\frac{s^n_{i+1j} - s^n_{i-1j}}{2h_1}\frac{s^n_{ij+1} - s^n_{ij-1}}{2h_2} \\
&+g^{\mathbf{s}}_{12}(\mathbf{s}^n)_{ij}\frac{s^n_{i+1j} - s^n_{i-1j}}{2h_1}\frac{v^n_{ij+1} - v^n_{ij-1}}{2h_2} \\
&+g^{\mathbf{s}}_{12}[\mathbf{s}^n]_{ij}\frac{s^n_{ij+1} - s^n_{ij-1}}{2h_2}\frac{v^n_{i+1j} - v^n_{i-1j}}{2h_1} \\
&+g^{\mathbf{s}}_{22}(\mathbf{s}^n)_{ij}\frac{v^n_{i+1j} - v^n_{i-1j}}{2h_1}\frac{v^n_{ij+1} - v^n_{ij-1}}{2h_2} ,
\end{aligned}
$$

$$
\begin{aligned}
g_{22}[\mathbf{s}^n]_{ij} &= g^{\mathbf{s}}_{11}(\mathbf{s}^n)_{ij}\left(\frac{s^n_{ij+1} - s^n_{ij-1}}{2h_2}\right)^2 \\
&+2g^{\mathbf{s}}_{12}(\mathbf{s}^n)_{ij}\frac{s^n_{ij+1} - s^n_{ij-1}}{2h_2}\frac{v^n_{ij+1} - v^n_{ij-1}}{2h_2} \\
&+g^{\mathbf{s}}_{22}(\mathbf{s}^n)_{ij}\left(\frac{v^n_{ij+1} - v^n_{ij-1}}{2h_2}\right)^2 ,
\end{aligned}
$$

$$1 \le i \le N_1 - 1 , \quad 1 \le j \le N_2 - 1 , \quad k, l = 1, 2 .$$

Algorithm for Computation

The solution of the scheme described above is obtained by applying successively formulas (7.6) and (7.7) for the solution of the reference difference problem (7.5). Namely, assuming

$$
\begin{aligned}
&A^{n+1/2}_{ij} = g_{22}(\mathbf{s}^n)_{ij} , \quad B^{n+1/2}_{ij} = g_{22}(\mathbf{s}^n)_{ij} , \\
&C^{n+1/2}_{ij} = 2g_{22}(\mathbf{s}^n)_{ij} + \theta_1, \quad \theta_1 = (h_1)^2/\tau , \\
&F^n_{ij} = \theta_1 s^n_{ij} - 2g_{12}(\mathbf{s}^n)_{ij}\mathbf{M}_{ij}(s^n) + g_{11}(\mathbf{s}^n)_{ij}\mathbf{S}_{ij}(s^n) + \mathbf{P}_{ij}(\mathbf{s}^n) ,
\end{aligned}
\tag{7.21}
$$

we obtain, in accordance with (7.6) and (7.7), a solution of (7.17) for each fixed number j, $0 < j < N_2$,

$$
\begin{aligned}
&s^{n+1/2}_{ij} = \alpha^{n+1/2}_{i+1j} s^{n+1/2}_{i+1j} + \beta^{n+1/2}_{i+1j} , \\
&i = 1, \dots , N_1 - 1 , \quad s^{n+1/2}_{N_1 j} = \varphi^1(1, jh_2) ,
\end{aligned}
\tag{7.22}
$$

where

$$\alpha_{i+1j}^{n+1/2} = \frac{B_{ij}^{n+1/2}}{C_{ij}^{n+1/2} - \alpha_{ij}^{n+1/2} A_{ij}^{n+1/2}} , \qquad (7.23)$$
$$i = 1, \dots , N - 1 , \quad \alpha_{1j}^{n+1/2} = 0 ,$$

$$\beta_{i+1j}^{n+1/2} = \frac{A_{ij}^{n+1/2} \beta_{ij}^{n+1/2} + F_{ij}^{n}}{C_{ij}^{n+1/2} - \alpha_{ij}^{n+1/2} A_{ij}^{n+1/2}} , \qquad (7.24)$$
$$i = 1, \dots , N_1 - 1 , \quad \beta_{1j}^{n+1/2} = \varphi^1(0, jh_2) .$$

Analogously the solution of (7.18), for each index j, $1 \le j \le N_2 - 1$, is expressed by the following recursive formula

$$v_{ij}^{n+1/2} = \alpha_{i+1j}^{n+1/2} v_{i+1j}^{n+1/2} + \beta_{i+1j}^{n+1/2} , \qquad (7.25)$$
$$i = 1, \dots , N_1 - 1 , \quad v_{N_1 j}^{n+1/2} = \varphi^2(1, jh_2) ,$$

where $\alpha_{i+1j}^{n+1/2}$ and $\beta_{i+1j}^{n+1/2}$ are computed by (7.24) and (7.25), respectively, with

$$\alpha_{1j}^{n+1/2} = 0 , \quad \beta_{1j}^{n+1/2} = \varphi_2(0, jh_2) ,$$

$$A_{ij}^{n+1/2} = g_{22}(\mathbf{s}_{ij}^{n})_{ij} , \quad B_{ij}^{n+1/2} = g_{22}(\mathbf{s}^{n})_{ij} , \qquad (7.26)$$
$$C_{ij}^{n+1/2} = 2g_{22}(\mathbf{s}^{n})_{ij} + \theta_1, \quad \theta_1 = (h_1)^2/\tau ,$$

$$F_{ij}^{n} = \theta_1 v_{ij}^{n} - 2g_{12}(\mathbf{s}^{n})_{ij}\mathbf{M}_{ij}(v^{n}) + g_{11}(\mathbf{s}^{n})_{ij}\mathbf{S}_{ij}(v^{n}) + \mathbf{Q}_{ij}(\mathbf{s}^{n}) .$$

In order to compute (7.19) by (7.6) we assume, for each fixed i, $1 \le i \le N_1 - 1$,

$$A_{ij}^{n+1} = g_{11}(\mathbf{s}^{n})_{ij} , \quad B_{ij}^{n+1} = g_{11}(\mathbf{s}^{n})_{ij} , \quad C_{ij}^{n+1} = 2g_{11}(\mathbf{s}^{n})_{ij} + \theta_2,$$
$$\theta_2 = (h_2)^2/\tau , \qquad (7.27)$$
$$F_{ij}^{n+1/2} = \theta_2 s_{ij}^{n+1/2} - g_{11}(\mathbf{s}^{n})_{ij}\mathbf{S}_{ij}(s^{n}) .$$

In accordance with (7.6) and (7.7) we find the solution of (7.19)

$$s_{ij}^{n+1} = \alpha_{ij+1}^{n+1} s_{ij+1}^{n+1} + \beta_{ij+1}^{n+1} , \qquad (7.28)$$
$$j = 1, \dots , N_2 - 1 , \quad s_{iN_2}^{n+1} = \varphi^1(ih_1, 1) ,$$

where

$$\alpha_{ij+1}^{n+1} = \frac{B_{ij}^{n+1}}{C_{ij}^{n+1} - \alpha_{ij}^{n+1} A_{ij}^{n+1}} , \qquad (7.29)$$
$$j = 1, \dots , N_2 - 1 , \quad \alpha_{i1}^{n+1} = 0 ,$$

$$\beta_{ij+1}^{n+1} = \frac{A_{ij}^{n+1}\beta_{ij}^{n+1} + F_{ij}^{n+1/2}}{C_{ij}^{n+1} - \alpha_{ij}^{n+1}A_{ij}^{n+1}}\,, \qquad (7.30)$$
$$j = 1, \dots, N_2 - 1\,, \quad \beta_{i1}^{n+1} = \varphi^1(ih_1, 0)\,.$$

Analogously the solution of (7.20) for each fixed index i, $1 < i < N_1 - 1$, is expressed by the recursive formula

$$v_{ij}^{n+1} = \alpha_{ij+1}^{n+1}v_{ij+1}^{n+1} + \beta_{ij+1}^{n+1}\,, \quad v_{iN_2}^{n+1} = \varphi^2(ih_1, 1), \qquad (7.31)$$

where α_{ij}^{n+1} and β_{ij}^{n+1} are computed by (7.30), and (7.31), respectively, with $\theta_2 = (h_2)^2/\tau$,

$$\alpha_{i1}^{n+1} = 0\,, \quad \beta_{i1}^{n+1} = \varphi^2(ih_1, 0)\,,$$
$$A_{ij}^{n+1} = g_{11}(\mathbf{s}^n)_{ij}\,, \quad B_{ij}^{n+1} = g_{11}(\mathbf{s}^n)_{ij}\,, \quad C_{ij}^{n+1} = 2g_{11}(\mathbf{s}^n)_{ij} + \theta_2\,, \qquad (7.32)$$
$$F_{ij}^{n+1/2} = \theta_2 v_{ij}^{n+1/2} - g_{11}(\mathbf{s}^n)_{ij}\mathbf{S}(v^n)\,.$$

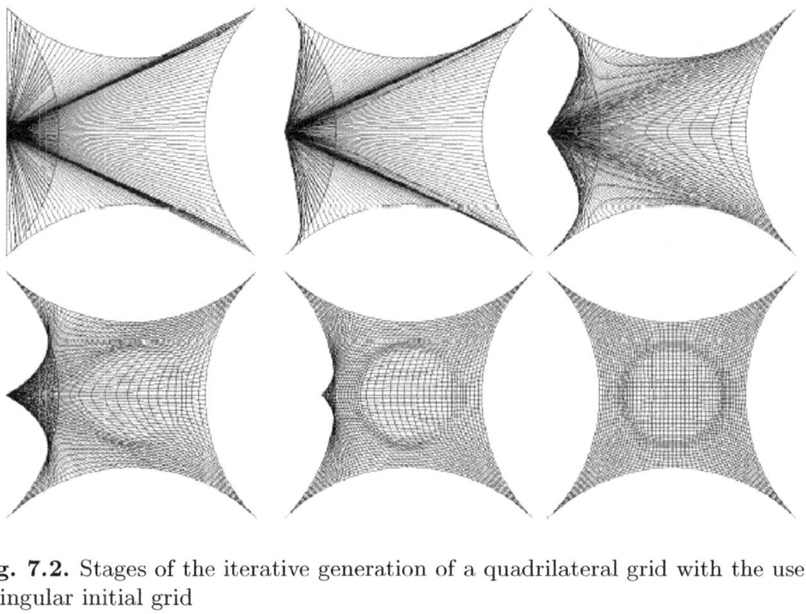

Fig. 7.2. Stages of the iterative generation of a quadrilateral grid with the use of a singular initial grid

An approximate solution of (7.13) is the solution $\mathbf{s}_{ij}^n$ at a step n such that

$$\max_{0\le i\le N_1, 0\le j\le N_2} \frac{1}{\tau}\sqrt{(s_{ij}^{n+1} - s_{ij}^n)^2 + (v_{ij}^{n+1} - v_{ij}^n)^2} \le \varepsilon\,, \qquad (7.33)$$

for some sufficiently small $\varepsilon > 0$.

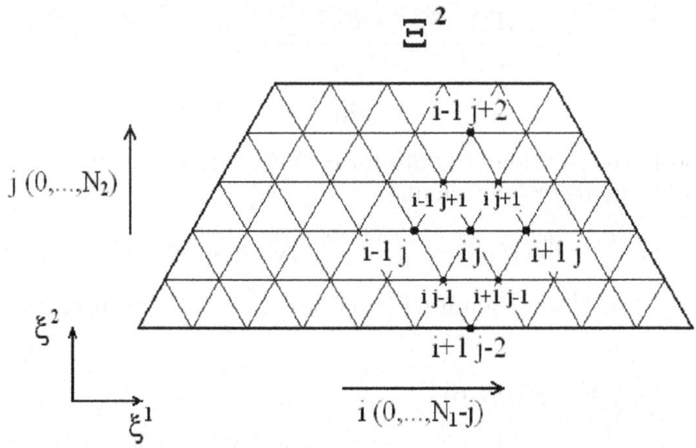

Fig. 7.3. Two-dimensional triangular stencil for finite differences

Step–by–Step Algorithm

The numerical algorithm described above is a sequential process. The details of the algorithm are presented here in a step-by-step manner.
Step 1.
Define an initial grid distribution, using the formulas (7.14) or (7.16), by introducing two difference functions s^0_{ij} and v^0_{ij}, $0 \le i \le N_1$, $0 \le j \le N_2$, such that

$$(s^0_{0j}, v^0_{0j}) = \mathbf{s}(0, jh_2) , \quad (s^0_{N_1 j}, v^0_{N_2 j}) = \mathbf{s}(1, jh_2) ,$$
$$(s^0_{i0}, v^0_{i0}) = \mathbf{s}(ih_1, 0) , \quad (s^0_{iN_2}, v^0_{iN_2}) = \mathbf{s}(ih_1, 1) .$$

Step 2.
Compute the functions $g_{11}(\mathbf{s}^0_{ij})$, $g_{12}(\mathbf{s}^0_{ij})$, $g_{22}(\mathbf{s}^0_{ij})$, $g(\mathbf{s}^0_{ij})$, $\mathbf{M}_{ij}(s^0)$, $\mathbf{M}_{ij}(v^0)$, $\mathbf{P}_{ij}(\mathbf{s}^0)$, $\mathbf{Q}_{ij}(\mathbf{s}^0)$.
Step 3.
Compute $s^{0+1/2}_{ij}$ using (7.22 – 7.24).
Step 4.
Compute $v^{0+1/2}_{ij}$ using (7.25 – 7.26).
Step 5.
Compute s^{0+1}_{ij} using (7.28 – 7.30).
Step 6.
Compute v^{0+1}_{ij} using (7.31 – 7.32).
Step 9.
Return to step 2 assuming $(s^0_{ij}, v^0_{ij}) = (s^1_{ij}, v^1_{ij})$, where s^1_{ij} and v^1_{ij} are taken from steps 5 and 6.

Continue until the tolerance (7.33) is observed.

This algorithm is readily transformed for the numerical solution of the grid equations (6.171), as well as of the inverted Beltrami equations simply by substituting in the above formulas $\sqrt{g^{\mathbf{s}}}$ for $w(\mathbf{s})$.

Figure 7.2 demonstrates some steps of the grid generation in a two-dimensional domain by the solution of the inverted Beltrami equations with the iterative algorithm described. The initial grid is singular (all its interior nodes merge into one node lying outside of the domain).

Generation of Triangular Grids

The numerical algorithms described for generating quadrilateral grids are naturally applied to the generation of triangular grids when the logical domain is a symmetric trapezoid. The scheme, similar to that in Fig. 7.1, is demonstrated in Fig. 7.3.

An example of a triangular domain grid generated by such algorithm is exhibited by Fig. 7.4. As well as in Fig. 7.2 the initial grid is singular. All its interior points are placed into three points, two of them lie outside of the domain.

Algorithm for Generating Grids on Two-Dimensional Surfaces

In the same way as for domains there are generated grids in a two-dimensional surface S^{x2} represented as

$$\mathbf{x}(\mathbf{s}): S^2 \to R^3\,, \quad \mathbf{x} = (x^1, x^2, x^3)\,, \quad \mathbf{s} = (s^1, s^2)\,, \tag{7.34}$$

by solving the boundary value problem for the inverted diffusion equations (6.170) and (6.171) as well as for the corresponding inverted Beltrami equations with respect to a monitor metric $g^{\mathbf{s}}_{ij}$ over S^{x2}.

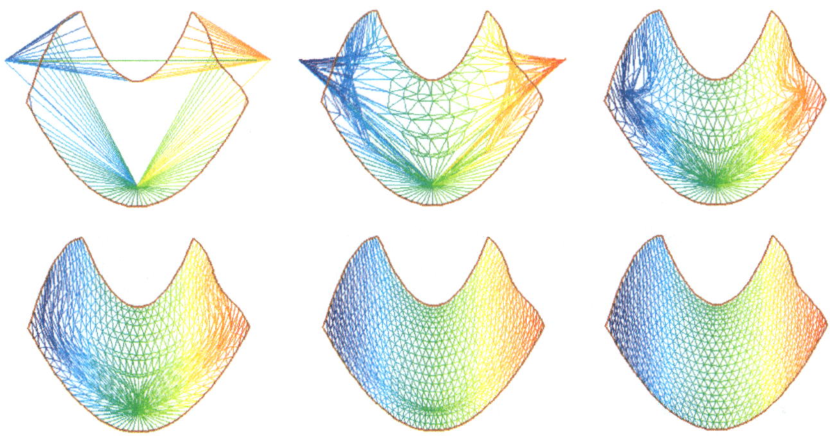

Fig. 7.4. Stages for generating a triangular grid by using a singular initial grid

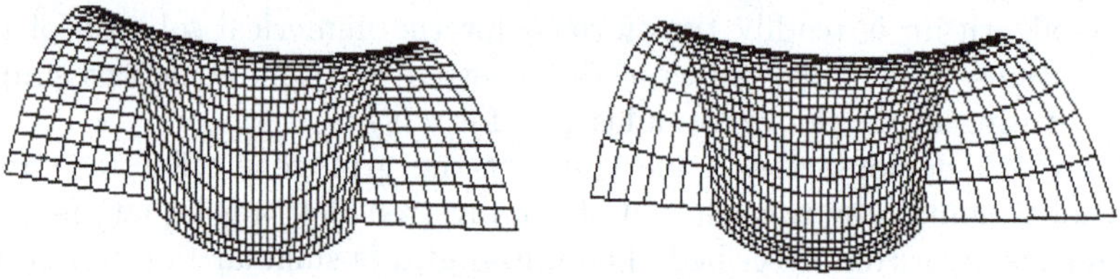

Fig. 7.5. Adaptive surface grids generated with the use of a monitor function. The boundary grid nodes in the left figure are generated without the use of the monitor function

Similarly to the case of a two-dimensional domain we also choose a square or trapezoid for the logical domain Ξ^2. We also assume that the boundary transformation

$$\boldsymbol{\varphi}(\boldsymbol{\xi}) : \partial\Xi^2 \to \partial S^2 , \quad \boldsymbol{\varphi} = (\varphi^1, \varphi^2) ,$$

which is continuous on $\partial\Xi^2$, has been specified on the boundary grid points of $\partial\Xi^2$, for example, by computing it through the algorithm described in Sect. 7.1.1.

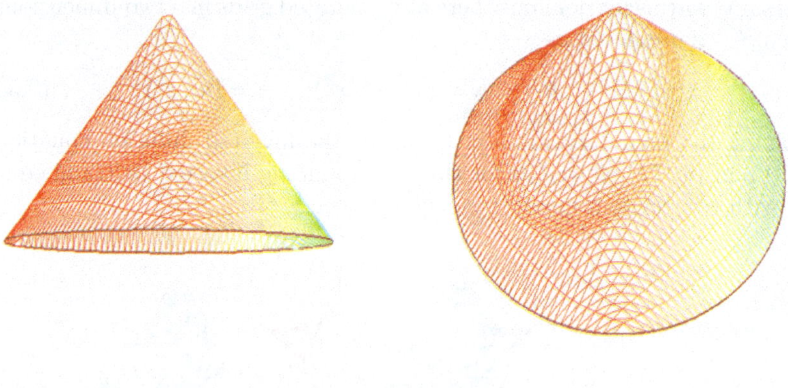

Fig. 7.6. A triangular adaptive grid on a conical surface

The grid on S^{x2} is obtained by mapping with $\mathbf{x}(\mathbf{s})$ the grid nodes computed in S^2 through the numerical solution of the Dirichlet problem with respect to $\mathbf{s}(\boldsymbol{\xi})$ for the inverted grid equations.

Figures 7.5 and 7.6 illustrate surface grids generated by the algorithm.

7.1.3 Three–Dimensional Equations

For generating grids in a three-dimensional domain $S^3 \subset R^3$ with the use of the inverted diffusion equations with respect to a monitor metric $g^{\mathbf{s}}_{ij}$, $i, j =$

1, 2, 3 we formulate the following boundary value problem for the equations (6.190)

$$w[\mathbf{s}(\boldsymbol{\xi})]L^3[s^i] = g^{\mathbf{s}} J^2 \frac{\partial}{\partial s^j}\left[w(\mathbf{s}) g^{ij}_{\mathbf{s}}\right], \quad i,j = 1,2,3\,, \qquad \mathbf{s}(\boldsymbol{\xi})\Big|_{\partial \Xi^3} = \boldsymbol{\varphi}(\boldsymbol{\xi})\,, \tag{7.35}$$

where

$$\begin{aligned}L^3[v] &= g^{\boldsymbol{\xi}} g^{ij}_{\boldsymbol{\xi}} \frac{\partial^2 v}{\partial \xi^i \partial \xi^j} \\ &= [g^{\boldsymbol{\xi}}_{22} g^{\boldsymbol{\xi}}_{33} - (g^{\boldsymbol{\xi}}_{23})^2] \frac{\partial^2 v}{\partial \xi^1 \partial \xi^1} + 2[g^{\boldsymbol{\xi}}_{23} g^{\boldsymbol{\xi}}_{13} - g^{\boldsymbol{\xi}}_{12} g^{\boldsymbol{\xi}}_{33}] \frac{\partial^2 v}{\partial \xi^1 \partial \xi^2} \\ &\quad + 2[g^{\boldsymbol{\xi}}_{12} g^{\boldsymbol{\xi}}_{23} - g^{\boldsymbol{\xi}}_{22} g^{\boldsymbol{\xi}}_{13}] \frac{\partial^2 v}{\partial \xi^1 \partial \xi^3} + [g^{\boldsymbol{\xi}}_{11} g^{\boldsymbol{\xi}}_{33} - (g^{\boldsymbol{\xi}}_{13})^2] \frac{\partial^2 v}{\partial \xi^2 \partial \xi^2} \\ &\quad + 2[g^{\boldsymbol{\xi}}_{12} g^{\boldsymbol{\xi}}_{13} - g^{\boldsymbol{\xi}}_{11} g^{\boldsymbol{\xi}}_{23}] \frac{\partial^2 v}{\partial \xi^2 \partial \xi^3} + [g^{\boldsymbol{\xi}}_{11} g^{\boldsymbol{\xi}}_{22} - (g^{\boldsymbol{\xi}}_{12})^2] \frac{\partial^2 v}{\partial \xi^3 \partial \xi^3}\,.\end{aligned}$$

Analogously to the solution of (7.13) we find a solution of (7.35) as a limit with $t \to \infty$ of the solution of the corresponding parabolic problem

$$\begin{aligned}&\frac{\partial s^i}{\partial t} = w[\mathbf{s}(\boldsymbol{\xi})]L^3[s^i] - g^{\mathbf{s}} J^2 \frac{\partial}{\partial s^j}\left[w(\mathbf{s}) g^{ij}_{\mathbf{s}}\right], \quad i,j = 1,2,3\,, \\ &\mathbf{s}(\boldsymbol{\xi}, t) = \boldsymbol{\varphi}(\boldsymbol{\xi})\,, \quad \boldsymbol{\xi} \in \partial \Xi^3\,, \quad t \geq 0\,, \\ &\mathbf{s}(\boldsymbol{\xi}, 0) = \mathbf{s}_0(\boldsymbol{\xi})\,, \quad \boldsymbol{\xi} \in \Xi^3\,.\end{aligned} \tag{7.36}$$

Initial Transformation

The initial transformation

$$\mathbf{s}(\boldsymbol{\xi}, 0) = \mathbf{s}_0(\boldsymbol{\xi}) : \Xi^3 \to S^3\,.$$

can be found by propagating the values of $\boldsymbol{\varphi}(\boldsymbol{\xi})$ into the interior of the unit cube Ξ^3, for example, through the formula of Lagrange transfinite interpolation. In particular, for the simplest expressions of the blending functions

$$\alpha^i_{0j}(s) = 1 - s\,, \quad \alpha^i_{1j}(s) = s\,,$$

we find from (1.10)

$$F_1^i(\xi^1,\xi^2,\xi^3) = (1-\xi^1)\varphi^i(0,\xi^2,\xi^3) + \xi^1\varphi^i(1,\xi^2,\xi^3)\ ,$$
$$F_2^i(\xi^1,\xi^2,\xi^3) = F_1^i(\xi^1,\xi^2,\xi^3) + (1-\xi^2)[\varphi^i(\xi^1,0,\xi^3)$$
$$-F_1^i(\xi^1,0,\xi^3)] + \xi^2[\varphi^i(\xi^1,1,\xi^3) - F_1^i(\xi^1,1,\xi^3)]\ , \tag{7.37}$$
$$x^i(\xi^1,\xi^2,\xi^3) = F_2^i(\xi^1,\xi^2,\xi^3) + (1-\xi^3)[\varphi^i(\xi^1,\xi^2,0)$$
$$-F_2^i(\xi^1,\xi^2,0)] + \xi^3[\varphi^i(\xi^1,\xi^2,1) - F_2^i(\xi^1,\xi^2,1)]\ ,\quad i=1,2,3\ ,$$

A numerical algorithm for solving the problem (7.36) is formulated analogously to the two-dimensional algorithm reviewed by formulas (7.17–7.20), namely, by splitting the process of the numerical solution into a series of one-dimensional algorithms. In particular assuming

$$s(\boldsymbol{\xi},t) = s^1(\boldsymbol{\xi},t),\quad v(\boldsymbol{\xi},t) = s^2(\boldsymbol{\xi},t),\quad p(\boldsymbol{\xi},t) = s^3(\boldsymbol{\xi},t),$$

the problem (7.36) with respect to the difference variables s^n_{ijl}, $0 \le i,j,l \le N$, $0 \le n$, representing $s(\boldsymbol{\xi},t)$ is approximated at the points $(ih, jh, lh, n\tau)$, $h = 1/N$, by the scheme:

$$\begin{aligned}\frac{s^{n+1/3}_{ijl} - s^n_{ijl}}{\tau} &= \frac{1}{h^2}[a^{11}\mathbf{D_{11}}(s^{n+1/2}_{ijl}) + 2a^{12}\mathbf{D_{12}}(s^n_{ijl})\\ &+a^{22}\mathbf{D_{22}}(s^n_{ijl}) + 2a^{23}\mathbf{D_{23}}\\ &+2a^{13}\mathbf{D_{13}}(s^n_{ijl}) + a^{33}\mathbf{D_{33}}(s^n_{ijl})] + \mathbf{P^1}(\mathbf{s}^n_{ijl}),\\ &1 \le i,j,l \le N-1\ ,\quad n \ge 0\ ,\quad i \text{ fixed}\ ,\end{aligned}$$

$$\begin{aligned}\frac{s^{n+2/3}_{ijl} - s^{n+1/3}_{ijl}}{\tau} &= \frac{1}{h^2}[a^{22}\mathbf{D_{22}}(s^{n+1}_{ijl} - s^n_{ijl})]\ ,\\ &1 \le i,j,l \le N-1\ ,\quad n \ge 0\ ,\quad j \text{ fixed}\ ,\end{aligned}$$

$$\begin{aligned}\frac{s^{n+1}_{ijl} - s^{n+2/3}_{ijl}}{\tau} &= \frac{1}{h^2}[a^{33}\mathbf{D_{33}}(s^{n+1}_{ijl} - s^n_{ijl})]\ ,\\ &1 \le i,j,l \le N-1\ ,\quad n \ge 0\ ,\quad l \text{ fixed}\ .\end{aligned}$$

where

$$\mathbf{D_{11}}(z_{ijl}) = z_{i+1jl} - 2z_{ijl} + z_{i-1jl}\,, \quad 1 \le i,j,l \le N-1\,,$$

$$\mathbf{D_{12}}(z_{ijl}) = \frac{1}{4}(z_{i+1j+1l} - z_{i-1j+1l} - z_{i+1j-1l} + z_{i-1j-1l})\,, \; 1 \le i,j,l \le N-1\,,$$

$$\mathbf{D_{22}}(z_{ijl}) = z_{ij+1l} - 2z_{ijl} + z_{ij-1l}\,, \quad 1 \le i,j,l \le N-1\,,$$

$$\mathbf{D_{33}}(z_{ijl}) = z_{ijl+1} - 2z_{ijl} + z_{ijl-1}\,, \quad 1 \le i,j,l \le N-1\,,$$

$$\mathbf{D_{13}}(z_{ijl}) = \frac{1}{4}(z_{i+1jl+1} - z_{i-1jl+1} - z_{i+1jl-1} + z_{i-1jl-1})\,, \; 1 \le i,j,l \le N-1\,,$$

$$\mathbf{D_{23}}(z_{ijl}) = \frac{1}{4}(z_{ij+1l+1} - z_{ij+1l-1} - z_{ij-1l+1} + z_{ij-1l-1})\,, \; 1 \le i,j,l \le N-1\,,$$

$$a^{km} = [w(\mathbf{s})g^{\boldsymbol{\xi}} g_{\boldsymbol{\xi}}^{km}]^n_{ijl}\,, \quad 1 \le i,j,l \le N-1\,, \quad k,m = 1,2,3\,,$$

$$\mathbf{P^1}(\mathbf{s}^n_{ijl}) = -[g^{\mathbf{s}} J^2 \frac{\partial}{\partial s^m}(w(\mathbf{s}) g_{\mathbf{s}}^{1m})]^n_{ijl}\,, \quad 1 \le i,j,l \le N-1\,, \quad m = 1,2,3\,.$$

Similar expressions are written out for computing the variables $v(\boldsymbol{\xi}, t)$ and $p(\boldsymbol{\xi}, t)$.

An example of a three-dimensional spatial grid generated with the use of this scheme is demonstrated in Fig. 7.7. The grid adaptation in the vicinity of a sphere is provided by the weight function $w(\mathbf{s})$, where

$$w(\mathbf{s}) = 0.8\; exp\Big[3\; exp(-\frac{R^2 - 0.0625}{0.0001})\Big],$$
$$R^2 = (s^1 - 0.5)^2 + (s^2 - 0.5)^2 + (s^3 - 0.5)^2$$

Analogously, assuming in (7.35) $w(\mathbf{s}) = \sqrt{g^{\mathbf{s}}}$, we get a numerical algorithm for solving the Dirichlet boundary value problem for the three-dimensional inverted Beltrami grid equations.

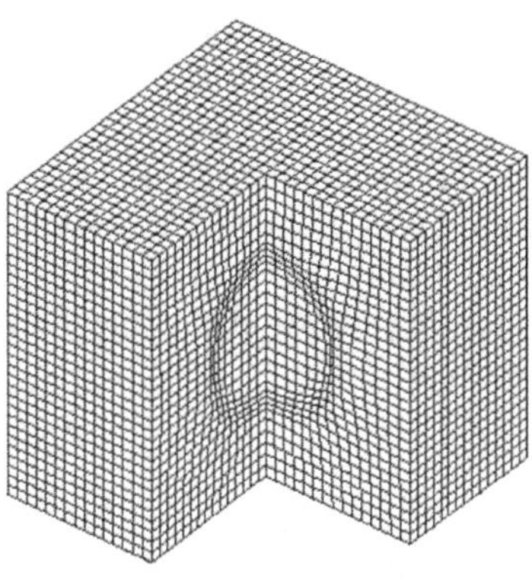

Fig. 7.7. Three-dimensional grid with node clustering in the vicinity of a sphere

7.2 Method of Minimization of Energy Functional

This section describes another finite-difference grid generation algorithm based of the minimization of the functional (5.32). Following to Charakhch'yan and Ivanenko (1988,1997) who originated it the algorithm is first expounded for the two-dimensional version of the functional in the Euler metric, i.e.

$$I[\mathbf{s}] = \int_{\Xi^2} \frac{(x_\xi)^2 + (x_\eta)^2 + (y_\xi)^2 + (y_\eta)^2}{J} d\xi d\eta, \tag{7.38}$$

where $J = x_\xi y_\eta - x_\eta y_\xi$, and then an explanation is given how it can be generalized to monitor metrics and other dimensions. Note the functional (5.32) in the Euler metric $g^{\mathbf{s}}_{ij} = \delta^i_j$, $i, j = 1, 2$, becomes the functional (7.38) when the following designations

$$\xi^1 = \xi, \quad \xi^2 = \eta, \quad \mathbf{s}(\xi, \eta) = [x(\xi, \eta), y(\xi, \eta)].$$

are assumed.

By the algorithm the functional (7.38) is approximated by a discrete functional $I^h[\ ]$. This is made by approximating the integrand of (7.38) at each grid cell of the logical domain Ξ^2 and then carrying out summation over all cells.

7.2.1 Generation of Fixed Grids

The problem of grid generation is treated as a discrete analog of the problem of finding the components $x(\xi, \eta)$ and $y(\xi, \eta)$ of the intermediate transformation $\mathbf{s}(\xi, \eta)$ producing one-to-one mapping of the logical square

$$0 < \xi < 1, \ 0 < \eta < 1$$

onto a physical domain X^2.

Instead of the logical square on the plane ξ, η the parametric rectangle

$$1 < \xi < N, \ 1 < \eta < M.$$

is introduced to simplify the computational formulas. This rectangle is associated with the square grid (ξ_i, η_j) on the plane ξ, η such that $\xi_i = i$, $\eta_j = j$, $i = 1, \dots, N$; $j = 1, \dots, M$.

It is readily shown that if a smooth mapping of one domain onto another with a one-to-one transformation between boundaries possesses a positive Jacobian, then such a mapping will be one-to-one. Hence, the grid coordinate system, generated in the domain X^2, will be non-degenerate if the Jacobian of the mapping $\mathbf{s}(\xi, \eta) = [x(\xi, \eta)\ ,\ y(\xi, \eta)]$ is positive:

$$J = x_\xi y_\eta - x_\eta y_\xi > 0. \tag{7.39}$$

Thus, the problem of the construction of the grid coordinates in the domain X^2 can be formulated as the problem of finding a smooth mapping of the parametric rectangle onto the domain X^2, which satisfies the condition of the Jacobian positiveness.

Formulation of Discrete Functional

Let the coordinates $(x,y)_{ij}$ of grid nodes be given. To construct the mapping $x^h(\xi,\eta)$, $y^h(\xi,\eta)$ of the parametric rectangle onto the domain X^2 such that $x^h(i,j) = x_{ij}$ and $y^h(i,j) = y_{ij}$ quadrilateral isoparametric finite elements are used. The square cell numbered as $i+1/2, j+1/2$ on the plane ξ,η is mapped onto the quadrilateral cell on the plane x,y , formed by the nodes with coordinates $(x,y)_{ij}$, $(x,y)_{ij+1}$, $(x,y)_{i+1j+1}$, $(x,y)_{i+1j}$. The cell vertices are numbered from 1 to 4 in the clockwise direction. The node (i,j) corresponds to the vertex 1, node $(i,j+1)$ to vertex 2 and so on. Each vertex is associated with a triangle: vertex 1 with $\triangle_{412}$, vertex 2 with $\triangle_{123}$ and so on. The doubled area J_k, $k=1,2,3,4$, of these triangles is introduced as follows

$$J_k = (x_{k-1} - x_k)(y_{k+1} - y_k) - (y_{k-1} - y_k)(x_{k+1} - x_k)$$

where one should put $k-1=4$ if $k=1$, $k+1=1$ if $k=4$.

The functions x^h , y^h for $i \le \xi \le i+1$, $j \le \eta \le j+1$ are represented in the form

$$\begin{aligned} x^h(\xi,\eta) &= x_1 + (x_4 - x_1)(\xi - i) + (x_2 - x_1)(\eta - j) \\ &\quad + (x_3 - x_4 - x_2 + x_1)(\xi - i)(\eta - j), \\ y^h(\xi,\eta) &= y_1 + (y_4 - y_1)(\xi - i) + (y_2 - y_1)(\eta - j) \\ &\quad + (y_3 - y_4 - y_2 + y_1)(\xi - i)(\eta - j). \end{aligned} \tag{7.40}$$

Each side of the square is linearly transformed onto the appropriate side of the quadrilateral. Consequently, the global transformation x^h , y^h is continuous on the cell boundaries. To check the one-to-one property of the transformation (7.40) we write out the expression for its Jacobian

$$J^h = x^h_\xi y^h_\eta - x^h_\eta y^h_\xi = \det\begin{pmatrix} x_4 - x_1 + A(\eta - j) & x_2 - x_1 + A(\xi - i) \\ y_4 - y_1 + B(\eta - j) & y_2 - y_1 + B(\xi - i) \end{pmatrix},$$

where $A = x_3 - x_4 - x_2 + x_1$, $B = y_3 - y_4 - y_2 + y_1$. The function J^h is linear, not bilinear, since the coefficient before $\xi\eta$ in this determinant is equal to zero. Consequently, if $J^h > 0$ at all corner points of the square, it does not vanish inside this square. At the corner node 1 ($\xi = i$, $\eta = j$) of the cell $i+1/2, j+1/2$ the Jacobian equals

$$J^h(i,j) = (x_4 - x_1)(y_2 - y_1) - (y_4 - y_1)(x_2 - x_1),$$

i.e. $J^h(i,j) = J_1$ is the doubled area of the triangle $\triangle_{412}$, introduced above. From this follows that the condition of the Jacobian positiveness $x^h_\xi y^h_\eta - x^h_\eta y^h_\xi > 0$ is equivalent to the system of inequalities

$$[J_k]_{i+1/2j+1/2} > 0, \quad k = 1,2,3,4; \; i = 1,\ldots,N-1; \; j = 1,\ldots,M-1. \tag{7.41}$$

If conditions (7.41) are satisfied, then all the grid cells are convex quadrilaterals. The set of grids satisfying these inequalities is called a convex grid set and denoted by D. This set belongs to the Euclidean space R^{N_1} , where $N_1 = 2(N-2)(M-2)$ is the total number of degrees of freedom of the grid equal to twice the number of its interior nodes.

Finaly the problem is formulated as follows. The convex grid, satisfying inequalities (7.41), must be generated in the domain X^2 for the given coordinates of the boundary nodes.

The mapping $x(\xi,\eta)$, $y(\xi,\eta)$ is approximated by functions $x^h(\xi,\eta)$, $y^h(\xi,\eta)$ introduced in (7.40). Substituting those expressions in (7.38) and replacing integrals over square cells by the quadrature formulas with nodes coinciding with the grid vertices on the plane ξ,η, the following discrete analog of the functional (7.38) is obtained:

$$I^h = \sum_{i=1}^{N-1}\sum_{j=1}^{M-1}\sum_{k=1}^{4}\frac{1}{4}[F_k]_{i+1/2\ j+1/2}, \tag{7.42}$$

where F_k is the integrand evaluated in the k - th grid node as

$$F_k = [(x_{k+1}-x_k)^2 + (x_k - x_{k-1})^2 + (y_{k+1}-y_k)^2 + (y_k - y_{k-1})^2]J_k^{-1}, \tag{7.43}$$

and J_k is the doubled area of the triangle introduced above.

Notice some properties of the function (7.42). For this purpose we introduce a parametric rectangle $0<\xi<1, 0<\eta<\alpha$, where $\alpha=(M-1)/(N-1)$ is the constant, instead of the unit logical square as a domain of integration in (7.38). In this case the continuous limit of the expression $I^h/(N-1)^2$ when $N,M\to\infty$ in such a way, that $(M-1)/(N-1)=\alpha=const$, will be the functional (7.38).

It is readily obtained the following identity

$$\begin{aligned} I &= \int_0^1\int_0^\alpha \frac{x_\xi^2+y_\xi^2+x_\eta^2+y_\eta^2-2(x_\xi y_\eta - x_\eta y_\xi)+2(x_\xi y_\eta - x_\eta y_\xi)}{J}d\xi d\eta \\ &= \int_0^1\int_0^\alpha \frac{(x_\xi - y_\eta)^2+(x_\eta - y_\xi)^2}{J}d\xi d\eta + 2\alpha. \end{aligned}$$

From this follows that the functional (7.38) has a lower bound equal to 2α. If this minimum is attained, the mapping $\mathbf{s}(\xi,\eta)$ is conformal:

$$x_\xi = y_\eta, \qquad x_\eta = -y_\xi.$$

To obtain the corresponding property of the discrete analog (7.42) of the functional (7.38) consider one term in (7.43) for $k=2$. We can assume that

$x_2 = 0$ and $y_2 = 0$ since (7.43) contains only finite differences of the grid node coordinates. In this case we obtain the following identity

$$F_2 = \frac{x_1^2 + y_1^2 + x_3^2 + y_3^2}{x_1 y_3 - x_3 y_1}$$
$$= \frac{x_1^2 + y_1^2 + x_3^2 + y_3^2 - 2(x_1 y_3 - x_3 y_1) + 2(x_1 y_3 - x_3 y_1)}{x_1 y_3 - x_3 y_1}$$
$$= \frac{(x_1 - y_3)^2 + (x_3 + y_1)^2}{x_1 y_3 - x_3 y_1} + 2.$$

From this follows that the function $I^h/(N-1)^2$ has on the set D a lower bound equal to $2(M-1)/(N-1)$. If this minimum is attained, the coordinates of the grid nodes satisfy a discrete analog of the conformal conditions

$$x_1 = y_3, \quad x_3 = -y_1.$$

If these conditions are satisfied for all cells, each grid cell will be a a square.

Note, that the function (7.42) is not convex and, in principle, multiple solutions may exist.

The function I^h possesses also the following very important property. If $G \to \partial D$ for $G \in D$, where ∂D is the boundary of the set of convex grids D, i.e. if at least one of the quantities J_k tends to zero for some cell while remaining positive, then $I^h(G) \to +\infty$. In fact, suppose that $J_k \to 0$ in (7.43) for some cell, but I^h does not tend to $+\infty$. Then the numerator in (7.43) must also tend to zero, i.e. the lengths of two sides of the cell tend to zero. Consequently, the areas of all triangles that contain these sides must also tend to zero. Repeating the argument as many times as necessary, we conclude that the lengths of the sides of all grid cells, including those at the boundary of the domain, must tend to zero, which is impossible.

Thus, if the set D is not empty, the system of algebraic equations

$$R_x = \frac{\partial I^h}{\partial x_{ij}} = 0, \; R_y = \frac{\partial I^h}{\partial y_{ij}} = 0, \quad i = 2, \ldots, N-1; \; j = 2, \ldots, M-1, \tag{7.44}$$

has at least one solution which is a convex grid. To find it, one must first find a certain initial grid $G_0 \in D$, and then use some method of unconstrained minimization. Since the function (7.42) has the infinite barrier on the boundary of the set D, each step of the method can be chosen so that the grid always remains convex. Note, that in the common case the discrete grid-generation equations (7.44) may have multiple solutions, but numerical experiments have not met such opportunity.

Method of Minimization

First there is considered a method for minimizing the function (7.42) assuming that the initial grid $G_0 \in D$ has been found. Suppose the grid at the

l-th step of the iterations is determined. For finding the grid nodes at the $(l+1)$-th step the quasi-Newtonian procedure for each interior node can be used:

$$\begin{aligned} \tau R_x + \frac{\partial R_x}{\partial x_{ij}}(x_{ij}^{l+1} - x_{ij}^l) + \frac{\partial R_x}{\partial y_{ij}}(y_{ij}^{l+1} - y_{ij}^l) = 0, \\ \tau R_y + \frac{\partial R_y}{\partial x_{ij}}(x_{ij}^{l+1} - x_{ij}^l) + \frac{\partial R_y}{\partial y_{ij}}(y_{ij}^{l+1} - y_{ij}^l) = 0 \end{aligned} \tag{7.45}$$

where τ is the iteration parameter. Note, that (7.45) is not the Newton - Raphson iteration because only a part of the second derivatives of (7.42) is taken into account. The rate of convergence for (7.45) is low by comparison. At the same time the Newton - Raphson method gives much more complex system of linear equations at each iteration.

Each of the derivatives in (7.45) is the sum of twelve terms, in accordance with the number of triangles containing the given node as a vertex. Rather than write out such cumbersome expressions, the first and second derivatives of the terms in (7.42) are considered:

$$\frac{\partial F_k}{\partial x_{k-1}} = 2\frac{x_{k-1} - x_k}{J_k} - F_k \frac{y_{k+1} - y_k}{J_k} \tag{7.46}$$

and so on. Arrays storing the derivatives of the function (7.42) were first cleared, and then all grid triangles were scanned and the appropriate derivatives added to the relevant elements of the arrays.

Now an algorithm is suggested for the choice of the iteration parameter τ in (7.45), which was used only for the problems with moving boundaries. Recall, that the minimized function (7.42) has the infinite barrier on the boundary of the set of convex grids D. Since if the initial grid G_0 is convex the iteration (7.45) gives, as a rule, a convex grid for any $\tau < 1$. But in extreme cases when G_0 is very close to the boundary of the set D, the grid $G(\tau)$ can cross the boundary of the set in the first iterations (7.45). Clearly, such condition is fatal for the method because the same barrier on the boundary of the set D does not allow the iterations to return into the set D in the following iterations. To avoid this, a certain basic parameter τ_0 is chosen so that $G(\tau_0/2) \in D$ and $G(\tau_0) \in D$. In the beginning $\tau_0 = 1$. If the above-mentioned conditions are violated, we put $\tau_0 = 1/4$ or $\tau_0 = 1/2$, depending on whether the grids $G(\tau_0/2)$ or $G(\tau_0)$ leave the set D, and so on.

In fixed boundary problems the simple choice $\tau = \text{const} \cdot \tau_0$ is used. For time-dependent problems with moving boundaries a version of the method of parabolas was developed. As the controlling quantity the squared residual of the equations (7.45)

$$W = \sum_{i,j}(R_x^2 + R_y^2)_{i,j}$$

was used. The parabola $W(\tau)$ is constructed from the grids obtained for $\tau = 0$, $\tau = \tau_0/2$ and $\tau = \tau_0$. The parameter τ is then chosen so that

$W(\tau) = \min$ in the interval $\theta\tau \le \tau \le \alpha\tau_0$. The parameter $\theta \sim 0.1$ is given a priori and bounds the value of τ away from zero. The parameter α bounds τ above, i.e., prevents a very large extrapolation along the parabola. If $\tau_0 = 1$, i.e., if the boundary of the set D is not crossed, we put $\alpha = 2$. If $\tau_0 < 1$, then $\alpha = 1$. Finally, if the algorithm gives $\tau < \tau_0/2$, the condition $I^h(\tau_0/2) < I^h(0)$ is checked. In the cases when this condition is found to be valid, $\tau = \tau_0/2$ was put.

For one iteration of the above method a measurement of the computational cost gives the value of about double (but not three times) the cost of the simple iteration. The reason is that the second derivatives of the function (7.42) are not used in calculating W while they are used in (7.45) to calculate the direction of minimization.

The algorithm described can be used only if the initial grid is convex. Otherwise, it is necessary either to obtain a convex grid by another algorithm as a preliminary stage of the method or to modify the computational formulas. The first approach is based on the minimization of the following function

$$I_D = \sum_{i=1}^{N-1}\sum_{j=1}^{M-1}\sum_{k=1}^{4}\left([\epsilon - J_k]_{i+1/2\ j+1/2}\right)_+^2, \quad (f)_+ = \max(0, f), \qquad (7.47)$$

for some given $\epsilon > 0$. This is accomplished by the gradient method with a suitable choice of the iteration parameter. The iterative process is broken off as soon as all inequalities (7.41) are satisfied. This method was used by Charakhch'yan (1993, 1994) for studying gas dynamics problems with moving boundaries when the initial interior grid nodes for minimizing (7.47) were taken from the previous time step. As a result, the initial grid is either convex or such that a convex grid is obtained after a few iterations.

In fixed boundary problems, the starting grid may be non-convex, containing numerous self-intersecting cells. In such case the preliminary stage of the method based on minimizing (7.47) can be unsuitable. Therefore another approach had been developed by Ivanenko (1988). The computational formulas (7.45) were modified so that the initial grid need not belong to the set D of convex grids. The quantities J_k appearing in the expressions for R_x, R_y and their derivatives are replaced with new quantities $\tilde{J}_k$

$$\tilde{J}_k = \begin{cases} J_k & \text{if } J_k > \epsilon, \\ \epsilon & \text{if } J_k \le \epsilon, \end{cases}$$

where $\epsilon > 0$ is some sufficiently small quantity.

It is quite important to choose an optimal value of ϵ so that the convex grid is constructed as fast as possible. The method used for specifying the value of ϵ is based on the computation of the absolute value of the average area of triangles with negative areas

$$\epsilon = \max[\alpha S/(N + 0.01), \epsilon_1],$$

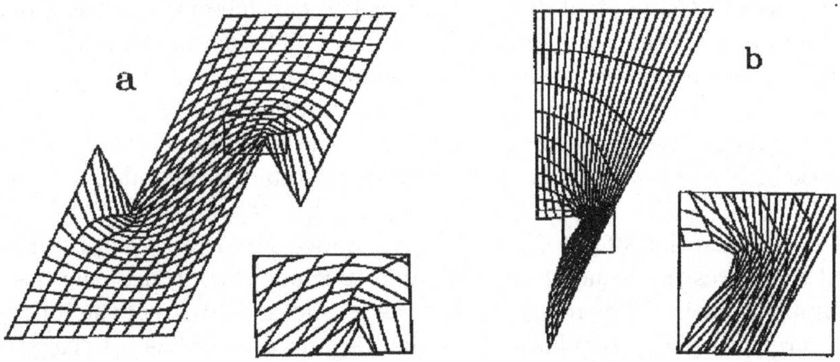

Fig. 7.8. Grids in a model domain (a) and for computing a cumulative jet (b).

where S is twice the absolute value of the total area of triangles with negative areas, and N the number of these triangles. The quantity $\epsilon_1 > 0$ sets a lower bound on ϵ to avoid very large values appearing in the computations. The coefficient α is chosen experimentally and is in the range $0.3 \leq \alpha \leq 0.7$.

In practical implementation, an arbitrary set of grid nodes can be marked as movable during iterations, while all other nodes are considered as stationary. All the terms in the function (7.42) which become independent on movable nodes are excluded from computations. Since the boundary nodes are always marked as stationary, four terms in (7.42) corresponding to "corner" triangles $\{(1,2);(1,1);(2,1)\}$, $\{(N-1,1);(N,1);(N,2)\}$, $\{(1,M-1);(1,M);(2,M)\}$, and $\{(N-1,M);(N,M);(N,M-1)\}$ are always excluded from computations. As a result, the method becomes applicable to those domains for which the angle between two intersecting boundaries is greater than or equal to π, despite the fact that the corresponding grid cell becomes non-convex regardless of the positions of interior nodes.

Examples of the grids generated by this method are exhibited in Figs. 7.8 and 7.9. Figure 7.9 demonstrates the application of the algorithm to generating a grid for computing a high-velocity impact of a thin foil (a) upon a conical target CD (Lomonosov, Frolova, and Charakhch'yan (1997)).

7.2.2 Adaptive Grid Generation

Numerical Algorithm

One approach of adaptive grid generation is based on the minimization of the functional (5.32) in the metric of a monitor surface.

Let the monitor surface be defined by a function $z = f(x, y)$ where $f \in C^1$. The expressions for the covariant elements and Jacobian of the monitor metric in the grid coordinates ξ, η are as follows:

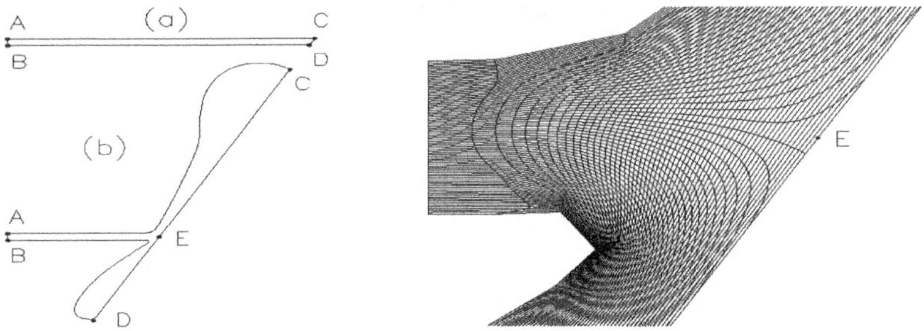

Fig. 7.9. A fragment of the grid (right-hand) in the vicinity of the point E (left-hand).

$$g_{11}^{\boldsymbol{\xi}} = g_{11}^{\mathrm{s}}\left(\frac{\partial x}{\partial \xi}\right)^2 + 2g_{12}^{\mathrm{s}}\frac{\partial x}{\partial \xi}\frac{\partial y}{\partial \xi} + g_{22}^{\mathrm{s}}\left(\frac{\partial y}{\partial \xi}\right)^2,$$

$$g_{22}^{\boldsymbol{\xi}} = g_{11}^{\mathrm{s}}\left(\frac{\partial x}{\partial \eta}\right)^2 + 2g_{12}^{\mathrm{s}}\frac{\partial x}{\partial \eta}\frac{\partial y}{\partial \eta} + g_{22}^{\mathrm{s}}\left(\frac{\partial y}{\partial \eta}\right)^2,$$

$$g^{\boldsymbol{\xi}} = (J)^2 g^{\mathrm{s}} = (J)^2[1 + (f_x)^2 + (f_y)^2],$$

where

$$g_{11}^{\mathrm{s}} = 1 + (f_x)^2, \quad g_{12}^{\mathrm{s}} = f_x f_y, \quad g_{22}^{\mathrm{s}} = 1 + (f_y)^2.$$

Substituting these expression in (5.32) for $n = 2$ we obtain the functional

$$I_a = \int_0^1\int_0^1 \frac{(x_\xi^2 + x_\eta^2)[1 + (f_x)^2] + (y_\xi^2 + y_\eta^2)[1 + (f_y)^2] + 2f_x f_y(x_\xi y_\xi + x_\eta y_\eta)}{J[1 + (f_x)^2 + (f_y)^2]^{1/2}} d\xi d\eta. \tag{7.48}$$

Now we again consider the grid $(x, y)_{ij}$, $i = 1, \dots, N;\ j = 1, \dots, M$ and, to simplify the computational formulas, the parametric rectangle $1 < \xi < N,\ 1 < \eta < M$ substitutes for the unit square $0 < \xi < 1,\ 0 < \eta < 1$. The functional I_a is approximated by the function

$$I_a^h = \sum_{i=1}^{N-1}\sum_{j=1}^{M-1}\sum_{k=1}^{4}\frac{1}{4}[F_k]_{i+1/2\ j+1/2}, \tag{7.49}$$

$$F_k = \frac{D_1[1 + (f_x)_k^2] + D_2[1 + (f_y)_k^2] + 2D_3(f_x)_k(f_y)_k}{J_k[1 + (f_x)_k^2 + (f_y)_k^2]^{1/2}}, \tag{7.50}$$

where

$$
\begin{aligned}
D_1 &= (x_{k-1} - x_k)^2 + (x_{k+1} - x_k)^2, \\
D_2 &= (y_{k-1} - y_k)^2 + (y_{k+1} - y_k)^2, \\
D_3 &= (x_{k-1} - x_k)(y_{k-1} - y_k) + (x_{k+1} - x_k)(y_{k+1} - y_k), \\
J_k &= (x_{k-1} - x_k)(y_{k+1} - y_k) - (x_{k+1} - x_k)(y_{k-1} - y_k).
\end{aligned}
$$

Derivatives $(f_x)_k$ and $(f_y)_k$ in the k-th cell vertex are equal to the corresponding values of derivatives, evaluated at the grid node ij

$$
\begin{aligned}
(f_x)_{ij} &= \frac{(f_{i+1j} - f_{i-1j})(y_{ij+1} - y_{ij-1}) - (f_{ij+1} - f_{ij-1})(y_{i+1j} - y_{i-1j})}{(x_{i+1j} - x_{i-1j})(y_{ij+1} - y_{ij-1}) - (x_{ij+1} - x_{ij-1})(y_{i+1j} - y_{i-1j})}, \\
(f_y)_{ij} &= \frac{(f_{i+1j} - f_{i-1j})(x_{ij+1} - x_{ij-1}) - (f_{ij+1} - f_{ij-1})(x_{i+1j} - x_{i-1j})}{(x_{i+1j} - x_{i-1j})(y_{ij+1} - y_{ij-1}) - (x_{ij+1} - x_{ij-1})(y_{i+1j} - y_{i-1j})}.
\end{aligned}
\tag{7.51}
$$

These formulas must be modified for the boundary nodes. Indices, "leaving" the computational domain must be replaced by the nearest boundary indices. For example, if $j = 1$, then $(i, j - 1)$ must be replaced by (i, j).

Function (7.49) possesses the same property as the function (7.42): $I_a^h(G) \to +\infty$ if $G \to \partial D$ for $G \in D$ where D is the set of convex grids, ∂D is the boundary of the set.

As before, equations (7.45) are used to minimize the function I_a^h. Quantities $(f_x)_{ij}$ and $(f_y)_{ij}$ are assumed to be parameters and therefore all their derivatives in (7.45) vanish. Note that if $(f_x)_{ij}$ and $(f_y)_{ij}$ vanish, the function I_a^h reduces to the function I^h (7.42).

The adaptive grid generation algorithm is formulated as follows:

1. Generate a grid for the given domain using unconstrained minimization algorithm described.

2. Compute the values of the control function at each grid node. The result is f_{ij}.

3. Evaluate derivatives $(f_x)_{ij}$ and $(f_y)_{ij}$ using the formulas (7.51).

4. Make one step in the minimization process for the function I_a^h using equations (7.45) and compute new values of x_{ij} and y_{ij}.

5. Repeat starting with Step 2 to convergency.

It is important that at each step of the iterative process the grid remains convex.

Redistribution of Boundary Nodes

There are several ways to redistribute the grid nodes along the boundary ∂X^2 of the domain X^2 during adaptation. The simplest one is a fixed position of every point on ∂X^2, referred to as "fixed position." However if some physical quantities are not smooth (e.g. shock waves), then some instability in the mesh generation and, consequently, in the physical problem solution near the points where the discontinuity joins ∂X^2 may arise. In some methods, called as "unconstrained minimization", the boundary nodes are treated as interior and the vectors of their shift are projected onto ∂X^2. This way can be used only if the discontinuity is nearly orthogonal to ∂X^2. If not, then, when condensing, the boundary nodes overlap, adjacent cells degenerate, and modeling breaks. The next method referred to as "1-D minimization" relies on using the 1-D functional along ∂X^2. This method is more robust than the two ones discussed above and can usually be used for adaptation. However, the 1-D and 2-D functionals are as a rule inconsistent. By this reason the parameters of adaptation for the interior and boundary nodes should be selected separately. It requires additional work when modeling unsteady flow problems.

In the method suggested by Azarenok (2002) instead of (7.49) there was minimized the function

$$\tilde{I}_a^h = \sum_{i=1}^{N-1}\sum_{j=1}^{M-1}\sum_{k=1}^{4}\frac{1}{4}\left[F_k\right]_{i+1/2j+1/2} + \sum_{ij\in\mathcal{L}}\lambda_{ij}G_{ij} = I_a^h + \sum_{ij\in\mathcal{L}}\lambda_{ij}G_{ij}, \quad (7.52)$$

where the constraints $G_{ij} = G(x_{ij}\,,\, y_{ij}) = 0$ define ∂X^2, λ_{ij} are the Lagrange multipliers, and $\mathcal{L}$ is the set of the boundary nodes. The function $G(x,y)$ is assumed piecewise differentiable, so the function $\tilde{I}_a^h$ holds the infinite barrier on the boundary of the set of convex grids as I_a^h does if $f \in C^1$.

If the set of convex grids is not empty, the system of the following algebraic equations

$$R_x = \frac{\partial I_a^h}{\partial x_{ij}} + \lambda_{ij}\frac{\partial G_{ij}}{\partial x_{ij}} = 0, \quad R_y = \frac{\partial I_a^h}{\partial y_{ij}} + \lambda_{ij}\frac{\partial G_{ij}}{\partial y_{ij}} = 0, \quad G_{ij} = 0, \quad (7.53)$$

has at least one solution that is a convex mesh. Here $\lambda_{ij} = 0$ if $ij \notin \mathcal{L}$ and the constraints are defined for the boundary nodes $ij \in \mathcal{L}$.

Consider the method of minimizing the function (7.52) assuming the grid to be convex at the lth step of the iterative procedure. The quasi-Newton procedure to find the coordinates x_{ij}^{l+1}, y_{ij}^{l+1} from the system (7.53) was used:

$$\begin{aligned}
&\tau R_x + \frac{\partial R_x}{\partial x_{ij}}(x_{ij}^{l+1} - x_{ij}^l) + \frac{\partial R_x}{\partial y_{ij}}(y_{ij}^{l+1} - y_{ij}^l) + \frac{\partial R_x}{\partial \lambda_{ij}}(\lambda_{ij}^{l+1} - \lambda_{ij}^l) = 0,\\
&\tau R_y + \frac{\partial R_y}{\partial x_{ij}}(x_{ij}^{l+1} - x_{ij}^l) + \frac{\partial R_y}{\partial y_{ij}}(y_{ij}^{l+1} - y_{ij}^l) + \frac{\partial R_y}{\partial \lambda_{ij}}(\lambda_{ij}^{l+1} - \lambda_{ij}^l) = 0,\\
&\tau G_{ij} + \frac{\partial G_{ij}}{\partial x_{ij}}(x_{ij}^{l+1} - x_{ij}^l) + \frac{\partial G_{ij}}{\partial y_{ij}}(y_{ij}^{l+1} - y_{ij}^l) = 0,
\end{aligned}$$

where

$$\begin{aligned}
&\frac{\partial R_x}{\partial x_{ij}} = \frac{\partial^2 I_a^h}{\partial x_{ij}^2} + \lambda_{ij}\frac{\partial^2 G_{ij}}{\partial x_{ij}^2}, \quad \frac{\partial R_x}{\partial y_{ij}} = \frac{\partial^2 I_a^h}{\partial x_{ij}\partial y_{ij}} + \lambda_{ij}\frac{\partial^2 G_{ij}}{\partial x_{ij}\partial y_{ij}},\\
&\frac{\partial R_y}{\partial x_{ij}} = \frac{\partial^2 I_a^h}{\partial x_{ij}\partial y_{ij}} + \lambda_{ij}\frac{\partial^2 G_{ij}}{\partial x_{ij}\partial y_{ij}}, \quad \frac{\partial R_y}{\partial y_{ij}} = \frac{\partial^2 I_a^h}{\partial y_{ij}^2} + \lambda_{ij}\frac{\partial^2 G_{ij}}{\partial y_{ij}^2},\\
&\frac{\partial R_x}{\partial \lambda_{ij}} = \frac{\partial G_{ij}}{\partial x_{ij}}, \quad \frac{\partial R_y}{\partial \lambda_{ij}} = \frac{\partial G_{ij}}{\partial y_{ij}}.
\end{aligned}$$

Resolving the last equation of (7.2.2) with respect to $y_{ij}^{l+1} - y_{ij}^l$ and substituting it in the two remaining equations, the system

$$\begin{pmatrix} a_{11} & a_{12} \\ a_{21} & a_{22} \end{pmatrix} \begin{pmatrix} x_{ij}^{l+1} - x_{ij}^l \\ \lambda_{ij}^{l+1} - \lambda_{ij}^l \end{pmatrix} = \begin{pmatrix} a_{13} \\ a_{23} \end{pmatrix},$$

is obtained, where

$$a_{11} = \frac{\partial R_x}{\partial x_{ij}} - \frac{\partial R_x}{\partial y_{ij}} \frac{\partial G_{ij}}{\partial x_{ij}} \Big/ \frac{\partial G_{ij}}{\partial y_{ij}},$$

$$a_{12} = \frac{\partial G_{ij}}{\partial x_{ij}},$$

$$a_{13} = \tau \left[\frac{\partial R_x}{\partial y_{ij}} G_{ij} \Big/ \frac{\partial G_{ij}}{\partial y_{ij}} - R_x \right],$$

$$a_{21} = \frac{\partial R_y}{\partial x_{ij}} - \frac{\partial R_y}{\partial y_{ij}} \frac{\partial G_{ij}}{\partial x_{ij}} \Big/ \frac{\partial G_{ij}}{\partial y_{ij}},$$

$$a_{22} = \frac{\partial G_{ij}}{\partial y_{ij}},$$

$$a_{23} = \tau \left[\frac{\partial R_y}{\partial y_{ij}} G_{ij} \Big/ \frac{\partial G_{ij}}{\partial y_{ij}} - R_y \right].$$

Denoting $\triangle = a_{11}a_{22} - a_{12}a_{21}$, $\triangle_1 = a_{13}a_{22} - a_{23}a_{12}$, $\triangle_2 = a_{11}a_{23} - a_{21}a_{13}$ (since $G_{ij} = 0$, the terms a_{13}, a_{23} are simplified), we obtain

$$x_{ij}^{l+1} = x_{ij}^l + \triangle_1/\triangle, \quad \lambda_{ij}^{l+1} = \lambda_{ij}^l + \triangle_2/\triangle, \tag{7.54}$$

while y_{ij}^{l+1} is determined from the third equation of (7.2.2). If the constraints are resolved in y in the form $G(x,y) = y - g(x) = 0$, then

$$\frac{\partial G_{ij}}{\partial x_{ij}} = -\frac{\partial g_{ij}}{\partial x_{ij}}, \quad \frac{\partial G_{ij}}{\partial y_{ij}} = 1,$$

and the upper formulas are simplified. Analogously the constrains may be resolved in x in the form $G(x,y) = x - \tilde{g}(y) = 0$. Note the equation $G(x,y) = 0$ can be locally resolved by one of these two forms.

If ∂X^2 is specified by parametric functions $x = x(t),\ y = y(t)$ or tabular values $(x,y)_{ij}$, the following algorithm can be used. Assume the index j is fixed and i is variable. When calculating the coordinates of the (ij)th node, in the interval (x_{i-1j}, x_{i+1j}) we construct an interpolating parabola $t = t(x)$ using the values in three nodes $(i-1j)$, (ij), and $(i+1j)$. From (7.54) we compute an intermediate value $\tilde{x}_{ij}^{l+1}$, further from the interpolation formula we determine $t_{ij} = t(\tilde{x}_{ij}^{l+1})$ and final values x_{ij}^{l+1}, y_{ij}^{l+1} from the parametric formulas.

Another way for redistributing the nodes along ∂X^2, given as parametric functions or by tabular values, employs an unconstrained minimization of the function in a parametric form and is based on solving the following system of algebraic equations, referred to as "parametric minimization,"

$$R_t = R_x \frac{\partial x_{ij}}{\partial t_{ij}} + R_y \frac{\partial y_{ij}}{\partial t_{ij}} = 0,$$

via the quasi-Newton procedure

$$\tau R_t + \frac{\partial R_t}{\partial t_{ij}} \left(t_{ij}^{l+1} - t_{ij}^l\right) = 0. \tag{7.55}$$

Here

$$\frac{\partial R_t}{\partial t_{ij}} = \frac{\partial R_x}{\partial x_{ij}} \left(\frac{\partial x_{ij}}{\partial t_{ij}}\right)^2 + \frac{\partial R_y}{\partial y_{ij}} \left(\frac{\partial y_{ij}}{\partial t_{ij}}\right)^2 + \left(\frac{\partial R_x}{\partial y_{ij}} + \frac{\partial R_y}{\partial x_{ij}}\right) \frac{\partial x_{ij}}{\partial t_{ij}} \frac{\partial y_{ij}}{\partial t_{ij}}$$

$$+ R_x \frac{\partial^2 x_{ij}}{\partial t_{ij}^2} + R_y \frac{\partial^2 y_{ij}}{\partial t_{ij}^2}, \qquad R_x = \frac{\partial I^h}{\partial x_{ij}}, \quad R_y = \frac{\partial I^h}{\partial y_{ij}}.$$

To the analytical control functions both the constrained and parametric minimization give similar results. Real-world 2-D flow computations have shown that it is better to perform adaptation along the boundary using constrained minimization (7.2.2) since the procedure (7.55) may not ensure consistent redistribution of the nodes in X^2 and on ∂X^2.

The use of the constrained minimization without adaptation (i.e. when f=*const.*) means that we seek the conformal mapping $x(\xi, \eta), y(\xi, \eta)$ of the parametric rectangle onto the domain X^2 with an additional parameter, the so-called conformal modulus.

Application to Unsteady Gas Dynamics Problems

The calculation of hydrodynamical problems on the adaptive moving meshes requires special conservative numerical schemes which update directly the flow parameters on the moving mesh at the new time level. Another way, when interpolation of parameters from the fixed mesh to the moving one is used at every time step, smears the singularities in the solution causing decrease of accuracy of modeling. We describe here a modification of the Godunov's scheme of the second-order accuracy in time and space on moving meshes, suggested by Azarenok (2000), to compute a two-dimensional gas flow in the Euler approach.

System of Equations

Two-dimensional equations of gas dynamics, namely, the laws of conservation of mass, momentum, and total energy are written in the integral form which can be derived by transformation of the volume integrals in the space (x, y, t) to the surface integrals by virtue of the Gauss's theorem:

$$\int_{\Omega} \left(\frac{\partial \boldsymbol{\sigma}}{\partial t} + \frac{\partial \boldsymbol{a}}{\partial x} + \frac{\partial \boldsymbol{b}}{\partial y} \right) d\Omega = \oiint \boldsymbol{\sigma} dxdy + \boldsymbol{a} dydt + \boldsymbol{b} dtdx = \boldsymbol{0}, \qquad (7.56)$$

where Ω is an arbitrary control volume in space (x, y, t), $\partial\Omega$ is the boundary of Ω,

$$\boldsymbol{\sigma} = \begin{bmatrix} \rho \\ \rho u \\ \rho v \\ E \end{bmatrix}, \quad \boldsymbol{a} = \begin{bmatrix} \rho u \\ \rho u^2 + p \\ \rho u v \\ u(E+p) \end{bmatrix}, \quad \boldsymbol{b} = \begin{bmatrix} \rho v \\ \rho u v \\ \rho v^2 + p \\ v(E+p) \end{bmatrix}.$$

Here u and v are the velocity components, p and ρ are the pressure and density, $E{=}\rho[e{+}0.5(u^2{+}v^2)]$ is the total energy, while e is the specific internal energy. The equation of state is $p{=}(\gamma{-}1)\rho e$ where γ is the ratio of specific heats. Denote the vector-valued unknown function as $\boldsymbol{f} = (u, v, p, \rho)^{\top}$. The conservation laws (7.56) hold for any functions $\boldsymbol{f}$ both smooth and discontinuous describing an ideal gas flow.

Numerical Scheme

Let a curvilinear moving grid in the x-y plane be introduced with the coordinate lines ξ, η, the $(i{+}1/2, j{+}1/2)$th cell of which at the time range (t^n, t^{n+1}) is shown in Fig. 7.10 by a domain Ω in R space (x, y, t), being a hexahedron with planar top and bottom faces. The bottom (top) face of the hexahedron Ω is the control volume at the time t^n (t^{n+1}).

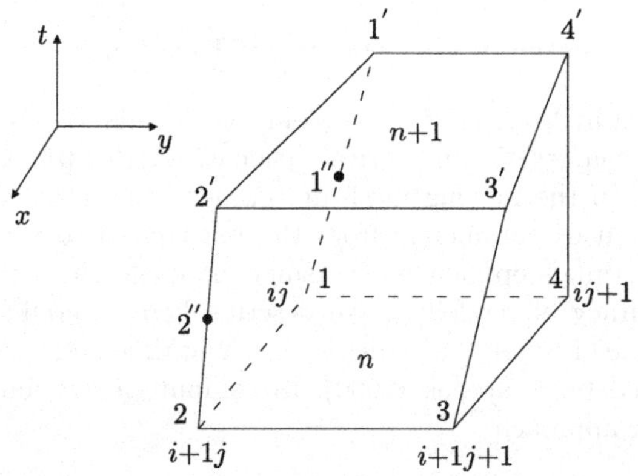

Fig. 7.10. Hexahedron Ω in R space with bottom (1234) and top (1'2'3'4') faces, being the cell of the 2D moving mesh at the time t^n and t^{n+1}, respectively.

Integrating (7.56) over the boundary $\partial\Omega$ of the hexahedron gives a cell-centered finite-volume approximation of the governing gas dynamics equations

$$\begin{aligned}&\boldsymbol{\sigma}^{n+1}_{i+1/2j+1/2}A_{1'2'3'4'}-\boldsymbol{\sigma}^{n}_{i+1/2j+1/2}A_{1234}\\&+\boldsymbol{Q}_{411'4'}+\boldsymbol{Q}_{233'2'}+\boldsymbol{Q}_{122'1'}+\boldsymbol{Q}_{344'3'}=0,\end{aligned}\tag{7.57}$$

where $\boldsymbol{\sigma}^{n+1}_{i+1/2j+1/2}$ and $\boldsymbol{\sigma}^{n}_{i+1/2j+1/2}$ are the average values of $\boldsymbol{\sigma}$ at the time t^{n+1} and t^n in the center of the top and bottom faces, respectively; $A_{1'2'3'4'}$ and A_{1234} are the areas of the corresponding faces. Each of the four vector values $\boldsymbol{Q}_{411'4'}$, $\boldsymbol{Q}_{233'2'}$, $\boldsymbol{Q}_{122'1'}$ and $\boldsymbol{Q}_{344'3'}$ is the amount of the mass, momentum, and energy which flows into and out the quadrilateral cell 1234 within time $\triangle t=t^{n+1}-t^n$ through the corresponding moving edges of the cell.

For example, the vector-valued quantity $\boldsymbol{Q}_{122'1'}$ that is the change of the parameters due to the flux through the edge 12 within time $\triangle t$ is given by

$$\boldsymbol{Q}_{122'1'}=\boldsymbol{\sigma}^{n+1/2}_{i+1/2j}A^{xy}_{122'1'}+\boldsymbol{a}^{n+1/2}_{i+1/2j}A^{yt}_{122'1'}+\boldsymbol{b}^{n+1/2}_{i+1/2j}A^{tx}_{122'1'},\tag{7.58}$$

where $\boldsymbol{\sigma}^{n+1/2}_{i+1/2j}$, $\boldsymbol{a}^{n+1/2}_{i+1/2j}$, and $\boldsymbol{b}^{n+1/2}_{i+1/2j}$ are calculated using the parameters $\boldsymbol{f}=(u,v,p,\rho)^\top$ in the center of the face $122'1'$, i.e. at the mid-point of edge 12 at the time $t^{n+1/2}$ (or at the mid-point of edge $1''2''$); $A^{xy}_{122'1'}$, $A^{yt}_{122'1'}$, $A^{tx}_{122'1'}$ are the areas of the projections of the face $122'1'$ onto the coordinate planes x-y, y-t, and t-x, respectively, given by

$$\begin{aligned}A^{xy}_{122'1'}&=\int\limits_{122'1'}dxdy=0.5[(x_{2'}-x_1)(y_{1'}-y_2)-(x_{1'}-x_2)(y_{2'}-y_1)],\\A^{yt}_{122'1'}&=\int\limits_{122'1'}dydt=0.5\triangle t(y_{2'}+y_2-y_1-y_{1'}),\\A^{tx}_{122'1'}&=\int\limits_{122'1'}dtdx=-0.5\triangle t(x_{2'}+x_2-x_1-x_{1'}).\end{aligned}$$

These expressions are obtained from the formula for area of the quadrangle 1234

$$\begin{aligned}A_{1234}&=A(x_1,y_1;x_2,y_2;x_3,y_3;x_4,y_4)\\&=0.5[(x_3-x_1)(y_4-y_2)-(x_4-x_2)(y_3-y_1)]\ ,\end{aligned}$$

when passing its contour in the anticlock-wise manner.

The values $\boldsymbol{f}^{n+1}_{i+1/2j+1/2}$ are updated by two stages using a predictor-corrector procedure. At the first stage (predictor) we compute the intermediate values $\bar{\boldsymbol{f}}^{n+1}_{i+1/2j+1/2}$ at the $(n{+}1)$th level by using (7.57).

Let us consider the curvilinear coordinate ξ. Assume the function $\boldsymbol{f}$ to be linear within the cell $(i+1/2, j+1/2)$ in the ξ-direction. The values $\boldsymbol{f}^n_{ij+1/2}$ and $\boldsymbol{f}^n_{i+1j+1/2}$, specified at the left and right ends of the segment $((i, j+1/2), (i+1, j+1/2))$ at the time t^n, are defined as

$$\begin{aligned}\boldsymbol{f}^n_{ij+1/2} &= \boldsymbol{f}^n_{i+1/2\ j+1/2} - 0.5\delta \boldsymbol{f}^n_{i+1/2} h^n_{i+1/2}\ ,\\ \boldsymbol{f}^n_{i+1j+1/2} &= \boldsymbol{f}^n_{i+1/2\ j+1/2} + 0.5\delta \boldsymbol{f}^n_{i+1/2} h^n_{i+1/2}\ .\end{aligned}$$

Here $\delta \boldsymbol{f}^n_{i+1/2}$ is the "effective" derivative in the ξ-direction, while the spacing $h^n_{i+1/2}$ is the length of the underlying segment. Note that $\delta \boldsymbol{f}^n_{i+1/2}$ and $h^n_{i+1/2}$ are the notations for $(\delta \boldsymbol{f}^n_\xi)_{i+1/2j+1/2}$ and $(h^n_\xi)_{i+1/2j+1/2}$, respectively. When determining $\delta \boldsymbol{f}^n_{i+1/2}$, to suppress spurious oscillations in the vicinity of discontinuities, the monotonicity algorithm should be applied. The spacing $h^n_{i+1/2}$ is given by

$$h^n_{i+1/2} = \\ 0.5\sqrt{(x^n_{i+1j} + x^n_{i+1j+1} - x^n_{ij} - x^n_{ij+1})^2 + (y^n_{i+1j} + y^n_{i+1j+1} - y^n_{ij} - y^n_{ij+1})^2}.$$

By analogy the values $\boldsymbol{f}^n_{i+1/2j}$ and $\boldsymbol{f}^n_{i+1/2j+1}$ are calculated at the left and right ends of the segment in the η-direction in the cell. Note, since we interpolate $\boldsymbol{f}$ along the curvilinear coordinate lines ξ and η the order of interpolation, in general, is less then 2 and equal 2 only if the mesh is rectangular and quasiuniform.

In order to find the values $\boldsymbol{Q}_{122'1'}$ we substitute in (7.58) the determined values of $\boldsymbol{f}$ at the mid-point of the lateral edge 12 of the quadrilateral 1234, i.e. at the time t^n instead of the ones at the time $t^{n+1/2}$. The values $\boldsymbol{Q}_{411'4'}$, $\boldsymbol{Q}_{233'2'}$, and $\boldsymbol{Q}_{344'3'}$ can be found in a similar way. Finally, from (7.57) we obtain the intermediate values $\bar{\boldsymbol{f}}^{n+1}_{i+1/2j+1/2}$ at the $(n+1)$th level.

We now discuss the second stage, corrector. For this purpose we set the effective derivatives at t^{n+1} equal to the ones at t^n, i.e. $\delta\bar{\boldsymbol{f}}^{n+1}_{i+1/2}=\delta \boldsymbol{f}^n_{i+1/2}$. Then the values in the center of the faces $122'1'$ and $344'3'$, namely, at the mid-point of the edges 12 and 34 at the time $t^{n+1/2}$ are

$$\begin{aligned}\boldsymbol{f}^{n+1/2}_{ij+1/2} &= 0.5[\boldsymbol{f}^n_{i+1/2j+1/2} + \bar{\boldsymbol{f}}^{n+1}_{i+1/2j+1/2} - 0.5\delta \boldsymbol{f}^n_{i+1/2}(h^n_{i+1/2} + h^{n+1}_{i+1/2})]\ ,\\ \boldsymbol{f}^{n+1/2}_{i+1j+1/2} &= 0.5[\boldsymbol{f}^n_{i+1/2j+1/2} + \bar{\boldsymbol{f}}^{n+1}_{i+1/2j+1/2} + 0.5\delta \boldsymbol{f}^n_{i+1/2}(h^n_{i+1/2} + h^{n+1}_{i+1/2})]\ .\end{aligned}$$

We can obtain $\boldsymbol{f}^{n+1/2}_{i+1/2j}$ and $\boldsymbol{f}^{n+1/2}_{i+1/2j+1}$ in a similar way. These four vector values are used as the pre-wave states in the center of the corresponding lateral faces of the hexahedron for the Riemann problem.

Let us consider the face $122'1'$. To get the postwave states $\boldsymbol{f}^{n+1/2}$ in the center of this face (for brevity we omit subscripts i, j), i.e. at the mid-point of the segment $(1'', 2'')$, we solve the Riemann problem with the pre-wave states $(r, p, \rho)^{n+1/2}$ at this point on the both sides of the face. One state $(r, p, \rho)_{+}^{n+1/2}$ relates to the underlying hexahedron and the other $(r, p, \rho)_{-}^{n+1/2}$ to the hexahedron adjacent to the face $122'1'$ (corresponding to the $(i+1/2, j-1/2)$th cell). Here $r^{n+1/2}$ is the normal component of the velocity to the segment $(1'', 2'')$. We also use the tangential components of the velocity $q^{n+1/2}$ on those sides. The normal and tangential components of the velocity are given by

$$r^{n+1/2} = n_x u^{n+1/2} + n_y v^{n+1/2} , \qquad q^{n+1/2} = n_y u^{n+1/2} - n_x v^{n+1/2} ,$$

where n_x, n_y are the components of the outward unit normal vector to the segment $(1'', 2'')$.

After solving the Riemann problem, the post-wave values $(r, p, \rho)_{R}^{n+1/2}$ in the face center are defined. The post-wave tangential component of the velocity $q_R^{n+1/2}$ is given by

$$q_R^{n+1/2} = \begin{cases} q_+^{n+1/2} & \text{if } \; w_{12} \leq d_{cont} , \\ q_-^{n+1/2} & \text{otherwise} , \end{cases} \tag{7.59}$$

where d_{cont} is the contact discontinuity speed in the Riemann problem, w_{12} is the velocity of the edge 12 in the normal direction to this edge, and $q_+^{n+1/2}, q_-^{n+1/2}$ are the pre-wave tangential components of the velocity in the underlying hexahedron and in the one adjacent to the face $122'1'$, respectively. This condition expresses the fact that the tangential component of the velocity is discontinuous across the tangential discontinuity. The velocity w_{12} can be derived from the equality

$$\Delta t l_{1''2''} w_{12} = A_{122'1'}^{xy} , \tag{7.60}$$

where $l_{1''2''}$ is the length of the segment $(1'', 2'')$. Next we restore the Cartesian components of the post-wave velocity in the center of the face $122'1'$

$$u_R^{n+1/2} = n_x r_R^{n+1/2} + n_y q_R^{n+1/2} , \qquad v_R^{n+1/2} = n_y r_R^{n+1/2} - n_x q_R^{n+1/2} .$$

Given the post-wave values $(u, v, p, \rho)_R^{n+1/2}$ in the center of the face $122'1'$, we calculate $\boldsymbol{Q}_{122'1'}$ via (7.58). Similarly we treat the Riemann problem in the center of the other three faces to obtain $\boldsymbol{Q}_{411'4'}$, $\boldsymbol{Q}_{233'2'}$, and $\boldsymbol{Q}_{344'3'}$.

The final values of $\boldsymbol{f}_{i+1/2j+1/2}^{n+1}$ at the time t^{n+1} are obtained by using (7.57). This scheme is of the second-order accuracy in the domains of smooth flow provided that the mesh is quasiuniform and close to rectangular.

Riemann Problem on the Moving Mesh

To demonstrate how to take into account the movement of grid nodes let us consider the midpoint of the segment $(1''2'')$ within the time interval

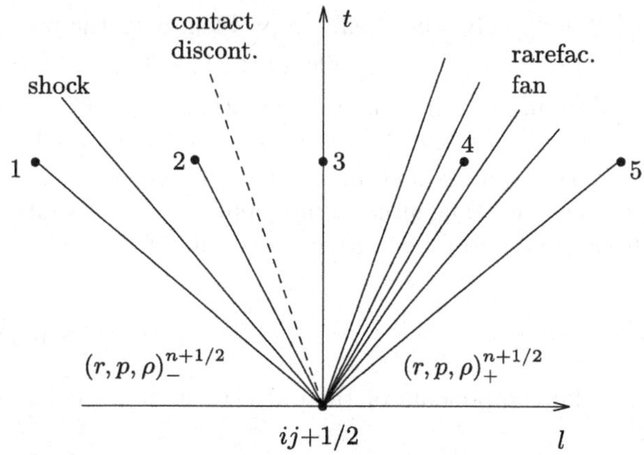

Fig. 7.11. Five possible cases of location of the segment $((x,y)^{n+1/2}_{i\ j+1/2},(x,y)^{n+1}_{i\ j+1/2})$ in the wave pattern. Points 1,...,5 indicate the location of the point $(ij{+}1/2)$ at t^{n+1}. The axis l with zero at the point $(ij{+}1/2)$ is aligned with the normal vector towards the segment $(1'',2'')$ (in the direction $\xi>0$) at the time $t^{n+1/2}$.

$(t^{n+1/2},t^{n+1})$ (see Fig. 7.11). Assume that after solving the Riemann problem at the point $(x,y)^{n+1/2}_{ij+1/2}$ we have the wave pattern depicted in Fig. 7.11. There are 5 cases of location of the segment $((x,y)^{n+1/2}_{ij+1/2},(x,y)^{n+1}_{ij+1/2})$ in the wave pattern depending on the velocity w_{12} of the edge 12. As the post-wave values $(r,p,\rho)^{n+1/2}_R$ we take:

1. $(r,p,\rho)^{n+1/2}_R=(r,p,\rho)^{n+1/2}_-$ if $w_{12}<d_{\text{sh}}$, where d_{sh} is the speed of the left shock in the l-axis direction.
2. $(r,p,\rho)^{n+1/2}_R=(r,p,\rho)^{n+1/2}_2$ if $d_{\text{sh}}<w_{12}<d_{\text{cont}}$, where the vector $(r,p,\rho)^{n+1/2}_2$ defines the flow parameters behind the shock, d_{cont} is the speed of the contact discontinuity which equals to the velocity u in that domain.
3. $(r,p,\rho)^{n+1/2}_R=(r,p,\rho)^{n+1/2}_3$ if $d_{\text{cont}}<w_{12}<d^{\text{lft}}_{\text{rar}}$, where the vector $(r,p,\rho)^{n+1/2}_3$ defines the parameters in the domain between the contact discontinuity and left characteristic of the rarefaction wave expanding with the speed $d^{\text{lft}}_{\text{rar}}$.
4. $(r,p,\rho)^{n+1/2}_R=\phi(\alpha)$ if $d^{\text{lft}}_{\text{rar}}<w_{12}<d^{\text{rght}}_{\text{rar}}$, i.e. we calculate the flow parameters in the rarefaction wave using the similarity variable $\alpha=l/(t-t^{n+1/2})$. Here $d^{\text{rght}}_{\text{rar}}$ is the speed of the right characteristic in the rarefaction fan.
5. $(r,p,\rho)^{n+1/2}_R=(r,p,\rho)^{n+1/2}_+$ if $w_{12}>d^{\text{rght}}_{\text{rar}}$.

Note that in the first-order Godunov's scheme the above algorithm is applied at the time t^n.

Stability condition

To demonstrate how the stability condition on moving mesh is obtained let us consider the one-dimensional case with the $(i+1/2)$th cell depicted in Fig. 7.12. At the nth time level in this cell, the local time step is determined by

$$\triangle t_{i+1/2} = \frac{h^n_{i+1/2}}{\max(d^{rght}_i - w_{i+1}\ ,\ -d^{lft}_{i+1} - w_i)}\ , \tag{7.61}$$

where d^{rght}_i and d^{lft}_{i+1} are the extreme right and left wave speeds at the points x^n_i and x^n_{i+1}, respectively, obtained by solving the Riemann problem at $t^{n+1/2}$, w_i is the velocity of the node x_i, i.e. the slope of the intercell boundary (x^n_i, x^{n+1}_i). The condition (7.61) implies that we estimate the time within which the left-going characteristic (in the linearized analysis this is a straight line), emanating from the $(i+1)$th node, arrives at the ith node moving with the velocity w_i, as well as the time within which the right characteristic, emanating from the ith node, arrives at the $(i+1)$th node moving with the velocity w_{i+1}. From these two time steps we take the minimal one. The resulting time step over the mesh is given by

$$\triangle t = c_{cfl}\ \min_i\ \triangle t_{i+1/2}\ . \tag{7.62}$$

The coefficient c_{cfl} is a correction to the non-linearity of the Eqs. (7.56). To calculate the node velocity w_i, on one hand it is necessary to know the time step $\triangle t$, and on the other hand to take into account that w_i participates in determining $\triangle t$. By these reasons at the time level $n+1$ we use $\triangle t$ obtained at the preceding level n. The coefficient $c_{\text{cfl}}<1$, usually about 0.5, may be corrected during the computation.

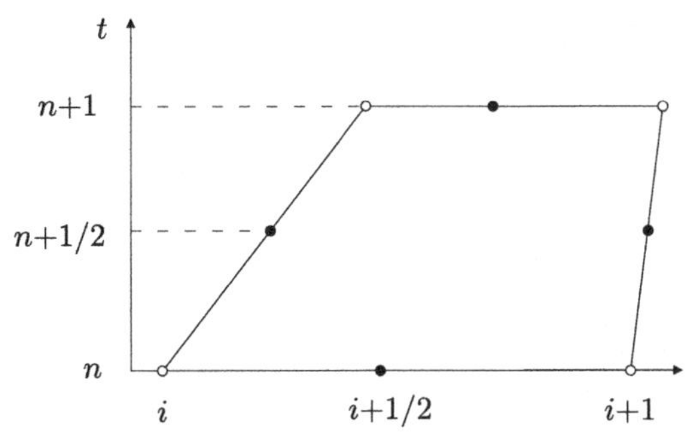

Fig. 7.12. Computing cell of 1D moving grid at the time t^n and t^{n+1}

In the 2D case the choice of the admissible step $\triangle t$ may be estimated in the energetic norm to the underlying Eqs. (7.56) written in a differential form as a t-hyperbolic by Friedrichs's system. The step Δt in the $(i{+}1/2, j{+}1/2)$th cell (see Fig. 7.10) is given by

$$\Delta t_{i+1/2 j+1/2} = \frac{\Delta t' \; \Delta t''}{\Delta t' + \Delta t''} , \qquad (7.63)$$

where

$$\begin{aligned} \Delta t' &= \frac{h'}{\max(d_{14}^{\mathrm{rght}} - w_{23}; -d_{23}^{\mathrm{lft}} + w_{14})} , \\ \Delta t'' &= \frac{h''}{\max(d_{12}^{\mathrm{rght}} - w_{34}; -d_{34}^{\mathrm{lft}} + w_{12})} , \\ h' &= \frac{A_{1234}}{0.5\sqrt{(x_4 + x_3 - x_1 - x_2)^2 + (y_4 + y_3 - y_1 - y_2)^2}} , \\ h'' &= \frac{A_{1234}}{0.5\sqrt{(x_3 + x_2 - x_4 - x_1)^2 + (y_3 + y_2 - y_4 - y_1)^2}} . \end{aligned} \qquad (7.64)$$

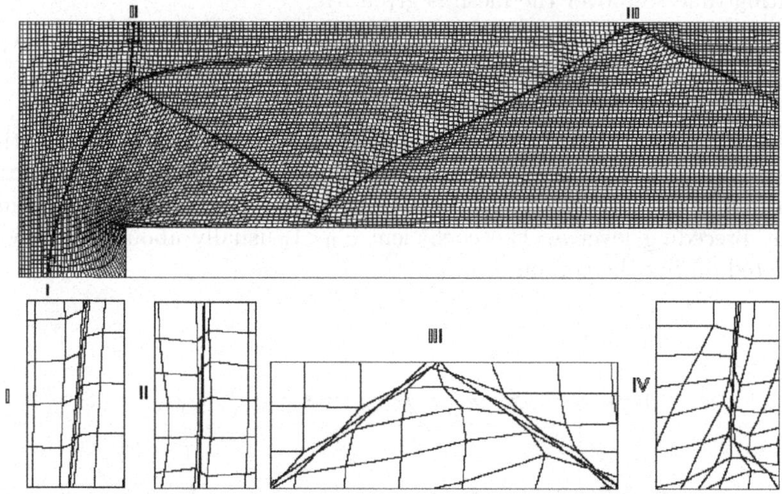

Fig. 7.13. Supersonic flow in the mach wind tunnel containing a step. The boundary nodes are distributed using constrained minimization. Adaptive mesh with fragments I, II, III near the boundary and IV comprising the triple point.

Here $\Delta t'$ and $\Delta t''$ are the admissible time steps to the one-dimensional scheme in the ξ and η-direction, respectively; h', h'' are the "average heights" of the bottom face 1234, and w is the velocity of the corresponding cell edge. For example, w_{12} is the velocity of the edge 12 in the normal direction determined via (7.60). Next, d_{12}^{rght} and d_{14}^{rght} are the "extreme right wave" speeds defined from solving the Riemann problem to the faces $122'1'$ and $11'4'4$, respectively; d_{23}^{lft} and d_{34}^{lft} are the "extreme left wave" speeds to the faces $233'2'$ and $433'4'$, respectively.

The resulting time step over the mesh is given by

$$\triangle t = c_{\text{cfl}} \min_{ij} \triangle t_{i+1/2j+1/2} \, .$$

7.2.3 Numerical Examples

Robustness of the adaptive mesh method is demonstrated in the two numerical examples.

The first is a test presented by Fig 7.13 of the planar unsteady supersonic flow in the wind tunnel containing a step (for details see Colella and Woodward (1984)). This test was performed by Azarenok and Ivanenko (2002) on the adaptive grids by applying the above flow solver when as a monitor function f in (7.48) there was used the modulus of velocity $|V|$. The boundary nodes were adapted by applying 1-D minimization. One of the main difficulties was to capture the triple point, caused by the irregular reflection of the bow shock from the top wall, with clustered grid lines that required especial efforts.

The use of constrained minimization for the boundary nodes allows both to eliminate the above difficulty connected with capturing the triple point (see fragment IV of the mesh in Fig. 7.13) and to perform robust node clustering in the domains where the shocks are attached to the boundary or reflected from it (see fragments $I - III$). The shock waves are smeared over 2 to 3 cells. There is also demonstrated compression of grid lines to the contact discontinuity emanating from the triple point.

The second example is related to modeling motion of a detonation wave. The adaptive mesh, obtained when modeling the unstable detonation wave motion (for details see Azarenok and Tang (2005)) is presented in Fig. 7.14. The pressure is used as a monitor function. To perform a stable mesh adaptation there was also employed the constrained minimization for the redistribution of boundary nodes. The figure exhibits clustering of the grid lines to the main incident shock and transverse waves.

The calculations related to Fig. 7.13 and 7.14 were carried out by B. Azarenok.

7.3 Generation of Multi-Block Grids

The numerical algorithms described above are formulated for generating a local single-block grid. This section reviews some approaches for extending the algorithms for generating multi-block grids.

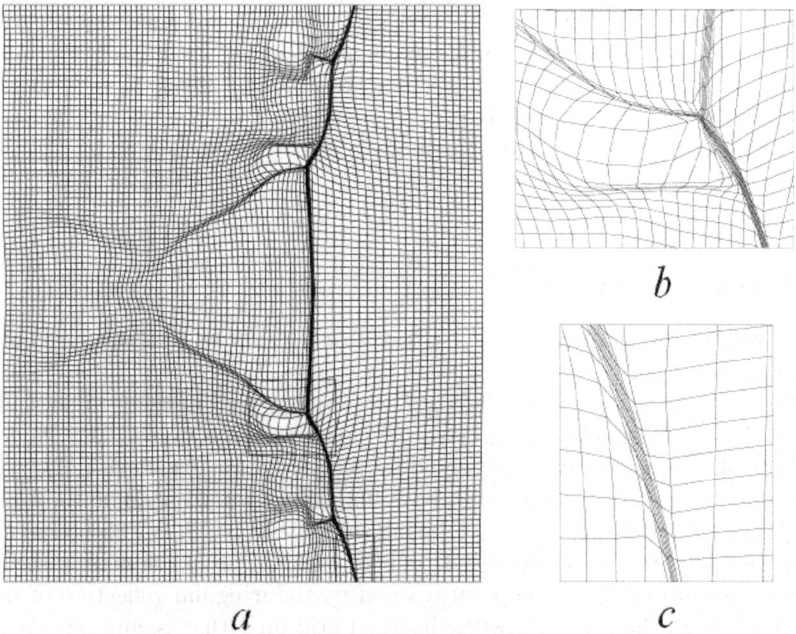

Fig. 7.14. Adaptive mesh (a) and its fragments (b), (c).

In the commonly applied block strategy, the physical geometry is divided into a few contiguous subgeometries referred to as blocks, which may be considered as the cells of a coarse, generally unstructured grid (see Fig. 7.15 for a tokamak-related domain). And then a separate structured or unstructured mesh is generated in each block. The union of these local grids constitutes a mesh referred to as a multi-block grid. The main reasons for using multi-block grids rather than single-block grids are that

(1) the physical geometry is exceedingly complicated, having a multiply connected boundary, cuts, narrow protuberances, cavities, etc.;
(2) the physical problem is heterogeneous relative to some of the physical quantities, so that different mathematical models are required in different zones of the geometry to adequately describe the physical phenomena;
(3) the solution of the problem behaves non-uniformly: zones of smooth and rapid variation of different scales may exist;
(4) opportunity to apply parallel algorithms.

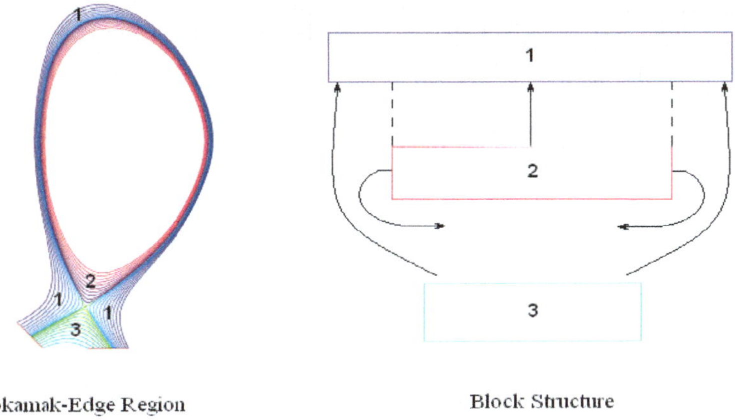

Fig. 7.15. Block structure for the tokamak-edge region.

7.3.1 Block-Structured Grids

If meshes in all blocks are structured then the multi-block grid can be considered as locally structured at the level of an individual block, but globally unstructured when viewed as a collection of blocks. Such grids are called block-structured grids. Thus a common idea with a block-structured grid technique is the use of different structured grids, or coordinate systems, in different zones of the physical geometry, allowing the most appropriate grid configuration to be used in each zone.

Block-structured grids are considerably more flexible in handling complex geometries than structured grids. Since these grids retain the simple regular connectivity pattern of a structured mesh on a local level, these block-structured grids maintain, in nearly the same manner as structured grids, compatibility with efficient finite-difference, finite-volume, or spectral element algorithms used to solve partial differential equations. However, the generation of block-structured grids may take a fair amount of user interaction and, therefore, requires the implementation of an automation technique to lay out the block topology.

The blocks of locally structured grids in a three-dimensional region are typically homeomorphic to a three-dimensional cube, thus having the shape of a curvilinear hexahedron. However, some domains can be more effectively partitioned with the use of cylindrical blocks as well. Cylindrical blocks are commonly applied to the numerical solution of problems in regions with holes and to the calculation of flows past aircraft or aircraft components (wings, fuselages, etc.). For many problems it is easier to take into account the shape of the physical geometry and the structure of the solution by using cylindrical

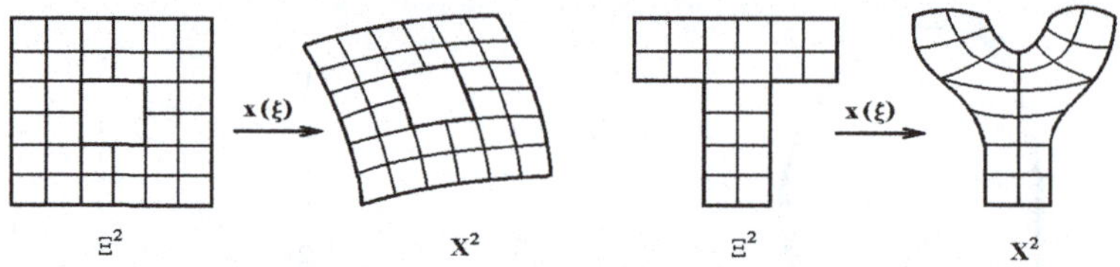

Fig. 7.16. Computational domains adjusted to the physical domains

blocks. Also, the total number of blocks and sections might be smaller than when using only blocks homeomorphic to a cube.

Topology of the Grid

The correct choice of the topology in a block, depending on the geometry of the logical domain and the qualitative type of the transformation of the region onto the block, has a considerable influence on the quality of the grid. There are two ways of specifying the computational domain for a block:

(1) as a complicated polyhedron which maintains the schematic form of the block subdomain (Fig. 7.16);
(2) simply as a solid cube or a cube with cuts (Fig. 7.17).

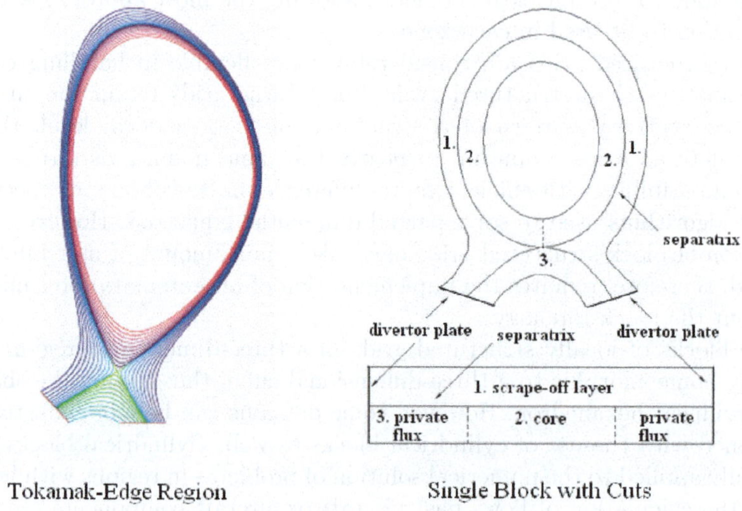

Fig. 7.17. A single-block topology for the tokamak-edge region.

With the first approach, the problem of constructing the grid transformation is simplified, and this method is often used to generate a single-blocked grid in a complicated domain. The second approach relies on a simplified geometry of the computational domain but requires sophisticated methods to derive suitable grid transformations.

In a block which is homeomorphic to a cylinder with thick walls, the grid topology is determined by the topology of the two-dimensional grids in the transverse sections. In applications, for sections of this kind, which are annular planes or surfaces with a hole, wide use is made of three basic grid topologies: H (Fig. 7.18), O (Fig. 7.19), and C (Fig. 7.20).

In H-type grids, the computational domain is a square with an interior cut which is opened by the construction of the coordinate transformation and mapped onto an interior boundary of the block. The outer boundary of the square is mapped onto the exterior of the block. The interior boundary has two points with singularities where one coordinate line splits. $H-$type grids are used, for instance, when calculating the flow past thin bodies (aircraft wings, turbine blades, etc.).

In O-type grids, the computational domain is a solid square. In this case the system of coordinates in the block with a hole is obtained by bending the square, sticking two opposite sides together and then deforming. The stuck sides determine the cut, called the fictive edge, in the block. An example of O-type grid is the nodes and cells of a polar system of coordinates. The O-type grid can be constructed without singularities when the boundary of the block is smooth. Grids of this kind are used when calculating the flow past bulky aircraft components (fuselages, gondolas, etc.) and, in combination with H-type grids, for multilayered block structures.

The computational domain is also a solid square in a C-type grid, but the mapping onto the block with a hole involves the identification of some segments of one of its sides and then deforming it. In the C-type grid, the coordinate lines of one family leave the outer boundary of the block, circle

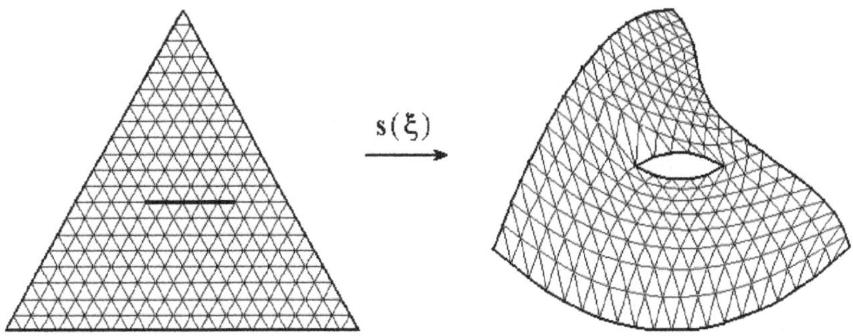

Fig. 7.18. H-type grid

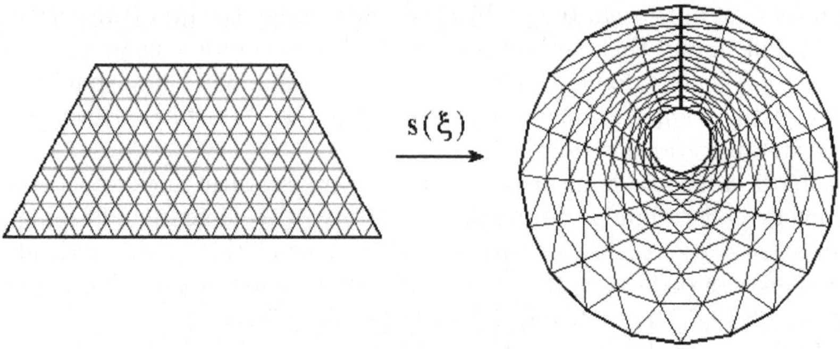

Fig. 7.19. O-type grid

the inner boundary and return again to the outer boundary. There is one point on the inner boundary which has the same type of singularity as in the H-type grid. The C-type grids are commonly used in regions with holes and long protuberances.

The O and C-type techniques in fact introduce artificial interior cuts in multiply connected physical geometries to generate single block-structured grids. The cuts are used to join the disconnected components of the boundary of the geometry in order to reduce their number. Theoretically, this operation can allow one to generate a single coordinate transformation in a multiply connected physical geometry.

The choice of the grid topology in a block depends on the structure of the solution, the shape of the physical geometry, and, in the case of continuous or smooth grid-line communication, on the topology of the grid in the adjacent block as well. For complicated physical geometries, such as those near aircraft surfaces or turbines with a large number of blades, it is difficult to choose the grid topology of the blocks, because each component of the system (wing,

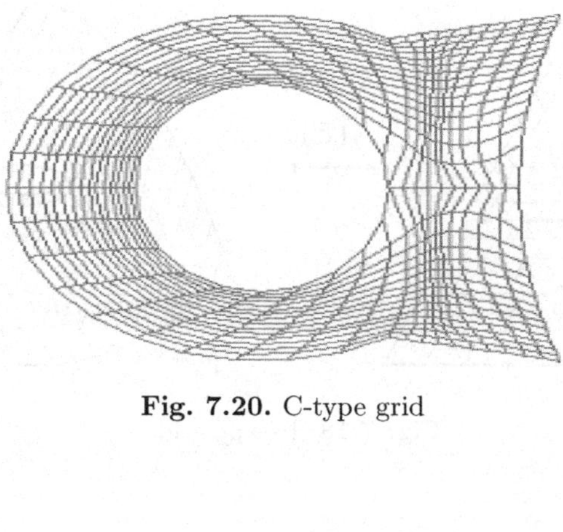

Fig. 7.20. C-type grid

fuselage, etc.) has its own natural type of grid topology, but these topologies are usually incompatible with each other.

Conditions Imposed on Grids in Blocks

A grid in a block must satisfy the conditions which are required to obtain an acceptable solution. In any specific case, these conditions are determined by features of the computer, the methods of grid generation available, the topology and conditions of interaction of the blocks, the numerical algorithms, and the type of data to be obtained.

One of the main requirements imposed on the grid is its adaptation to the solution. Multidimensional computations are likely to be very costly without the application of adaptive grid techniques. The basic aim of adaptation is to enhance the efficiency of numerical algorithms for solving physical problems by a special nonuniform distribution of grid nodes. The appropriate adaptive displacement of the nodes, depending on the physical solution, can increase the accuracy and rate of convergence and reduce oscillations and the interpolation error.

In addition to adaptation, the construction of locally structured grids often requires the coordinate lines to cross the boundary of the domain or the surface in an orthogonal or nearly orthogonal fashion. The orthogonality at the boundary can greatly simplify the specification of boundary conditions. Also, a more accurate representation of algebraic models of turbulence, the equations of a boundary layer, and parabolic Navier–Stokes equations is possible in this case. If for grids of O and C-type the coordinate lines are orthogonal to the boundary of each block and its interior cuts, the global block-structured grid will be smooth. It is also desirable for the coordinate lines to be orthogonal or nearly orthogonal inside the blocks. This will improve the convergence of the difference algorithms, and the equations, if written in orthogonal variables, will have a simpler form.

For unsteady gas-dynamics problems, some coordinates in the entire domain or on the boundary are required to have Lagrange or nearly Lagrange properties. With Lagrangian coordinates the computational region remains fixed in time and simpler expressions for the equations can be obtained in this case.

It is also important that the grid cells do not collapse, the changes in the steps are not too abrupt, the lengths of the cell sides are not very different, and the cells are finer in any zone of high gradient, large error, or slow convergence. Requirements of this kind are taken into account by introducing quantitative and qualitative characteristics of the grid, both with the help of coordinate transformations and by using the sizes of cell edges, faces, angles, and volumes. The characteristics used include the deviation from orthogonality, the Lagrange properties, the values of the transformation Jacobian or cell volume, and the smoothness and adaptivity of the transformation. For cell faces, the deviation from a parallelogram, rectangle, or square, as well as the ratio of the area of the face to its perimeter, is also used. A detailed review of these grid characteristics is given in the monograph of Liseikin (1999).

Communication of Adjacent Coordinate Lines

The requirement of mutual positioning or "communication" of grids in the vicinity of adjacent grid blocks and fictive edges or faces can also have a considerable influence on the construction of locally structured grids and on the efficiency of the numerical calculations. The coordinate lines defining the grid nodes of two adjacent blocks, in general, need not have points in common, and can join smoothly or nonsmoothly (Fig. 7.21). If all adjacent grid blocks join smoothly, interpolation is not required. If the coordinate lines do not join, then during the calculation the solution values at the nodes of one block must be transferred to those of the adjacent block in the neighborhood of their intersection. This is done by interpolation or (in mechanics) using conservation laws.

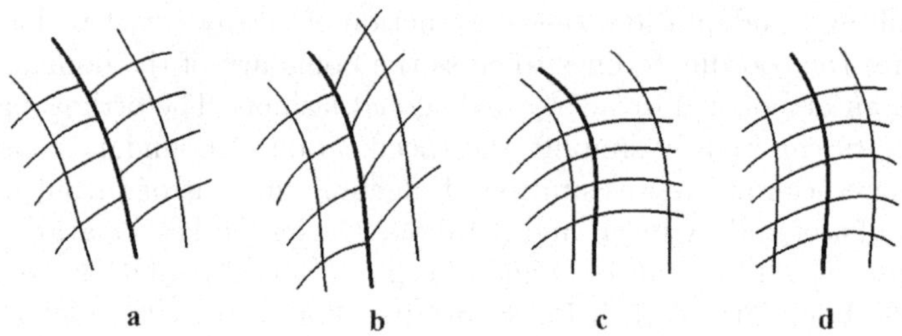

Fig. 7.21. Types of interface of grid lines between contiguous blocks (**a** discontinuous; **b, c** nonsmooth; **d** smooth)

The types of interaction between adjacent grid blocks and in the vicinity of fictive edges or faces are selected on the basis of the features of the physical quantities in the region of their intersection. If the gradient of the physical solution is not high in these zones and interpolation can, therefore, be performed with high accuracy, the coordinate lines do not need to join. This greatly simplifies the algorithm for constructing the grid. If there are high gradients of the solution near the intersection of two blocks or in the neighborhood of fictive edges or faces, a smooth matching is usually performed between the coordinate lines. The problem of smooth matching was typically overcome by an algebraic technique using Hermitian interpolation, or by elliptic methods, involving a choice of control functions. A combination of Laplace and Poisson equations, yielding equations of fourth or even sixth order, was also used for this purpose. A distinctive feature of these approaches is that the boundaries of joint blocks or fictive intersections as well as grid distribution on these geometries remain fixed.

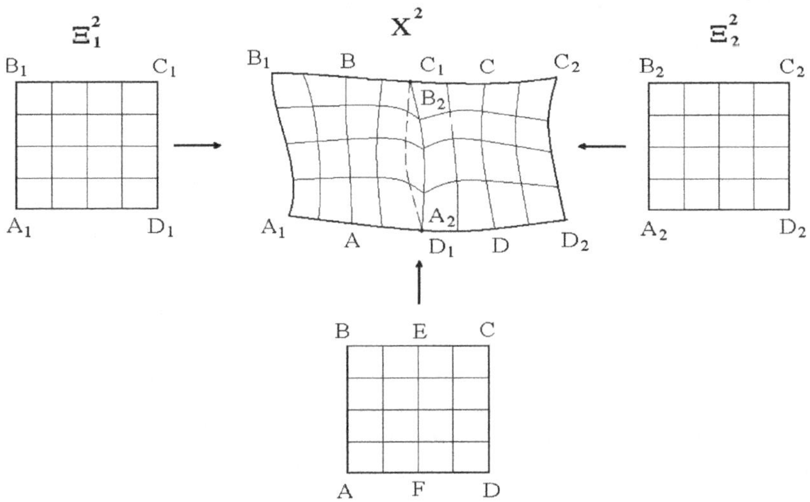

Fig. 7.22. Scheme for generating smooth grids by computing the common points of the adjacent blocks through grid equations

Approaches to Smoothing Grids

In this section we describe two novel iterative approaches for providing smooth matching of grid lines across adjacent blocks. These approaches are readily applied to generating smooth grids in the vicinity of fictive edges of faces as in the case of O- or C-types of meshes. The essential difference from the previous methods for generating smooth block-structured grids is that the current approaches are based on the computation of both the position of the joint boundary segments of the blocks and grid distribution in these segments. The position of the segments and grid distribution are found 1) through the numerical solution of grid equations and 2) by the interpolation from the nodes of the grid hypersurfaces neighboring the joint boundary segment.

Computation Through Grid Equations.

The idea of this version of the approach is demonstrated in Fig. 7.22 representing a two-block structured scheme for generating smooth quadrilateral grids in a domain X^2. The left-hand block of the domain is bounded by

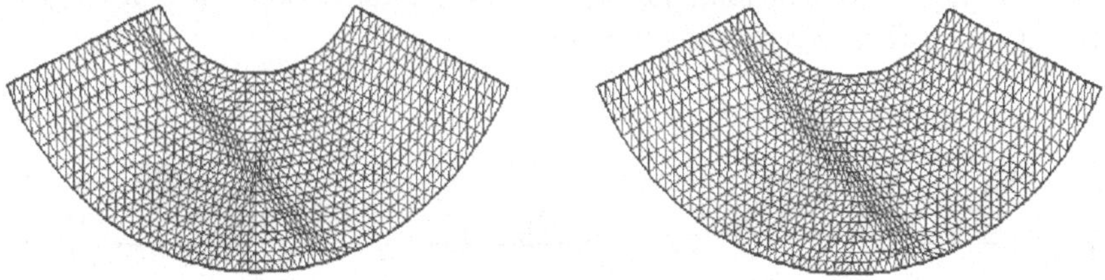

Fig. 7.23. Examples of nonsmooth and smooth triangular grids.

the curves A_1B_1, B_1C_1, C_1D_1, and D_1A_1. Similarly the right-hand block is bounded by the segments A_2B_2, B_2C_2, C_2D_2, and D_2A_2. The curves C_1D_1 and B_2A_2 are identical presenting the joint boundary of the blocks. By the iterations the grid in the domain X^2 is computed independently at each block via solving numerically by the methods described in Sects. 7.1-7.3 the Dirichlet boundary value problem for grid equations at the nodes of the corresponding logical domains Ξ_1^2 and Ξ_2^2 with the identical boundary node distribution at joint segments in X^2. These segments and their grid points are found in the process of iterations by solving the boundary value problem at the points of a new logical domain $ABCD$.

Thus, during the first iteration we specify the joint boundary segment $A_2B_2 = D_1C_1$ in X^2, the grid nodes at this segment, and the transformation at this nodes from the grid points of the segments C_1D_1 and A_2B_2 of the corresponding logical domains Ξ_1^2 and Ξ_2^2. Then we compute independently the grid nodes in the both blocks of X^2 through the grid equations. Having done this we choose in the both blocks the corresponding grid lines AB and CD, for example, neighboring the joint boundary line $C_1D_1 = B_2A_2$. After this we solve numerically the boundary value problem for the grid equations at the points of a new logical domain $ABCD$. The number of the points at

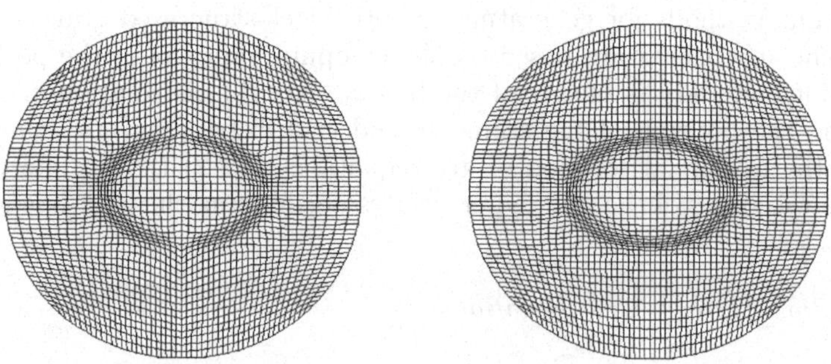

Fig. 7.24. Examples of nonsmooth and smooth quadrilateral grids.

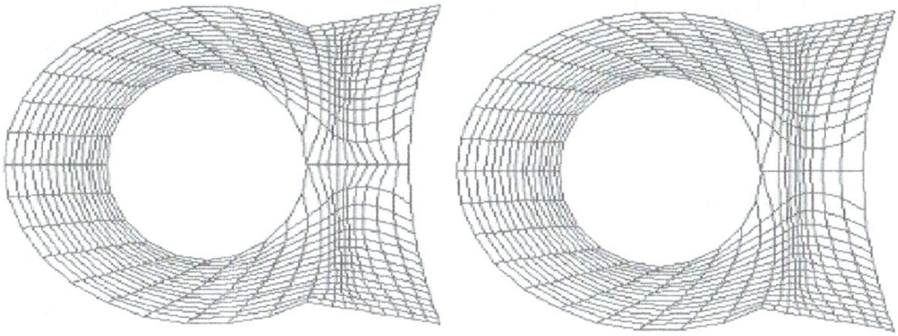

Fig. 7.25. Examples of nonsmooth and smooth C-type grids.

the segments AD in the domain X^2 and in this new logical domain coincides. The boundary points at the segments AB and CD in the logical domain are mapped at the computed points of the corresponding segments AB and CD in X^2. The transformation of the boundary points at the segments AD and BC coincides with the initial boundary transformation from the corresponding points of the segments A_1D_1, A_2B_2 and B_1C_1, B_2C_2. In particular, the points E and F are mapped at the points C_1 and D_1, respectively. By the computation the points of the segment EF in the new logical domain are transformed at the points of a new joint boundary segment (dotted line) of two new blocks. Then the iterations continue up to satisfy a tolerance condition. Similarly, there are generated smooth block-structured triangular grids.

Figures 7.23, 7.24, and 7.25 demonstrate nonsmooth grids (left-hand) computed in two-dimensional domains without the use of the smoothing algorithm and smooth grids (right-hand) found by the application of the algorithm.

Analogous procedure is formulated for generating smooth block-structured surface and three-dimensional domain grids. Figure 7.26 exhibits nonsmooth (left-hand) and smooth (right-hand) grids on a surface.

Fig. 7.26. Examples of nonsmooth and smooth surface grids.

An essential feature of this approach is that the equations for generating grids in the blocks and for computing new joint boundary segments of the blocks are the same. So such a smoothing process does not breach the properties of the grids realized by particular monitor metrics.

Computation by Interpolation

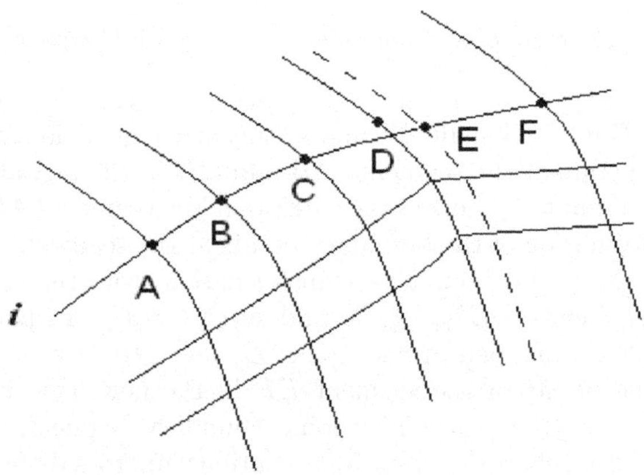

Fig. 7.27. Scheme for generating smooth grids by interpolation.

In this approach the grid point at the i-th level of a new joint boundary segment (dotted line in Fig.7.27) is computed by the following procedure: after computing the grid points at both blocks having a joint boundary line with a grid node D we choose at the i-th level of grid lines two points neighboring the previous joint segment point D of one block (the points B and C in Fig. 7.27) and one neighboring a point of another block (point F in Fig. 7.27). Through these points we draw a smooth line, for example, a circle segment. The i-th level of a new joint boundary point E will lie on this segment. Its position at the segment is defined from the relation $\overline{CE}/\overline{EF} = \overline{AB}/\overline{BC}$, where A is the third grid point of the first block computed at the i-th level (overline means the distance between the corresponding points). When all grid points of the new joint boundary segment are found by such way then the grid points in the blocks are found independently by solving grid equations. The process continues up to satisfy a required tolerance.

Figure 7.28 illustrates an O-type nonsmooth (left-hand) and smooth (right-hand) surface grids generated by this approach.

Similarly there are generated smooth block-structured three-dimensional domain grids.

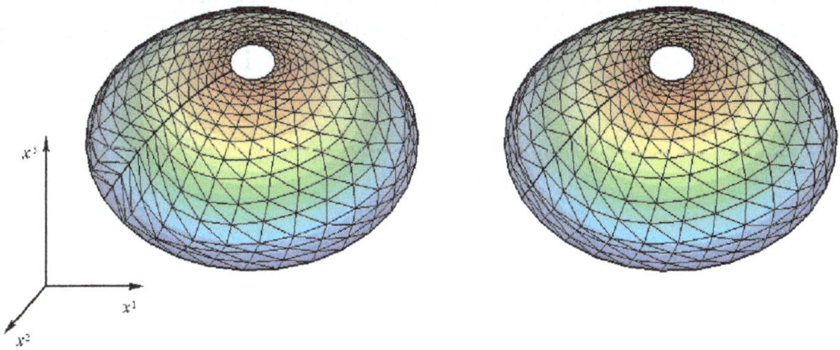

Fig. 7.28. Examples of nonsmooth and smooth surface grids.

7.4 Application of Layer-Type Functions to Grid Codes

Formulations of general monitor metrics (5.58) as well as of the metrics (5.73) and (5.101) for generating field-aligned and balanced grids rely on some weight functions whose role is both to specify metrics and to determine the influence of individual metrics on the resulting mesh. This section reviews the application of some basic layer-type univariate mappings to defining the weight functions.

7.4.1 Specification of Basic Functions

A suitable set of weight functions can be formulated by using three basic functions $\varphi_i(x,\varepsilon), i = 1,2,3, x \geq 0,\ 0 < \varepsilon \ll 1$ which model locally the qualitative behavior of solutions to singularly perturbed problems along a coordinate transverse to the layers of their rapid variation.

The first function is the familiar exponential layer-type mapping

$$\varphi_1(x,\varepsilon) = \exp(-bx/\varepsilon^k)\,, \qquad k > 0\,, \qquad b > 0\,, \tag{7.65}$$

representing a layer-type function of the first order. The following mapping is a power function, namely,

$$\varphi_2(x,\varepsilon) = \frac{\varepsilon^{kb}}{(\varepsilon^k + x)^b}\,, \qquad k > 0\,, \qquad b > 0\,, \tag{7.66}$$

The third mapping is a logarithmic map

$$\varphi_3(x,\varepsilon) = \frac{\ln(1 + x\varepsilon^{-k})}{\ln(1 + \varepsilon^{-k})}\,, \qquad k > 0\,. \tag{7.67}$$

The interval where any function $\varphi_i(x,\varepsilon)$, $i = 1,2,3$, provides a rapid stretching of the coordinate x coincides with the interval where the first derivative with respect to x of this function is large. The first derivatives of the basic functions are

$$\frac{d\varphi_1}{dx}(x,\varepsilon) = -b\varepsilon^{-k}\exp(-bx/\varepsilon^k)\,, \qquad k>0\,,\quad b>0\,,$$

$$\frac{d\varphi_2}{dx}(x,\varepsilon) = -\frac{b\varepsilon^{kb}}{(\varepsilon^k+x)^{b+1}}\,, \qquad k>0\,,\quad b>0\,,$$

$$\frac{d\varphi_3}{dx}(x,\varepsilon) = \frac{1}{\ln(1+\varepsilon^{-k})(\varepsilon^k+x)}\,, \qquad k>0\,,$$

Thus the lengths of the intervals of the rapid transition of the functions equals to

$$x_1 = C_1\varepsilon^k ln(\varepsilon^{-1}), \quad x_2 = C_2\varepsilon^{kb/(b+1)}, \quad x_3 = C_3/ln(1+\varepsilon^{-k}),$$

respectively. The intervals $[0,x_1]$, $[0,x_2]$, and $[0,x_3]$ are referred to as layers of singularity of corresponding functions $\varphi_i(x,\varepsilon)$, $i = 1,2,3$.

The quantities k and b in the expressions for the functions $\varphi_i(x,\varepsilon)$ are positive constants that control some characteristics of the functions and the layers of their singularity. In particular, the number k exhibits the scale of a layer. The constant b controls the type of stretching nonuniformity and the width of the layer. The parameter ε provides the major contribution to determining the slopes of the functions in the vicinity of the point $x = 0$.

The basic functions $\varphi_i(x,\varepsilon)$, $i = 1,2,3$, have the boundary layers of rapid variation near the point $x = 0$. It is evident that the procedures of scaling, shifting, and matching can yield layer-type functions with arbitrary boundary and interior layers as well. These procedures are described in detail in the monographs by Liseikin (1999, 2001a).

Originally the layer-type functions $\varphi_i(x,\varepsilon)$ were used in the so called stretching method for specifying grid node clustering in the zones of boundary and interior layers for the numerical solution of singularly perturbed problems (see Bakhvalov (1969) and Liseikin (2001a)).

7.4.2 Numerical Grids Aligned to Vector-Fields

Application to Formulation of Field-Aligned Monitor Metrics

For generating numerical grids in a domain X^n, aligned to a vector-field $\mathbf{B}$, there, in accordance with Sect. 5.3.3, can be used the contravariant metric components (5.77). For specifying these components two vector-fields $\mathbf{B}_1$ and $\mathbf{B}_2$ may be chosen by

$$\mathbf{B}_1 = \mathbf{B}, \quad \mathbf{B}_2 = k\mathbf{D},$$

where $\mathbf{D}$ is orthogonal to the field $\mathbf{B}$, k is a small positive function. Thus the contravariant components of the monitor metric for generating field-aligned grids in the domain X^n are as follows:

$$g^{ij}_{\mathbf{s}} = \epsilon(\mathbf{s})\delta^i_j + B^iB^j + k^2D^iD^j, \quad i,j = 1,\dots,n. \tag{7.68}$$

Since the vector field $\mathbf{B}$ is orthogonal to $\mathbf{D}$ and $|\mathbf{B}| = |\mathbf{D}|$ we find from (5.79)

$$g_{\mathbf{s}} = \det(g^{ij}_{\mathbf{s}}) = [\epsilon(\mathbf{s}) + |\mathbf{B}|^2][\epsilon(\mathbf{s}) + k^2|\mathbf{B}|^2]. \tag{7.69}$$

Taking into account that for $n = 2$

$$B^iB^j = |\mathbf{B}|^2\delta^i_j - D^iD^j, \quad i,j = 1,2,$$

the contravariant metric components (7.68), for $n = 2$ and $l = 2$, are as follows:

$$\begin{aligned} g^{ij}_{\mathbf{s}} &= [\epsilon(\mathbf{s}) + |\mathbf{B}|^2]\delta^i_j + (k^2 - 1)D^iD^j \\ &= [\epsilon(\mathbf{s}) + k^2|\mathbf{B}|^2]\delta^i_j + (1 - k^2)B^iB^j, \quad i,j = 1,2, \end{aligned} \tag{7.70}$$

i.e. of the form (5.77) for $l = 1$. So, availing us of (5.81) yields that the covariant components of the monitor metric in the Cartesian coordinates s^1, s^2 are expressed in both forms

$$\begin{aligned} g^{\mathbf{s}}_{ij} &= \frac{1}{\epsilon(\mathbf{s}) + |\mathbf{B}|^2}\Big[\delta^i_j - \frac{k^2 - 1}{\epsilon(\mathbf{s}) + k^2|\mathbf{B}|^2}D^iD^j\Big], \quad i,j = 1,2, \\ g^{\mathbf{s}}_{ij} &= \frac{1}{\epsilon(\mathbf{s}) + k^2|\mathbf{B}|^2}\Big[\delta^i_j - \frac{1 - k^2}{\epsilon(\mathbf{s}) + |\mathbf{B}|^2}B^iB^j\Big], \quad i,j = 1,2, \end{aligned} \tag{7.71}$$

Thus the covariant metric components in the grid coordinates ξ^1, ξ^2 are as

$$\begin{aligned} g^{\boldsymbol{\xi}}_{ij} &= \frac{1}{\epsilon(\mathbf{s}) + |\mathbf{B}|^2}\Big[\frac{\partial \mathbf{s}}{\partial \xi^i} \cdot \frac{\partial \mathbf{s}}{\partial \xi^j} - \frac{k^2 - 1}{\epsilon(\mathbf{s}) + k^2|\mathbf{B}|^2}D^m\frac{\partial s^m}{\partial \xi^i}D^p\frac{\partial s^p}{\partial \xi^j}\Big] \\ &= \frac{1}{\epsilon(\mathbf{s}) + k^2|\mathbf{B}|^2}\Big[\frac{\partial \mathbf{s}}{\partial \xi^i} \cdot \frac{\partial \mathbf{s}}{\partial \xi^j} - \frac{1 - k^2}{\epsilon(\mathbf{s}) + |\mathbf{B}|^2}B^m\frac{\partial s^m}{\partial \xi^i}B^p\frac{\partial s^p}{\partial \xi^j}\Big], \end{aligned} \tag{7.72}$$

$i, j, m, p = 1, 2.$

For generating a grid with the requirement that grid coordinates are aligned with the vector field $\mathbf{B}$, we assume $k \sim 0.01 - 0.1$ and $\epsilon(\mathbf{s})$ as a function with small positive values when $|\mathbf{B}| \sim 1$, while $\epsilon(\mathbf{s}) \sim 1$ when $|\mathbf{B}| = 0$. The second condition for $\epsilon(\mathbf{s})$ is stipulated by the effect of the solution of inverted diffusive equations: the grid cells become very small at the points where all elements $g^{ij}_{\mathbf{s}}$ are small. The functions $\epsilon(\mathbf{s})$ satisfying these properties are formulated through the boundary layer-type functions (7.65)–(7.67) assuming $\epsilon(\mathbf{s}) = \varphi(|\mathbf{B}(\mathbf{s})|^2, \delta)$, where

$$\varphi(x,\delta) = \begin{cases} Mexp(-x/\delta), \\ M\delta^\alpha/(\delta + x)^\alpha, \quad \alpha > 0, \\ M\ln(\delta + x)/\ln\delta, \end{cases}$$

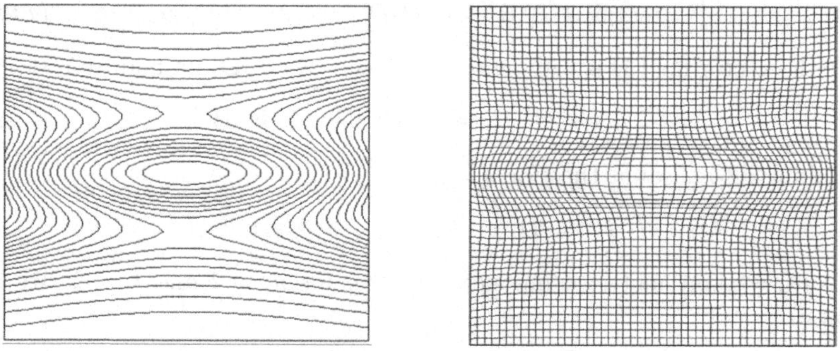

Fig. 7.29. Field-aligned grid (right-hand) for a symmetric vector field.

for $x \geq 0, \quad 0 < \delta << 1, \ M = const$. These boundary layer-type functions help solutions of grid equations switch from one mode to another.

In the case of a two-dimensional domain X^2 and a square logical domain Ξ^2, the boundary conditions at the points of the boundary segments $\xi^2 = 0; 1$ of the logical square Ξ^2, are found iteratively to satisfy the requirement of orthogonality

$$\frac{\partial \mathbf{s}}{\partial \xi^1} \cdot \frac{\partial \mathbf{s}}{\partial \xi^2} = 0.$$

The conditions at the boundary points of the other coordinate family can be either specified as fixed or they can be specified at one segment $\xi^1 = const$ and computed iteratively at the points of the another segment to satisfy at these points the requirement of grid lines alignment with the vector field

$$\mathbf{B}_3 = [B^1 + sgnB^1\varphi(|B^1|^2, \delta), \ (1 - \varphi(|B^1|^2, \delta))B^2].$$

This vector field is introduced to rule out, at the corresponding boundary points, the direction $(0, B^2)$ parallel to the boundary segment $\xi^1 = const$ for a coordinate curve $\xi^2 = const$ emanating from this segment.

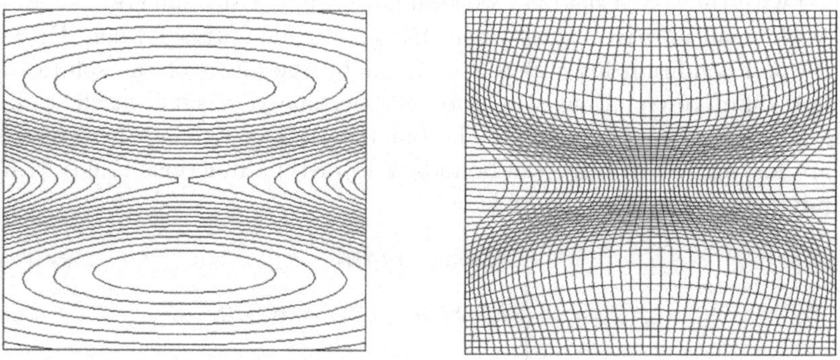

Fig. 7.30. Field-aligned grid (right-hand) for a vector field with two islands.

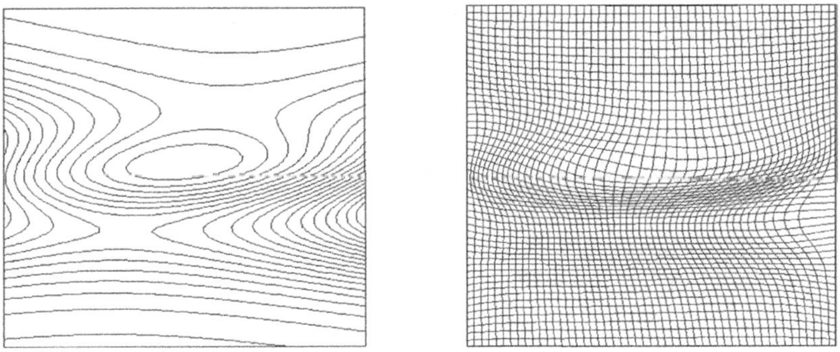

Fig. 7.31. Field-aligned grid (right-hand) for a nonsymmetric vector field.

Numerical Experiments

Figures 7.29–7.32 demonstrate isocontours (left-hand) of a vector field $\mathbf{B}$ in a two-dimensional domain X^2 and pictures (right-hand) of the corresponding grids. The vector field $\mathbf{B}$ is formed by a model function $\psi(\mathbf{s})$ via the relation

$$\mathbf{B} = (-\frac{\partial\psi}{\partial s^2}, \frac{\partial\psi}{\partial s^1}). \tag{7.73}$$

This vector field is subject to the natural equation for magnetic fields $div\mathbf{B} = 0$. In addition, for such a vector field, $\mathbf{D} = grad\ \psi$. Thus, in this case, the equations (7.72) become

$$g^{\boldsymbol{\xi}}_{ij} = \frac{1}{\epsilon(\mathbf{s}) + |\mathbf{B}|^2}\Big[\frac{\partial\mathbf{s}}{\partial\xi^i}\cdot\frac{\partial\mathbf{s}}{\partial\xi^j} - \frac{k^2-1}{\epsilon(\mathbf{s}) + k^2|\mathbf{B}|^2}\frac{\partial\psi}{\partial\xi^i}\frac{\partial\psi}{\partial\xi^j}\Big], \quad i,j = 1,2.$$

With these covariant components the two-dimensional diffusion equations (7.11) have the following form

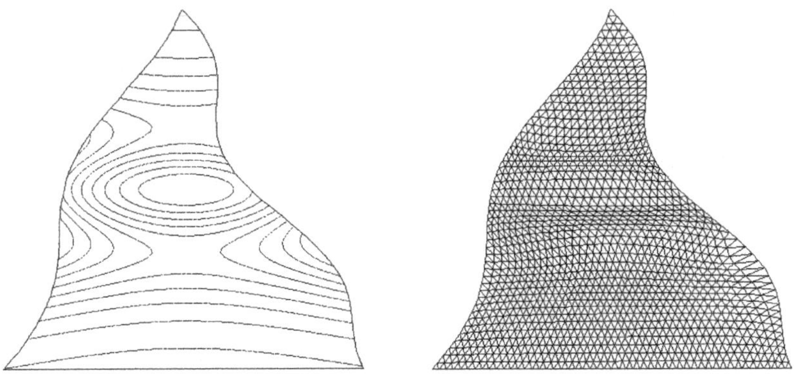

Fig. 7.32. Example of a triangular field-aligned grid.

$$w[\mathbf{s}(\boldsymbol{\xi})]\Big(g^{\boldsymbol{\xi}}_{22}\frac{\partial^2 s^i}{\partial\xi^1\partial\xi^1}-2g^{\boldsymbol{\xi}}_{12}\frac{\partial^2 s^i}{\partial\xi^1\partial\xi^2}+g^{\boldsymbol{\xi}}_{11}\frac{\partial^2 s^i}{\partial\xi^2\partial\xi^2}\Big)=K^i, \qquad (7.74)$$

where

$$\begin{aligned}K^i &= \frac{(J)^2}{g_{\mathbf{s}}}\frac{\partial}{\partial s^j}[w(\mathbf{s})g^{ij}_{\mathbf{s}}]\\ &= \frac{J}{g_{\mathbf{s}}}\Big\{\frac{\partial s^2}{\partial\xi^2}\frac{\partial}{\partial\xi^1}[w(\mathbf{s})g^{i1}_{\mathbf{s}}]-\frac{\partial s^2}{\partial\xi^1}\frac{\partial}{\partial\xi^2}[w(\mathbf{s})g^{i1}_{\mathbf{s}}]\\ &\quad -\frac{\partial s^1}{\partial\xi^2}\frac{\partial}{\partial\xi^1}[w(\mathbf{s})g^{i2}_{\mathbf{s}}]+\frac{\partial s^1}{\partial\xi^1}\frac{\partial}{\partial\xi^2}[w(\mathbf{s})g^{i2}_{\mathbf{s}}]\Big\}, \quad i,j=1,2.\end{aligned}$$

Equations (7.74) become the inverted Beltrami equations after substituting $\sqrt{g^{\mathbf{s}}}=1/\sqrt{g_{\mathbf{s}}}$ for $w(\mathbf{s})$.

Analogously there are written out the inverted diffusion grid equations in the form (6.171) and corresponding inverted Beltrami grid equations.

The equations (7.74) were solved by the numerical algorithms reviewed in Sect. 7.1.2. There were used the following expressions for $\epsilon(\mathbf{s})$ and $\psi(\mathbf{s})$:

$$\epsilon(\mathbf{s})=\begin{cases}0.05\ \exp(-|\mathbf{B}|^2/0.1), & \text{Fig. 7.29;}\\ 0.05\Big(\dfrac{0.3}{0.3+|\mathbf{B}|^2}\Big)^2, & \text{Fig. 7.30;}\\ 0.05\ln(0.005+|\mathbf{B}|^2)/\ln(0.005), & \text{Fig. 7.31;}\\ 0.1exp\Big(-\dfrac{|\mathbf{B}|}{0.07}\Big), & \text{Fig. 7.32.}\end{cases}$$

$$\psi(\mathbf{s})=\begin{cases}\phi(s^2)(1-\phi(s^2))[(s^1-0.5)^2+2(\phi(s^2)-0.5)^2], & \text{Fig. 7.29;}\\ \phi(s^2)(1-\phi(s^2))[(s^1-0.5)^2-6(\phi(s^2)-0.5)^2], & \text{Fig. 7.30;}\\ \phi(s^2)(1-\phi(s^2))[(s^1-0.5)^2+2(\phi(s^2) & \\ -0.5-0.2s^1)^2], & \text{Fig. 7.31;}\\ \phi(s^2)(1-\phi(s^2))[(s^1-0.5)^2+1.5(\phi(s^2)-0.5)^2], & \text{Fig. 7.32.}\end{cases}$$

where

$$\phi(s^2)=0.5\Big[1+tanh\Big(\frac{s^2-0.5}{0.2}\Big)\Big].$$

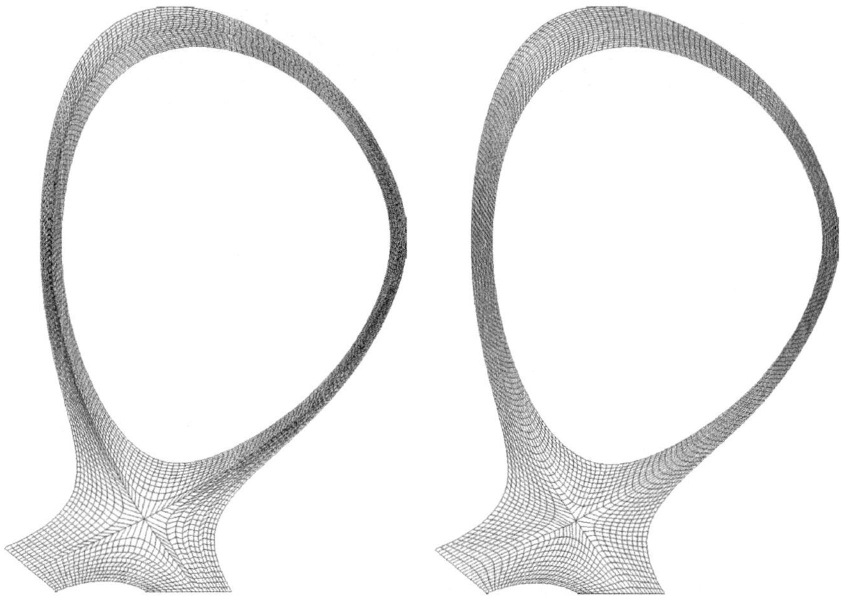

Fig. 7.33. Grids for the tokamak–edge region.

An example of a field-aligned grid for the tokamak edge region is exhibited in Fig. 7.33 (right-hand). The left-hand picture demonstrates an initial grid presented by A. Glasser. Originally grids for the tokamak edge region were performed by Petravic (1987) and Rognlien, Xu, and Hinmarsh (2002).

7.4.3 Application to Grid Clustering

The basic layer-type functions $\phi_i(x, \varepsilon)$, $i = 1, 2, 3$, can also be used to produce grid clustering in the vicinity of a hypersurface defined by the equation $\phi(\mathbf{s}) = 0$. For this purpose one can specify a monitor metric in the form (5.95) proposed for generating grids adapting to the gradient of a function $f(\mathbf{s})$ that has large variation near the hypersurface. One of such functions is defined by the formula $f(\mathbf{s}) = tanh[\phi(\mathbf{s})/\delta]$, that includes the layer-type map $\phi_1(x, \delta)$. Figure 7.34 (a) demonstrates a grid with node clustering near the curve $\phi(\mathbf{s}) = 0$. Figure 7.34 (b) shows grid clustering near two curves $\phi_1(\mathbf{s}) = 0$ and $\phi_2(\mathbf{s}) = 0$.

$$f(\mathbf{s}) = \begin{cases} 0.05\ tanh\Big(\dfrac{\phi(\mathbf{s})}{0.05}\Big), & (a); \\[2ex] 0.06\ tanh\Big(\dfrac{\phi_1(\mathbf{s})}{0.05}\Big) + 0.08\ tanh\Big(\dfrac{\phi_2(\mathbf{s})}{0.1}\Big), & (b); \end{cases}$$

where

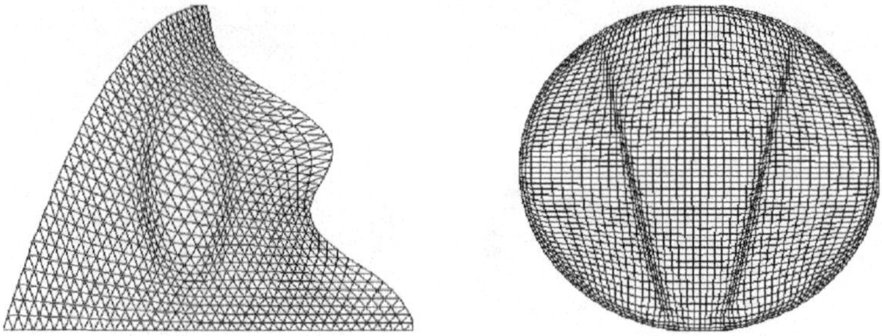

a *b*

Fig. 7.34. Triangular adaptive grids.

$$\phi(\mathbf{s}) = 100(s^1 - 0.5)^2 + 16(s^2 - 0.5)^2 - 1,$$
$$\phi_1(\mathbf{s}) = (s^1 - 0.5)^2 + (s^2 - 0.5)^2 - 0.5,$$
$$\phi_2(\mathbf{s}) = s^2 - 0.5 - 0.8\ sin(6(s^1 + 0.3)).$$

Another metric for providing grid clusstering is defined by the formula (5.99):

$$g_{\mathbf{s}}^{ij} = \omega[z(\phi)]\delta_j^i, \quad i,j = 1,\dots,n,$$

where the function $z(\phi)$ is formulated by the basic layer-type mappings $\phi_i(x,\varepsilon)$, $i = 1,2,3$. Figures 7.35 and 7.36 exhibit domain grids generated through such metric by the numerical solution of the inverted diffusion grid equations. Remind, in accordance with Sect. 5.1.5 the three-dimensional diffusion equations are the Beltrami equations in a modified metric.

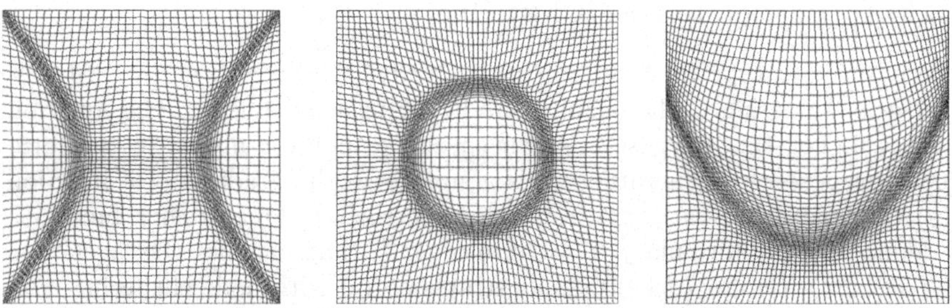

a *b* *c*

Fig. 7.35. Examples of quadrilateral adaptive grids.

For Fig. 7.35 there was assumed

$$\omega[z(\phi)] = \begin{cases} \exp(-z^2/0.5), \quad z(\mathbf{s}) = \exp\{-[\phi(\mathbf{s})]^2/0.001\}, & (a); \\ \left(\dfrac{0.5}{0.5+z^2}\right)^2, \quad z(\mathbf{s}) = \exp\{-[\phi(\mathbf{s})]^2/0.0005\}, & (b); \\ 2\exp(-z^2/0.3), \quad z(\mathbf{s}) = \dfrac{\ln\{0.0005 + [\phi(\mathbf{s})]^2\}}{\ln(0.0005)}, & (c); \end{cases}$$

$$\phi(\mathbf{s}) = \begin{cases} (s^1-0.5)^2 - 0.8(s^2-0.5)^2 - 0.05, & (a); \\ (s^1-0.5)^2 + (s^2-0.5)^2 - 0.0625, & (b); \\ s^2 - 2(s^1-0.5)^2 - 0.2, & (c). \end{cases}$$

Figure 7.36 demonstrates two-dimensional and three-dimensional adaptive grids in the case

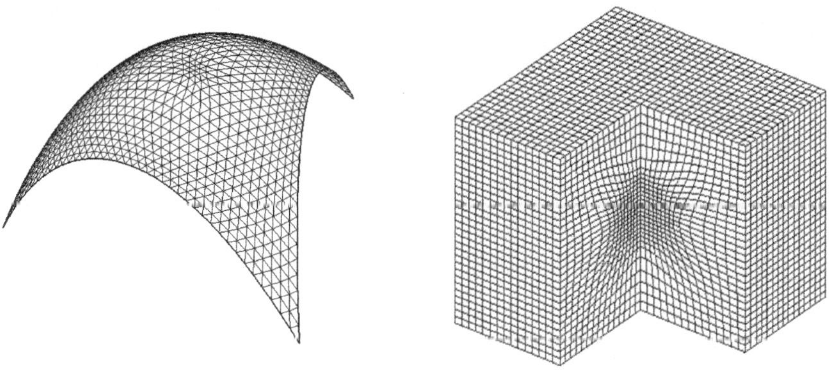

Fig. 7.36. Examples of triangular and hexahedral adaptive grids.

$$\omega[z(\phi)] = 0.5\ exp\Big[2\ exp(-\frac{\phi(\mathbf{s})^2}{0.001})\Big],$$

$$\phi(\mathbf{s}) = (s^1-0.5)^2 + (s^2-0.5)^2 + (s^3-0.5)^2,$$

which provides node clustering near central points of the geometries.

7.4.4 Application to Formulation of Weight Functions for Generating Balanced Grids

The basic layer-type functions (7.65)–(7.67) are changing rapidly in a very narrow zone when the parameter ε is small. Therefore they allow one to formulate weight functions for localizing the contribution of the corresponding terms of the metric elements (5.100) and (5.101) on the resulting quality of the balanced grids, obtained by the numerical solution of the inverted Beltrami or diffusion grid equations.

For computing the balanced numerical grids, that are field-aligned and adaptive to the values of one function and to the variations of another function, there was used formula (5.101) for the contravariant metric elements of the monitor metric, written in the following form

$$g^{ij}(\mathbf{s}) = (1-\alpha)g^{ij}_{al} + \alpha\Big((1-\beta)g^{ij}_{adg} + \beta g^{ij}_{adv}\Big), \quad i,j = 1,\dots,n.$$

$$g^{ij}_{al} = \delta^i_j \varepsilon(\mathbf{s}) + B^i B^j, \quad g^{ij}_{adv} = \delta^i_j f_1(\varphi_1),$$

$$g^{ij}_{adg} = \delta^i_j - \frac{1}{1 + |grad\ f_2(\varphi_2)|^2} \frac{\partial f_2(\varphi_2)}{\partial s^i} \frac{\partial f_2(\varphi_2)}{\partial s^j}.$$

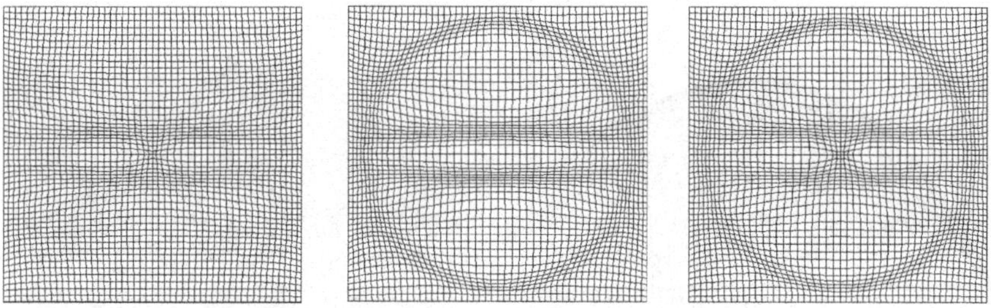

Fig. 7.37. Examples of balanced numerical grids.

Some two-dimensional balanced grids in a square domain X^2 are shown in figure 7.37. These grids were generated through the solution of equations (7.11) by the finite-difference algorithm described in Sect. 7.12. First picture of the figure demonstrates the grid aligned to a vector-field $\mathbf{B}$ formulated by (7.73) and adapted to the values of a function $\varphi_1(\mathbf{s})$. Second picture demonstrates the grid aligned to the same vector-field and adapted to the gradients of a function $f_2[\varphi_2(\mathbf{s})]$. Third picture demonstrates the grid aligned to the same vector-field and adapted to the values of one function and the gradients of another. These grids were generated with the help of the following functions and parameters

$$f_1(\varphi_1) = \Big(\frac{0.6}{0.6+\varphi_1}\Big)^3, \quad \varphi_1(\mathbf{s}) = \Big(\frac{0.01}{0.01+R^2}\Big)^5,$$

$$f_2(\varphi_2) = 0.05\tanh\Big(\frac{\varphi_2}{0.03}\Big), \quad \varphi_2(\mathbf{s}) = R^2 - 0.2,$$

$$R^2 = (s^1-0.5)^2 + (s^2-0.5)^2,$$

$$\psi(\mathbf{s}) = v(s^2)(1-v(s^2))[(s^1-0.5)^2 + 2(v(s^2)-0.5)^2],$$

$$v(s^2) = 0.5\Big[1 + tanh\Big(\frac{s^2-0.5}{0.2}\Big)\Big],$$

$$1) \quad \alpha = \varepsilon(\mathbf{s}) = \Big(\frac{0.3}{0.3+|\mathbf{B}|^2}\Big)^6, \quad \beta = 1,$$

$$2) \quad \alpha = \varepsilon(\mathbf{s}) = \Big(\frac{0.3}{0.3+|\mathbf{B}|^2}\Big)^8, \quad \beta = 0,$$

$$3) \quad \alpha = \varepsilon(\mathbf{s}) = \Big(\frac{0.3}{0.3+|\mathbf{B}|^2}\Big)^5, \quad \beta = exp(-(f_1)^2/0.1).$$

References

1. Alalykin, G.B., Godunov, S.K., Kireyeva, L.L., Pliner, L.A. (1970): *On Solution of One-Dimensional Problems of Gas Dynamics in Moving Grids.* Nauka, Moscow (Russian)
2. Albone, C.M. (1992): Embedded meshes of controllable quality synthesised from elementary geometric features. AIAA Paper 92-0633
3. Albone, C.M., Joyce, M.G. (1990): Feature-associated mesh embedding for complex configurations. AGARD Conference Proceedings 464.13
4. Allwright, S. (1989): Multiblock topology specification and grid generation for complete aircraft configurations. In Schmidt, W. (ed.): *AGARD Conference Proceedings 464, Applications of Mesh Generation to Complex 3-D Configurations.* Loen, Norway. Advisory Group for Aerospace Research and Development, NATO
5. Amsden, A.A., Hirt, C.W. (1973): A simple scheme for generating general curvilinear grids. J. Comput. Phys. 11, 348–359
6. Anderson, D.A. (1983): Adaptive grid methods for partial differential equations. In Ghia, K.N., Ghia U. (eds.): *Advances in Grid Generation.* ASME, Houston, pp. 1–15
7. Andrews, A.E. (1988): Progress and challenges in the application of artificial intelligence to computational fluid dynamics. AIAA Journal 26, 40–46
8. Arina, R., Casella, M. (1991): A harmonic grid generation technique for surfaces and three-dimensional regions. In Arcilla, A.S., Hauser, J., Eiseman, P.R., Thompson, J.F. (eds.): *Numerical Grid Generation in Computational Fluid Dynamics and Related Fields.* North-Holland, New York, pp. 935–946
9. Atta, E.H., Vadyak, J. (1982): A grid interfacing zonal algorithm for three-dimensional transonic flows about aircraft configurations. AIAA Paper 82-1017
10. Atta, E.H., Birchelbaw, L., Hall, K.A. (1987): A zonal grid generation method for complex configurations. AIAA Paper 87-0276
11. Azarenok, B.N. (2000): Realization of second-order Godunov's scheme. Comp. Meth. in Appl. Mech. and Engin., 189, pp. 1031–1052.
12. Azarenok, B.N. (2002): Variational barrier method of adaptive grid generation in hyperbolic problems of gas dynamics. SIAM J. Numer. Anal., 40(40), pp. 651–682.
13. Azarenok, B.N., Tang, T. (2005): Second-order Godunov-type scheme for reactive flow calculations on moving meshes. J. Comput. Phys., 206, pp. 48–80.

14. Baker, T.J. (1995): Prospects and expectations for unstructured methods. In: *Proceedings of the Surface Modeling, Grid Generation and Related Issues in Computational Fluid Dynamics Workshop.* NASA Conference Publication 3291, NASA Lewis Research Center, Cleveland, OH, pp. 273–287
15. Baker, T.J. (1997): Mesh adaptation strategies for problems in fluid dynamics. Finite Elements Anal. Design. 25, 243–273
16. Bakhvalov, N.S. (1969): On optimization of the methods of the numerical solution of boundary-value problems with boundary layers. J. Comput. Math. Math. Phys. 9(4), 842–859 (Russian) [English transl.: USSR Comput. Math. and Math. Phys. 9 (1969)]
17. Barfield, W.D. (1970): An optimal mesh generator for Lagrangian hydrodynamic calculations in two space dimensions. J. Comput. Phys. 6, 417–429
18. Belinsky, P.P., Godunov, S.K., Ivanov, Yu.V., Yanenko I.K. (1975): The use of one class of quasiconformal mappings to generate numerical grids in regions with curvilinear boundaries. Zh. Vychisl. Maths. Math. Fiz. 15, 1499–1511 (Russian)
19. Belk, D.M., Whitefield, D.L. (1987): Three-dimensional Euler solutions on blocked grids using an implicit two-pass algorithms. AIAA Paper 87-0450
20. Bell, J.B., Shubin, G.R. (1983): An adaptive grid finite-difference method for conservation laws. J. Comput. Phys. 52, 569–591
21. Benek, J.A., Buning, P.G., Steger, J.L. (1985): A 3-d chimera grid embedding technique. AIAA Paper 85-1523
22. Benek, J.A., Steger, J.L., Dougherty, F.C. (1983): A flexible grid embedding technique with application to the Euler equations. AIAA Paper 83-1944
23. Berger, M.J., Oliger, J. (1983): Adaptive mesh refinement for hyperbolic partial differential equations. Manuscript NA-83-02, Stanford University, March
24. Brackbill, J.U. (1993): An adaptive grid with directional control. J. Comput. Phys. 108, 38–50
25. Brackbill, J.U., Saltzman, J. (1982): Adaptive zoning for singular problems in two directions. J. Comput. Phys. 46, 342–368
26. Carey, G.F. (1997): *Computational Grids. Generation, Adaptation, and Solution Strategies.* Taylor and Francis, London
27. Chan, W.M., Buning, P.G. (1995): Surface grid generation methods for overset grids. Computer and Fluids. 24(5), 509–522
28. Chan, W.M., Steger, L.G. (1992): Enhancement of a three-dimensional hyperbolic grid generation scheme. Appl. Maths. Comput. 51(1), 181–205
29. Charakch'yan, A.A. (1993): Almost conservative difference schems for the equations of gas dynamics. Comput. Math. Math. Phys. 33, 1473
30. Charakch'yan, A.A. (1994): Compound difference schems for time-dependent equations on non-uniform nets. Commun. Numer. Methods. Eng. 10, 93
31. Charakch'yan, A.A., Ivanenko, S.A. (1997): A variational form of the Winslow grid generator. J. Comput. Phys. 136, 385–398

32. Chiba, N., Nishigaki, I., Yamashita, Y., Takizawa, C., Fujishiro, K. (1998): A flexible automatic hexahedral mesh generation by boundary-fit method. Comput. Methods Appl. Mech Engng. 161, 145–154
33. Chu, W.H. (1971): Development of a general finite difference approximation for a general domain. J. Comput. Phys. 8, 392–408
34. Chumakov, G.A., Chumakov, S.G. (1998): A method for the 2-D quasi-isometric regular grid generation, J. Comput. Phys. 133, 1–28
35. Colella, P., Woodward, P.R. (1984): The numerical simulation of two-dimensional fluid flow with strong shocks, J. Comput. Phys. 54, 115–173
36. Cordova, J.Q., Barth, T.J. (1988): Grid generation for general 2-D regions using hyperbolic equations, AIAA Paper 88-0520
37. Crowley, W.P. (1962): An equipotential zoner on a quadrilateral mesh. Memo, Lawrence Livermore National Lab., 5 July 1962
38. Danaev, N.T., Liseikin, V.D., Yanenko, N.N. (1980): Numerical solution on a moving curvilinear grid of viscous heat-conducting flow about a body of revolution. Chisl. Metody Mekhan. Sploshnoi Sredy 11(1), 51–61 (Russian)
39. Dannenhoffer, J.F. (1995): Automatic blocking for complex three-dimensional configurations. In: *Proceedings of the Surface Modeling, Grid Generation, and Related Issues in Computational Fluid Dynamics Workshop.* NASA Lewis Research Center, Cleveland OH, May, p. 123
40. Dvinsky, A.S. (1991): Adaptive Grid Generation from Harmonic Maps on Riemannian Manifolds. J. Comput. Phys. 95, 450–476
41. Edwards, T.A. (1985): Noniterative three-dimensional grid generation using parabolic partial differential equations. AIAA Paper 85-0485
42. Eells, J., Lemaire, L. (1988): Another report on harmonic maps. Bull. London Math. Soc. 20(5), 385–524
43. Eiseman, P.R. (1980): Geometric methods in computational fluid dynamics. ICASE Report 80-11 and Von Karman Institute for Fluid Dynamics Lecture Series Notes
44. Eiseman, P.R. (1985): Grid generation for fluid mechanics computations. Ann. Rev. Fluid Mech. 17, 487–522
45. Eiseman, P.R. (1987): Adaptive grid generation. Comput. Methods. Appl. Mech. Engng. 64, 321–376
46. Eriksson, L.E. (1982): Generation of boundary-conforming grids around wing-body configurations using transfinite interpolation. AIAA Journal 20, 1313–1320
47. Eriksson, L.E. (1983): Practical three-dimensional mesh generation using transfinite interpolation. Lecture Series Notes 1983-04, von Karman Institute for Fluid Dynamics, Brussels
48. Farrel, F.T., Jones, L.E. (1996): Some non-homeomorphic harmonic homotopy equivalences. Ball. London Math. Soc., 28, 177–182
49. Field, D.A. (1995): The legacy of automatic mesh generation from solid modeling. Comp. Aided Geom. Design 12, 651–673
50. Fletcher C.A.J. (1997): *Computational Techniques for Fluid Dynamics 1: Fundamental and General Techniques.* Springer, Berlin
51. Fomenko, A.T., Thi, D.T. (1991): *Minimal surfaces, stratified multivarifolds, and the Plateau problem.* AMS, New York

52. Georgala, J.M., Shaw, J.A. (1989): A discussion on issues relating to multiblock grid generation. In Schmidt, W. (ed.): *AGARD Conference Proceedings 464, Applications of Mesh Generation to Complex 3-D Configurations.* Loen, Norway. Advisory Group for Aerospace Research and Development, NATO
53. George, P.L., Borouchaki, H. (1998): *Delaunay Triangulation and Meshing.* Editions Hermes, Paris
54. Ghia, K.N., Ghia, U., Shin, C.T. (1983): Adaptive grid generation for flows with local high gradient regions. In Ghia, K.N., Ghia, U. (eds.): *Advances in Grid Generation.* ASME, Houston TX, pp. 35–47
55. Giannakopoulos, A.E., Engel, A.J. (1988): Directional control in grid generation. J. Comput. Phys. 74, 422–439
56. Glasser, A.H., Liseikin, V.D., Kitaeva, I.A. (2005): Control of grid properties with the help of monitor metrics. Comput. Math. Math. Phys., 45(8), 1416–1432
57. Glasser, A.H., Kitaeva, I.A., Yu.V. Likhanova, Liseikin, V.D., Lukin, V.S. (2005): Specification of monitor metrics for generating balanced numerical grids. Joint Bulletin of NCC and IIS, Numerical Analysis, 13, 1–13
58. Glasser, A.H., Tang, X.Z. (2004): The SEL macroscopic modeling code. Comp. Phys. Comm., 164, 237–243
59. Godunov, S.K., Prokopov, G.P. (1967): Calculation of conformal mappings in the construction of numerical grids. J. Comput. Maths. Math. Phys. 7, 1031–1059 (Russian)
60. Godunov, S.K., Prokopov, G.P. (1972): On utilization of moving grids in gasdynamics computations. J. Vychisl. Matem. Matem. Phys. 12, 429–440 (Russian) [English transl.: USSR Comput. Math. and Math. Phys. 12 (1972), 182–195]
61. Godunov, S.K., Romenskii, E.I., Chumakov, G.A. (1990): Grid generation in complex domains by means of quasi-conformal mappings. Proc. Institute of Mathematics, Novosibirsk, Nauka, 18, 75–84 (Russian)
62. Gordon, W.J., Hall, C.A. (1973): Construction of curvilinear coordinate systems and applications to mesh generation. Int. J. Numer. Meth. Engng. 7, 461–477
63. Gordon, W.J., Thiel, L.C. (1982): Transfinite mappings and their application to grid generation. In Thompson, J.F. (ed.): *Numerical Grid Generation.* North-Holland, New York, pp. 171–192
64. Hawken, D.F., Gottlieb, J.J., Hansen, J.S. (1991): Review of some adaptive node-movement techniques in finite-element and finite-difference solutions of partial differential equations. J. Comput. Phys. 95, 254–302
65. Hedstrom, G.W., Rodrigue C.M. (1982): Adaptive-grid methods for time-dependent partial differential equations. Lect. Notes Math. 960, 474–484
66. Ho-Le (1988): Finite element mesh generation methods: a review and classification. Computer-Aided Design 20, 27–38
67. Holcomb, J.F. (1987): Development of a grid generator to support 3-dimensional multizone Navier–Stokes analysis. AIAA Paper 87-0203
68. Huang, W. (2001): Variational mesh adaptation: isotropy and equidistribution. J. Comput. Phys. 174, 903–924

69. Huang, W., Ren, Y., Russel, R.D. (1994): Moving mesh PDEs based on the equidistribution principle. SIAM J. Numer. Anal. 31, 709–730
70. Ivanenko, S.A. (1988): Generation of non-degenerate meshes. USSR Comput. Math. Math. Phys. 28(5), 141
71. Ivanenko, S.A., Charakch'yan, A.A. (1988): An algorithm for constructing curvilinear grids consisting of convex quadrangles. Soviet. Math. Dokl. 36(1), 51
72. Jacquotte, O.-P. (1987): A mechanical model for a new grid generation method in computational fluid dynamics. Comput. Meth. Appl. Mech. Engng. 66, 323–338
73. Jeng, Y.N., Shu, Y.-L. (1995): Grid combination method for hyperbolic grid solver in regions with enclosed boundaries. AIAA Journal 33(6), 1152–1154
74. Khakimzyanov, G.S., Shokin, Yu.I., Barakhnin, V.B., Shokin, N.Yu. (2001): *Numerical Modeling of Flows with Surface Waves.* Novosibirsk, Sibirian Division of the Russian Academy of Sciences
75. Khamayseh, A., Mastin, C.W. (1996): Computational conformal mapping for surface grid generation. J. Comput. Phys. 123, 394–401
76. Kim, B., Eberhardt, S.D. (1995): Automatic multiblok grid generation for high-lift configuration wings. *Proceedings of the Surface Modeling, Grid Generation, and Related Issues in Computational Fluid Dynamics Workshop.* NASA, Lewis Research Center, Cleveland OH, May, p. 671
77. Knupp, P., Steinberg, S. (1993): *Fundamentals of Grid Generation.* CRC Press, Boca Raton
78. Kovenya, V.M., Tarnavskii, G.A., Chernyi S.G. (1990): *Application of a Splitting Method to Fluid Problems.* Nauka, Novosibirsk (in Russian)
79. Krugljakova, L.V., Neledova, A.V., Tishkin, V.F., Filatov, A.Yu. (1998): Unstructured adaptive grids for problems of mathematical physics (survey). Math. Modeling 10(3), 93–116 (Russian)
80. Langtangen (2003): *Computational Partial Differential Equations. Numerical Methods and Diffpack Programming.* Springer, Berlin
81. Lee, K.D., Loellbach, J.M. (1989): Geometry-adaptive surface grid generation using a parametric projection. J. Aircraft 2, 162–167
82. Lee, K.D., Huang, M., Yu, N.J., Rubbert, P.E. (1980): Grid generation for general three-dimensional confugurations. In Smith, R.E. (ed.): *Proc. NASA Langley Workshop on Numerical Grid Generation Techniques.* Oct., p. 355
83. Liao, G. (1991): On harmonic maps. In Castilio, J.E. (ed.): *Mathematical Aspects of Numerical Grid Generation.* Frontiers in Applied Mathematics, 8. SIAM, Philadelphia, pp. 123–130
84. Lin, K.L., Shaw, H.J. (1991): Two-dimensional orthogoal grid generation techniques. Comput. Struct. 41(4), 569–585
85. Liseikin, V.D. (1991a): On generation of regular grids on n-dimensional surfaces. J. Comput. Math. Math. Phys. 31, 1670–1689 (Russian). [English transl.: USSR Comput. Math. Math. Phys. 31(11) (1991), 47–57]
86. Liseikin, V.D. (1991b): Techniques for generating three-dimensional grids in aerodynamics (review). Problems Atomic Sci. Technology. Ser. Math. Model. Phys. Process 3, 31–45 (Russian)
87. Liseikin, V.D. (1992): On a variational method of generating adaptive grids on n-dimensional surfaces. Soviet Math. Docl. 44(1), 149–152

88. Liseikin, V.D. (1996): Construction of structured adaptive grids – a review. Comput. Math. Math. Phys., 36(1), 1–32
89. Liseikin, V.D. (1998a): Algebraic adaptation based on stretching functions. Russ. J. Numer. Anal. Math. Modeling 13(4), 307–324
90. Liseikin, V.D. (1998b): A method of algebraic adaptation. Comput. Math. Math. Phys., 38(10), 1624–1640
91. Liseikin, V.D. (1999): *Grid Generation Methods.* Springer, Berlin
92. Liseikin, V.D. (2001a): *Layer Resolving Grids and Transformations for Singular Perturbation Problems.* VSP, Utrecht
93. Liseikin, V.D. (2001b): Application of notions and relations of differential geometry to grid generation. Russ. J. Numer. Anal. Math. Modeling 16(1), 57–75
94. Liseikin, V.D. (2002a): Analysis of grids derived by a comprehensive grid generator. Russ. J. Numer. Anal. Math. Modeling 17(2), 183–202
95. Liseikin, V.D. (2002b): On geometric analysis of grid properties. Russ. Docl. Academ. Nauk 383(2), 167–170
96. Liseikin, V.D. (2003): On analysis of clustering of numerical grids produced by elliptic models. Russ. J. Numer. Anal. Math. Modeling 18(2), 159–183
97. Liseikin, V.D. (2004): *A Computational Differential Geometry Approach to Grid Generation.* Springer, Berlin
98. Liseikin, V.D. (2005): On a universal monitor metric for numerical grid generation. Doklady Mathematics 71(1), 15–19
99. Liseikin, V.D., Yanenko N.N. (1977): Selection of optimal numerical grids. Chisl. Metody Mekhan. Sploshnoi Sredy 8(7), 100–104 (Russian)
100. Lomonosov, I.V., Frolova, A.A., Charakhch'yan, A.A., (1997): Computation of high-velocity impact of thin foil upon conical target (survey). Math. Modeling 9(5), 48–60 (Russian)
101. Mastin, C.W. (1992): Linear variational methods and adaptive grids. Computers Math. Applic. 24(5/6), 51–56
102. McNally, D. (1972): FORTRAN program for generating a two-dimensional orthogonal mesh between two arbitrary boundaries. NASA, TN D-6766, May
103. Miki, K., Takagi, T. (1984): A domain decomposition and overlapping method for the generation of three-dimensional boundary-fitted coordinate systems. J. Comput. Phys. 53, 319–330
104. Nakamura, S. (1982): Marching grid generation using parabolic partial differential equations. Appl. Math. Comput. 10(11), 775–786
105. Nakamura, S., Suzuki M. (1987): Noniterative three-dimensional grid generation using a parabolic-hyperbolic hybrid scheme. AIAA Paper 87-0277
106. Noack, R.W. (1985): Inviscid flow field analysis of maneuvering hypersonic vehicles using the SCM formulation and parabolic grid generation. AIAA Paper 85-1682
107. Noack, R.W., Anderson D.A. (1990): Solution adaptive grid generation using parabolic partial differential equations: AIAA Journal 28(6), 1016–1023
108. Petravic, M. (1987): Orthogonal grid construction for modeling of transport in Tokamaks. J. Comput. Phys. 73, 125–130

109. Reed, C.W., Hsu, C.C., Shiau, N.H. (1988): An adaptive grid generation technique for viscous transonic flow problems. AIAA Paper 88-0313
110. Rizk, Y.M., Ben-Shmuel, S. (1985): Computation of the viscous flow around the shuttle orbiter at low supersonic speeds. AIAA Paper 85-0168
111. Rognlien, T.D., Xu, X.Q., Hinmarsh A.C. (2002): Application of parallel implicit methods to edge-plasma numerical simulations. J. Comput. Phys., 175, 249–268
112. Rubbert, P.E., Lee. K.D. (1982): Patched coordinate systems. In Thompson, J.F. (ed.): *Numerical Grid Generation*, North-Holland, New York, p. 235
113. Ryskin, G., Leal., L.G. (1983): Orthogonal mapping. J. Comput. Phys. 50(3), 71–100
114. Schonfeld, T., Weinerfelt, P., Jenssen, C.B. (1995): Algorithms for the automatic generation of 2d structured multiblock grids. *Proceedings of the Surface Modeling, Grid Generation, and Related Issues in Computational Fluid Dynamics Workshop.* NASA, Lewis Research Center, Cleveland OH, May, p. 561
115. Shaw, J.A., Weatherill, N.P. (1992): Automatic topology generation for multiblock grids. Appl. Math. Comput. 52, 355–388
116. Shephard, M.S., Grice, K.R., Lot, J.A., Schroeder, W.J. (1988): Trends in automatic three-dimensional mesh generation. Comput. Strict. 30(1/2), 421–429
117. Smith, R.E. (1981): Two-boundary grid generation for the solution of the three-dimensional Navier–Stokes equations. NASA TM-83123
118. Smith, R.E. (1982): Algebraic grid generation. In Thompson, J.F. (ed.): *Numerical Grid Generation.* North-Holland, New York, pp. 137–170
119. Smith, R.E., Eriksson, L.E. (1987): Algebraic grid generation. Comp. Meth. Appl. Mech. Eng. 64, 285–300
120. Sorenson, R.L. (1986): Three-dimensional elliptic grid generation about fighter aircraft for zonal finite-difference computations. AIAA Paper 86-0429
121. Sparis, P.D. (1985): A method for generating boundary-orthogonal curvilinear coordinate systems using the biharmonic equation. J. Comput. Phys. 61(3), 445–462
122. Starius, G. (1977): Constructing orthogonal curvilinear meshes by solving initial value problems. Numer. Math. 28, 25–48
123. Steger, J.L. (1991): Grid generation with hyperbolic partial differential equations for application to complex configurations. In Arcilla, A.S., Hauser, J., Eiseman, P.R., Thompson, J.F. (eds.): *Numerical Grid Generation in Computational Fluid Dynamics and Related Fields.* North-Holland, New York, pp. 871–886
124. Steger, J.L., Chaussee, D.S. (1980): Generation of body fitted coordinates using hyperbolic differential equations. SIAM. J. Sci. Stat. Comput. 1(4), 431–437
125. Steger, J.L., Rizk, Y.M. (1985): Generation of three-dimensional body-fitted coordinates using hyperbolic partial differential equations. NASA, TM 86753, June

126. Steger, J.L., Sorenson, R.L. (1979): Automatic mesh-point clustering near a boundary in grid generation with elliptic partial differential equations. J. Comput. Phys. 33, 405–410
127. Steinberg, S., Roache, P.J. (1986): Variational grid generation. Numer. Meth. Partial Differential Equations 2, 71–96
128. Steinbrenner, J.P., Chawner, J.R., Fouts, C.L. (1990): Multiple block grid generation in the interactive environment. AIAA Paper 90-1602
129. Stewart, M.E.M. (1992): Domain decomposition algorithm applied to multielement airfoil grids. AIAA Journal 30(6), 1457
130. Tai, C.H., Chiang, D.C., Su, Y.P. (1996): Three-dimensional hyperbolic grid generation with inherent dissipation and Laplacian smoothing. AIAA Journal 34(9), 1801–1806
131. Takagi, T., Miki, K., Chen, B.C.J., Sha, W.T. (1985): Numerical generation of boundary-fitted curvilinear coordinate systems for arbitrarily curved surfaces. J. Comput. Phys. 58, 69–79
132. Takahashi, H., Shimizu, H. (1991): A general purpose automatic mesh generation using shape recognition technique. Comput. Engng. ASME 1, 519–526
133. Tamamidis, P., Assanis, D.N. (1991): Generation of orthogonal grids with control of spacing. J. Comput. Phys. 94, 437–453
134. Thacker, W.C. (1980): A brief review of techniques for generating irregular computational grids. Int. J. Numer. Meth. Engng. 15(9), 1335–1341
135. Thoman, D.C., Szewczyk, A.A. (1969): Time-dependent viscous flow over a circular cylinder. Phys. Fluids Suppl. II, 76
136. Thomas, P.D. (1982): Composite three-dimensional grids generated by elliptic systems. AIAA Journal 20(9), 1195–1202
137. Thomas, P.D., Middlecoff, J.F. (1980): Direct control of the grid point distribution in meshes generated by elliptic equations. AIAA Journal 18(6), 652–656
138. Thomas, M.E., Bache, G.E., Blumenthal, R.F. (1990): Structured grid generation with PATRAN. AIAA Paper 90-2244
139. Thompson, J.F. (1984a): Grid generation techniques in computational fluid dynamics. AIAA Journal 22(11), 1505–1523
140. Thompson, J.F. (1984b): A survey of dynamically-adaptive grids in the numerical solution of partial differential equations. AIAA Paper 84-1606
141. Thompson, J.F. (1985): A survey of dynamically-adaptive grids in the numerical solution of partial differential equations. Appl. Numer. Math. 1, 3–27
142. Thompson, J.F. (1987): A general three-dimensional elliptic grid generation system on a composite block structure. Comput. Meth. Appl. Mech. Engng. 64, 377–411
143. Thompson, J.F. (1996): A reflection on grid generation in the 90s: trends, needs influences. In Soni, B.K., Thompson, J.F., Hauser, J., Eiseman, P.R. (eds.): *Numerical Grid Generation in CFD.* Mississippi State University, 1, pp. 1029–1110
144. Thompson, J.F., Weatherill, N.P. (1993): Aspects of numerical grid generation: current science and art. AIAA Paper 93-3539

145. Thompson, J.F., Thames, F.C., Mastin, C.W. (1974): Automatic numerical generation of body-fitted curvilinear coordinate system for field containing any number of arbitrary two-dimensional bodies. J. Comput. Phys. 15, 299–319
146. Thompson, J.F., Warsi, Z.U.A., Mastin C.W. (1982): Boundary-fitted coordinate systems for numerical solution of partial differential equations – a review. J. Comput. Phys. 47, 1–108
147. Thompson, J.F., Warsi, Z.U.A., Mastin C.W. (1985): *Numerical Grid Generation. Foundations and Applications.* North-Holland, New York
148. Visbal, M., Knight, D. (1982): Generation of orthogonal and nearly orthogonal coordinates with grid control near boundaries. AIAA Journal 20(3), 305–306
149. Vogel, A.A. (1990): Automated domain decomposition for computational fluid dynamics. Computers and Fluids 18(4), 329–346
150. Warsi, Z.U.A. (1981):*Tensors and Differential Geometry Applied to Analytic and Numerical Coordinate Generation.* MSSU-EIRS-81-1, Aerospace Engineering, Mississippi State University
151. Warsi, Z.U.A. (1982): Basic differential models for coordinate generation. In Thompson, J.F. (ed.): *Numerical Grid Generation.* North-Holland, New York, pp. 41–78
152. Warsi, Z.U.A. (1986): Numerical grid generation in arbitrary surfaces through a second-order differential-geometric model. J. Comput. Phys. 64, 82–96
153. Warsi, Z.U.A. (1990): Theoretical foundation of the equations for the generation of surface coordinates. AIAA Journal 28(6), 1140–1142
154. Warsi, Z.U.A., Thompson, J.F. (1990): Application of variational methods in the fixed and adaptive grid generation. Comput. Math. Appl. 19(8/9), 31–41
155. Weatherill, N.P., Forsey, C.R. (1984): Grid generation and flow calculations for complex aircraft geometries using a multi-block scheme. AIAA Paper 84-1665
156. White, A.B. (1990): Elliptic grid generation with orthogonality and spacing control on an arbitrary number of boundaries. AIAA Paper 90-1568
157. Widhopf, G.D., Boyd, C.N., Shiba, J.K., Than, P.T., Oliphant, P.H., Huang, S-C., Swedberg, G.D., Visich, M. (1990): RAMBO-4G: An interactive general multiblock grid generation and graphics package for complex multibody CFD applications. AIAA Paper 90-0328
158. Winslow, A.M. (1967): Equipotential zoning of two-dimensional meshes. J. Comput. Phys. 1, 149–172
159. Winslow, A.M. (1981): Adaptive mesh zoning by the equipotential method. UCID-19062, Lawrence Livermore National Laboratories
160. Wulf, A., Akrag, V. (1995): Tuned grid generation with ICEM CFD. *Proceedings of the Surface Modeling, Grid Generation, and Related Issues in Computational Fluid Dynamics Workshop.* NASA, Lewis Research Center, Cleveland OH, May, p. 477
161. Yanenko, N.N. (1971): *The Method of Fractional Steps. The Solution of Problems of Mathematiacal Physics in Several Variables.* Springer, Berlin.

162. Yanenko, N.N., Danaev, N.T., Liseikin, V.D. (1977): A variational method for grid generation. Chisl. Metody Mekhan. Sploshnoi Sredy 8(4), 157–163 (Russian)

163. Zegeling, P.A. (1993): *Moving-Grid Methods for Time-Dependent Partial Differential Equations.* CWI Tract 94, Centrum voor Wiskund en Informatica, Amsterdam

Index

Scientific Computation

A Computational Method in Plasma Physics
F. Bauer, O. Betancourt, P. Garabechan

Implementation of Finite Element Methods for Navier–Stokes Equations
F. Thomasset

Finite-Different Techniques for Vectorized Fluid Dynamics Calculations
Edited by D. Book

Unsteady Viscous Flows
D. P. Telionis

Computational Methods for Fluid Flow
R. Peyret, T. D. Taylor

Computational Methods in Bifurcation Theory and Dissipative Structures
M. Kubicek, M. Marek

Optimal Shape Design for Elliptic Systems
O. Pironneau

The Method of Differential Approximation
Yu. I. Shokin

Computational Galerkin Methods
C. A. J. Fletcher

Numerical Methods for Nonlinear Variational Problems
R. Glowinski

Numerical Methods in Fluid Dynamics
Second Edition M. Holt

Computer Studies of Phase Transitions and Critical Phenomena O. G. Mouritsen

Finite Element Methods in Linear Ideal Magnetohydrodynamics
R. Gruber, J. Rappaz

Numerical Simulation of Plasmas
Y. N. Dnestrovskii, D. P. Kostomarov

Computational Methods for Kinetic Models of Magnetically Confined Plasmas
J. Killeen, G. D. Kerbel, M. C. McCoy, A. A. Mirin

Spectral Methods in Fluid Dynamics
Second Edition
C. Canuto, M. Y. Hussaini, A. Quarteroni, T. A. Zang

Computational Techniques for Fluid Dynamics 1
Fundamental and General Techniques
Second Edition
C. A. J. Fletcher

Computational Techniques for Fluid Dynamics 2
Specific Techniques for Different Flow Categories
Second Edition
C. A. J. Fletcher

Methods for the Localization of Singularities in Numerical Solutions of Gas Dynamics Problems
E. V. Vorozhtsov, N. N. Yanenko

Classical Orthogonal Polynomials of a Discrete Variable
A. F. Nikiforov, S. K. Suslov, V. B. Uvarov

Flux Coordinates and Magnetic Filed Structure: A Guide to a Fundamental Tool of Plasma Theory
W. D. D'haeseleer, W. N. G. Hitchon, J. D. Callen, J. L. Shohet

Monte Carlo Methods in Boundary Value Problems
K. K. Sabelfeld

The Least-Squares Finite Element Method
Theory and Applications in Computational Fluid Dynamics and Electromagnetics
Bo-nan Jiang

Computer Simulation of Dynamic Phenomena
M. L. Wilkins

Grid Generation Methods
V. D. Liseikin

Radiation in Enclosures
A. Mbiock, R. Weber

Higher-Order Numerical Methods for Transient Wave Equations
G. C. Cohen

Fundamentals of Computational Fluid Dynamics
H. Lomax, T. H. Pulliam, D. W. Zingg

The Hybrid Multiscale Simulation Technology
An Introduction with Application to Astrophysical and Laboratory Plasmas A. S. Lipatov

Computational Aerodynamics and Fluid Dynamics
An Introduction J.-J. Chattot

Nonclassical Thermoelastic Problems in Nonlinear Dynamics of Shells Applications of the Bubnov–Galerkin and Finite Difference Numerical Methods
J. Awrejcewicz, V. A. Krys'ko

Scientific Computation

A Computational Differential Geometry Approach to Grid Generation Second Edition V. D. Liseikin

Stochastic Numerics for Mathematical Physics
G. N. Milstein, M. V. Tretyakov

Conjugate Gradient Algorithms and Finite Element Methods
M. Křížek, P. Neittaanmäki, R. Glowinski, S. Korotov (Eds.)

Finite Element Methods and Their Applications Z. Chen

Mathematics of Large Eddy Simulation of Turbulent Flows
L. C. Berselli, T. Iliescu, W. J. Layton

Large Eddy Simulation for Incompressible Flows
An Introduction Third Edition
P. Sagaut

Spectral Methods Fundamentals in Single Domains
C. Canuto, M. Y. Hussaini, A. Quarteroni, T. A. Zang

Stochastic Optimization
J. J. Schneider, S. Kirkpatrick

The manufacturer's authorised representative in the EU is Springer Nature Customer Service Centre GmbH, Europaplatz 3, 69115 Heidelberg, Germany. If you have any concerns regarding our products, please contact ProductSafety@springernature.com

Printed and bound by CPI Group (UK) Ltd, Croydon, CR0 4YY
07/01/2026
02030963-0007